PIC Microcontroller Projects in C

PIC Microcontroller Projects in C

Basic to Advanced

Dogan Ibrahim

AMSTERDAM • BOSTON • HEIDELBERG • LONDON
NEW YORK • OXFORD • PARIS • SAN DIEGO
SAN FRANCISCO • SINGAPORE • SYDNEY • TOKYO

Newnes is an imprint of Elsevier

Newnes

Newnes is an imprint of Elsevier
The Boulevard, Langford Lane, Kidlington, Oxford OX5 1GB, UK
225 Wyman Street, Waltham, MA 02451, USA

First edition 2008

Notice
No responsibility is assumed by the publisher for any injury and/or damage to persons or property as a matter
of products liability, negligence or otherwise, or from any use or operation of any methods, products,
instructions or ideas contained in the material herein. Because of rapid advances in the medical sciences, in
particular, independent verification of diagnoses and drug dosages should be made.

British Library Cataloguing in Publication Data
A catalogue record for this book is available from the British Library

Library of Congress Cataloging-in-Publication Data
A catalog record for this book is available from the Library of Congress

ISBN-13: 978-0-08-099924-1

For information on all Newnes publications visit our
website at http://store.elsevier.com/

Printed and bound in the UK
14 15 16 17 18   10 9 8 7 6 5 4 3 2 1

Working together
to grow libraries in
developing countries

www.elsevier.com • www.bookaid.org

Contents

Preface

A microcontroller is a single chip microprocessor system that contains data and program memory, serial and parallel input—output, timers, external and internal interrupts, all integrated into a single chip that can be purchased for as little as $2.00. About 40% of microcontroller applications are in office automation, such as PCs, laser printers, fax machines, intelligent telephones, and so forth. About one-third of microcontrollers are found in consumer electronic goods. Products like CD players, hi-fi equipment, video games, washing machines, cookers, and so on fall into this category. The communications market, automotive market, and the military share the rest of the application areas.

There are many different types of microcontrollers available from many manufacturers. This book is about the PIC18F family of high-end 8-bit microcontrollers, developed and manufactured by Microchip Inc. The highly popular PIC18F45K22 microcontroller is used in the projects in this book. Many simple, intermediate level, and advanced projects are given in the book. Most projects are developed using the highly popular mikroC Pro for PIC compiler as well as the MPLAB XC8 compiler. All the projects are fully documented where the following is given for each project: project description, project hardware (and project block diagram where appropriate), project PDL, project program, and for some projects suggestions are given for possible modifications and improvements. All the projects have been tested and are working.

Knowledge of the C programming language will be useful. Also, familiarity with at least one member of the PIC16F series of microcontrollers will be an advantage. The knowledge of assembly language programming is not required because all the projects in the book are based on using the C language.

This book is written for students, for practicing engineers, and for hobbyists interested in developing microcontroller-based projects using the PIC series of microcontrollers. Attempt has been made to include as many projects as possible, limited only by the size of the book.

Chapter 1 presents the basic features of microcontrollers.

Chapter 2 provides a short tutorial on the C language and then examines the features of the highly popular mikroC Pro for PIC programming language and compiler used in projects in this book.

Chapter 3 is about the MPLB X IDE and the XC8 programming language and compiler. Both the mikroC Pro and the XC8 program listings are given for most projects in the book. The reader should be able to convert easily from one language to the other.

Chapter 4 describes the commonly used program development tools, such as the PDL and flowcharts. Examples are given for both tools.

Chapter 5 gives simple projects using the PIC18F45K22 microcontroller. In this chapter, the projects range from simple LEDs, 7-segment LED displays, LCD displays, sound projects, and so on.

Chapter 6 provides intermediate level projects. The projects in this chapter range from using the interrupts, using a keypad, generating waveforms in real time, serial communications, GPS data decoding, various bus systems, and so on.

Chapter 7 provides more advanced projects. Some of the projects covered in this chapter are using the Bluetooth communication, RFid, real-time clock, using graphics LCDs, SD cards, Ethernet-based projects, using the CAN bus, multitasking in microcontroller systems, stepping motors, and DC motors. Although the projects on motors are not advanced, they are given in this chapter for completeness.

Dogan Ibrahim
London, 2014

Acknowledgments

The following material is reproduced in this book with the kind permission of the respective copyright holders and may not be reprinted, or reproduced in any way, without their prior consent.

Figures 3.1 and 3.12 are taken from Microchip Technology Inc. Data Sheet PIC18(L)F2X/4XK22 (DS41412F) and Microchip technology web site www.microchip.com.

Figure 6.48 is taken from the web site of Parallax Inc.

Figures 6.8, 7.13, 7.28, 7.30, 7.63 and 7.113 are taken from the web site of mikroElektronica.

PIC®, PICSTART®, and MPLAB® are all trademarks of Microchip Technology Inc.

Microcomputer Systems

Chapter Outline

1.1 Introduction

The term microcomputer is used to describe a system that includes a minimum of a microprocessor, program memory, data memory, and input—output (I/O) module. Some microcomputer systems include additional components such as timers, counters, interrupt processing modules, analog-to-digital converters, serial communication modules, USB modules, and so on. Thus, a microcomputer system can be anything from a large system having hard disks, keyboard, monitor, floppy disks, and printers to a single chip embedded controller.

In this book, we are going to consider only the type of microcomputers that consists of a single silicon chip. Such microcomputer systems are also called microcontrollers and they are used in many everyday household goods such as personal computers, digital watches, microwave ovens, digital TV sets, TV remote control units (CUs), cookers, hi-fi equipment, CD players, personal computers, fridges, etc.

There are a large number of different types of microcontrollers available in the market, developed and manufactured by many companies. In this book, we shall be looking at the programming and system design using the highly popular 8-bit programmable interface controller (PIC) series of microcontrollers manufactured by Microchip Technology Inc (www.microchip.com).

1.2 Microcontroller Systems

A microcontroller is a single chip computer. *Micro* suggests that the device is small, and *controller* suggests that the device can be used in control applications. Another term used for microcontrollers is *embedded controller*, since most of the microcontrollers in industrial, commercial, and domestic applications are built into (or embedded in) the devices they control.

A microprocessor differs from a microcontroller in many ways. The main difference is that a microprocessor requires several other external components for its operation as a computer, such as program memory and data memory, I/O module, and external clock module. A microcontroller on the other hand has all these support chips incorporated inside the same chip. In addition, because of the multiple chip concept, microprocessor-based systems consume considerably more power than the microcontroller-based systems. Another advantage of microcontroller-based systems is that their overall cost is much less than microprocessor-based systems.

All microcontrollers (and microprocessors) operate on a set of instructions (or the user program) stored in their program memories. A microcontroller fetches these instructions from its program memory one by one, decodes these instructions, and then carries out the required operations.

Microcontrollers have traditionally been programmed using the assembly language of the target device. Although the assembly language is fast, it has several disadvantages. An assembly program consists of mnemonics and in general it is difficult to learn and maintain a program written using the assembly language. Also, microcontrollers manufactured by different firms have different assembly languages and the user is required to learn a new language every time a new microcontroller is to be used.

Microcontrollers can also be programmed using high-level languages, such as BASIC, PASCAL, and C. High-level languages have the advantage that it is much easier to learn a high-level language than an assembler language. Also, very large and complex programs can easily be developed using a high-level language. In this book, we shall be learning the programming of high-end 8-bit PIC microcontrollers using two popular C programming languages: the mikroC Pro for PIC, developed by mikroElektronika (www.mikroe.com), and the MPLAB X IDE, developed by Microchip (www.microchip.com).

In general, a single chip is all that is required to have a running microcontroller-based computer system. In practical applications, additional components may be required to allow a microcomputer to interface to its environment. With the advent of the PIC family of microcontrollers, the development time of an electronic project has reduced to several months, weeks, or even hours.

Basically, a microcontroller (or a microprocessor) executes a user program that is loaded in its program memory. Under the control of this program, data are received from external devices (inputs), manipulated, and then sent to external devices (outputs).

For example, in a microcontroller-based fluid level control system, the aim is to control the level of the fluid at a given point. Here, the fluid level is read by the microcomputer via a level sensor device. The program running inside the microcontroller then actuates the pump and the valve and attempts to control the fluid level at the required value. If the fluid level is low, the microcomputer operates the pump to draw more fluid from the reservoir. In practice, the pump is controlled continuously in order to keep the fluid at the required level. Figure 1.1 shows the block diagram of our simple fluid level control system.

The system shown in Figure 1.1 is a very simplified fluid level control system. In a more sophisticated system we may have a keypad to set the required fluid level, and an LCD to display the current fluid level in the tank. Figure 1.2 shows the block diagram of this more sophisticated fluid level control system.

We can make our design even more sophisticated (see Figure 1.3) by adding an audible alarm to inform us if the fluid level is outside the required point. Also, the actual level at any time can be sent to a PC every second for archiving and further processing. For example, a graph of the daily fluid level changes can be plotted on the PC. Wireless interface (e.g. Bluetooth or RF) or internet connectivity can be added to the system so that the fluid level can be monitored or controlled remotely. Figure 1.4 shows the block diagram with a Bluetooth module attached to the microcontroller.

As you can see, because the microcontrollers are programmable, it is very easy to make the final system as simple or as complicated as we like.

Another example of a microcontroller-based system is the speed control of a direct current (DC) motor. Figure 1.5 shows the block diagram of such a system. Here, a speed sensor device reads current speed of the motor and this is compared with the desired speed

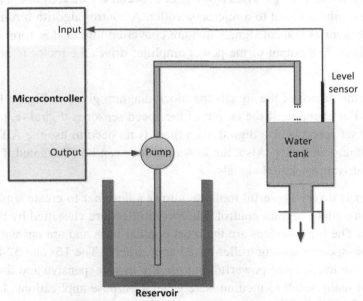

Figure 1.1: Microcontroller-Based Fluid Level Control System.

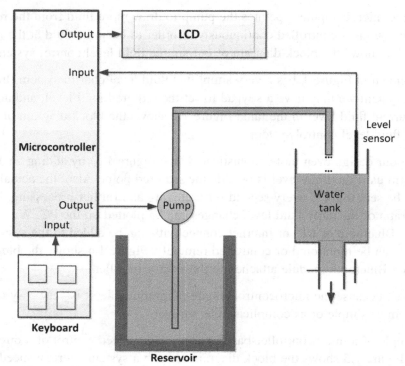

Figure 1.2: Fluid Level Control System with a Keypad and LCD.

(which is assumed to be analog). The error signal between the desired and the actual speed is converted into digital and fed to a microcontroller. A control algorithm running on the microcontroller generates control signals that are converted into analog form and are fed to a power amplifier. The output of the power amplifier drives the motor to achieve the desired speed.

Depending upon the nature of the signals the block diagram given in Figure 1.5 can take different shapes. For example, if the output of the speed sensor is digital (e.g. optical encoder) and the set speed is also digital, then there is no need to use the A/D converter at the input of the microcontroller. Also, the D/A converter can be eliminated if the power amplifier can be driven by digital signals.

A microcontroller is a very powerful tool that allows a designer to create sophisticated I/O data manipulation under program control. Microcontrollers are classified by the number of bits they process. The 8-bit devices are the most popular ones and are currently used in most low-cost low-speed microcontroller-based applications. The 16- and 32-bit microcontrollers are much more powerful, but usually more expensive and their use may not be justified in many small to medium-size general purpose applications. In this book, we will be using 8-bit PIC18F series of microcontrollers.

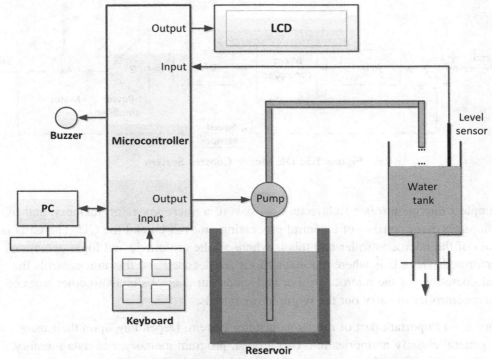

Figure 1.3: More Sophisticated Fluid Level Controller.

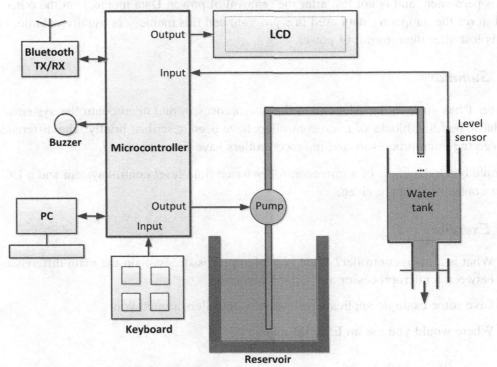

Figure 1.4: Using Bluetooth for Remote Monitoring and Control.

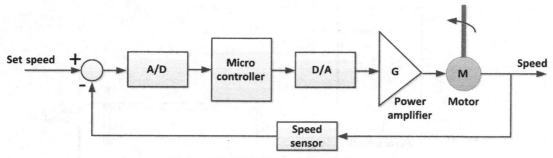

Figure 1.5: DC Motor Control System.

The simplest microcontroller architecture consists of a microprocessor, memory, and I/O. The microprocessor consists of a central processing unit (CPU) and the CU. The CPU is the brain of the microcontroller and this is where all the arithmetic and logic operations are performed. The CU is where the instructions are decoded and this unit controls the internal operations of the microcontroller and sends out control signals to other parts of the microcontroller to carry out the required operations.

Memory is an important part of a microcontroller system. Depending upon their usage, we can in general classify memories into two groups: program memory, and data memory. Program memory stores the user programs and this memory is usually nonvolatile, i.e. data is permanent and is not lost after the removal of power. Data memory on the other hand stores the temporary data used in a program and this memory is usually volatile, i.e. data is lost after the removal of power.

1.3 Summary

Chapter 1 has given an introduction to the microprocessor and microcontroller systems. The basic building blocks of microcontrollers have been described briefly. The differences between the microprocessors and microcontrollers have been explained.

Example block diagrams of a microcontroller-based fluid level control system and a DC motor control system are given.

1.4 Exercises

1. What is a microcontroller? What is a microprocessor? Explain the main differences between a microprocessor and a microcontroller.

2. Give some example applications of microcontrollers around you.

3. Where would you use an EPROM memory?

4. Where would you use a RAM memory?

5. Explain what type of memories are usually used in microcontrollers.

6. What is an I/O port?

7. What is an analog-to-digital converter? Give an example use for this converter.

8. Explain why a watchdog timer could be useful in a real-time system.

9. Why is the current sinking/sourcing important in the specification of an output port pin?

10. It is required to control the temperature in an oven using a microcontroller. Assuming that we have available an analog temperature sensor, an analog heater, and a fan, draw a block diagram to show how the system may be configured.

11. Repeat Exercise 10 by assuming that the temperature sensor gives digital output.

12. Repeat Exercise 10 by assuming that the heater can be controlled digitally.

13. It is required to monitor the temperature of an oven remotely and to display the temperature on a PC screen. Assuming that we have available a digital temperature sensor, and a Bluetooth transmitter—receiver module, draw a block diagram to show how the system may be configured.

mikroC Pro for PIC Programming Language

Chapter Outline

Some of the popular C compilers used in the development of commercial, industrial, and educational programmable interface controller (PIC) 18 microcontroller applications are

- mikroC Pro for PIC C compiler
- PICC18 C compiler
- MPLAB C18 C compiler
- MPLAB XC8 C Compiler
- CCS C compiler

mikroC Pro for PIC C compiler has been developed by MikroElektronika (web site: www.microe.com) and is one of the easy-to-learn compilers with rich resources, such as a large number of library functions and an integrated development environment with built-in simulator, and an in-circuit debugger (e.g. mikroICD). A demo version of the compiler with a 2 K program limit is available from MikroElektronika.

PICC18 C compiler is another popular C compiler, developed by Hi-Tech Software (web site: www.htsoft.com). This compiler has two versions: the standard compiler and the professional version. A powerful simulator and an integrated development environment

(Hi-Tide) is provided by the company. PICC18 is supported by the PROTEUS simulator (www.labcenter.co.uk) that can be used to simulate PIC microcontroller-based systems.

MPLAB C18 C compiler is a product of Microchip Inc. (web site: www.microchip.com). A limited-period demo version and a limited functionality version with no time limit of C18 are available from the Microchip web site. C18 includes a simulator and supports hardware and software development tools.

MPLAB XC8 C compiler is the latest C compiler from Microchip Inc. that supports all their 8-bit family of microcontrollers. The compiler is available for download free of charge.

CCS C compiler has been developed by Custom Computer Systems Inc. (web site: www. ccsinfo.com). The company provides a limited-period demo version of their compiler. CCS compiler provides a large number of built-in functions and supports an in-circuit debugger.

In this book, we shall be using the two popular C languages: mikroC Pro for PIC and MPLAB XC8. The details of mikroC Pro for PIC are given in this chapter. MPLAB XC8 is covered in detail in the next chapter.

2.1 Structure of a mikroC Pro for PIC Program

Figure 2.1 shows the simplest structure of a mikroC Pro for PIC program. This program flashes a light-emitting diode (LED) connected to port RB0 (bit 0 of PORTB) of a PIC

```
/*********************************************************************

                        LED FLASHING PROGRAM
                        ************************

        This program flashes an LED connected to port pin RB0 of PORTB with one
        second intervals.

        Programmer  : D. Ibrahim
        File        : LED.C
        Date        : July, 2013
        Micro       : PIC18F45K22
        *********************************************************************/

        void main()
        {
            for(;;)                          // Endless loop
            {
                ANSELB = 0;                  // Configure PORTB digital
                TRISB = 0;                   // Configure PORTB as output
                PORTB.0 = 0;                 // RB0 = 0
                Delay_Ms(1000);              // Wait 1 s
                PORTB.0 = 1;                 // RB0 = 1
                Delay_Ms(1000);              // Wait 1 s
            }                                // End of loop
        }
```

Figure 2.1: Structure of a Simple mikroC Pro for PIC Program.

microcontroller with 1 s intervals. Do not worry if you do not understand the operation of the program at this stage as all will be clear as we progress through this chapter. Some of the programming statements used in Figure 2.1 are described below in detail.

Comments are used by programmers to clarify the operation of the program or a programming statement. Two types of comments can be used in mikroC Pro for PIC programs: long comments extending several lines and short comments occupying only a single line. As shown in Figure 2.1, long comments start with characters "/*" and terminate with characters "*/". Similarly, short comments start with characters "//" and there is no need to terminate short comments.

In general, C language is case sensitive and variables with lower case names are different from those with upper case names. Currently, mikroC Pro for PIC variables are not case sensitive. The only exception is that identifiers **main** and **interrupt** must be written in lower case in mikroC Pro for PIC. In this book, we shall be assuming that the variables are case sensitive for compatibility with other C compilers and variables with same names but different cases shall not be used.

In C language, variable names can begin with an alphabetical character or with the underscore character. In essence, variable names can be any of the characters a−z and A−Z, the digits 0−9, and the underscore character "_". Each variable name should be unique within the first 31 characters of its name. Variable names can contain upper case and lower case characters and numeric characters can be used inside a variable name. Some names are reserved for the compiler itself and they cannot be used as variable names in our programs.

mikroC Pro for PIC language supports the variable types shown in Table 2.1.

Table 2.1: mikroC Pro for PIC Variable Types

Type	Size (Bits)	Range
unsigned char	8	0 to 255
unsigned short int	8	0 to 255
unsigned int	16	0 to 65535
unsigned long int	32	0 to 4294967295
signed char	8	−128 to 127
signed short int	8	−128 to 127
signed int	16	−32768 to 32767
signed long int	32	−2147483648 to 2147483647
float	32	±1.17549435082E-38 to ±6.80564774407E38
double	32	±1.17549435082E-38 to ±6.80564774407E38
long double	32	±1.17549435082E-38 to ±6.80564774407E38

Constants represent fixed values (numeric or character) in programs that cannot be changed. Constants are stored in the flash program memory of the PIC microcontroller; thus, the valuable and limited random-access memory (RAM) is not wasted.

2.2 Arrays

An array is declared by specifying its type, name, and the number of elements it will store. For example,

```
unsigned int Total[5];
```

Creates an array of type unsigned int, with name Total, and having five elements. The first element of an array is indexed with 0. Thus, in the above example, Total[0] refers to the first element of this. The array Total is stored in memory in five consecutive locations as follows:

```
Total[0]
Total[1]
Total[2]
Total[3]
Total[4]
```

Data can be stored in the array by specifying the array name and index. For example, to store 25 in the second element of the array, we have to write:

```
Total[1] = 25;
```

Similarly, the contents of an array can be read by specifying the array name and its index. For example, to copy the third array element to a variable called temp we have to write:

```
Temp = Total[2];
```

The contents of an array can be initialized during its declaration. An example is given below where array months has 12 elements and months[0] = 31, months[1] = 28, and so on.

```
unsigned char months[12] = {31,28,31,30,31,30,31,31,30,31,30,31};
```

The above array can also be declared without specifying the size of the array:

```
unsigned char months[ ] = {31,28,31,30,31,30,31,31,30,31,30,31};
```

Character arrays can be declared similarly. In the following example, a character array named Hex_Letters is declared with six elements:

```
unsigned char Hex_Letters[ ] = {'A', 'B', 'C', 'D', 'E', 'F'};
```

Strings are character arrays with a null terminator. Strings can either be declared by enclosing the string in double quotes, or each character of the array can be specified within single quotes, and then terminated with a null character:

```
unsigned char Mystring[ ] = "COMP";
```

And

```
unsigned char Mystring[ ] = {'C', 'O', 'M', 'P', '\0'};
```

In C programming language, we can also declare arrays with multiple dimensions. In the following example, a two-dimensional array named P is created having three rows and four columns. Altogether, the array has 12 elements. The first element of the array is P[0][0], and the last element is P[2][3]. The structure of this array is shown below:

P[0][0]	P[0][1]	P[0][2]	P[0][3]
P[1][0]	P[1][1]	P[1][2]	P[1][3]
P[2][0]	P[2][1]	P[2][2]	P[2][3]

2.3 Pointers

Pointers are an important part of the C language and they hold the memory addresses of variables. Pointers are declared same as any other variables, but with the character ("*") in front of the variable name. In general, pointers can be created to point to (or hold the addresses of) character variables, integer variables, long variables, floating point variables, or functions.

In the following example, an unsigned character pointer named pnt is declared:

```
unsigned char *pnt;
```

When a new pointer is created, its content is initially unspecified and it does not hold the address of any variable. We can assign the address of a variable to a pointer using the ("&") character:

```
pnt = &Count;
```

Now pnt holds the address of variable Count. Variable Count can be set to a value by using the character ("*") in front of its pointer. For example, Count can be set to 10 using its pointer:

```
*pnt = 10;      //Count = 10
```

Which is same as

```
Count = 10;     //Count = 10
```

Or, the value of Count can be copied to variable Cnt using its pointer:

```
Cnt = *pnt;      //Cnt = Count
```

2.4 Structures

A structure is created by using the keyword **struct**, followed by a structure name, and a list of member declarations. Optionally, variables of the same type as the structure can be declared at the end of the structure.

The following example declares a structure named Person:

```
struct Person
{
  unsigned char name[20];
  unsigned char surname[20];
  unsigned char nationality[20];
  unsigned char age;
}
```

Declaring a structure does not occupy any space in memory, but the compiler creates a template describing the names and types of the data objects or member elements that will eventually be stored within such a structure variable. It is only when variables of the same type as the structure are created that these variables occupy space in memory. For example, two variables **Me** and **You** of type Person can be created by the statement:

```
struct Person Me, You;
```

Variables of type Person can also be created during the declaration of the structure as shown below:

```
struct Person
{
  unsigned char name[20];
  unsigned char surname[20];
  unsigned char nationality[20];
  unsigned char age;
} Me, You;
```

We can assign values to members of a structure by specifying the name of the structure, followed by a dot ("."), and the name of the member. In the following example, the **age** of structure variable **Me** is set to 25, and variable **M** is assigned to the value of **age** in structure variable **You**:

```
Me.age = 25;
M = You.age;
```

Structure members can be initialized during the declaration of the structure. In the following example, the radius and height of structure Cylinder are initialized to 1.2 and 2.5, respectively:

```
struct Cylinder
{
  float radius;
  float height;
} MyCylinder = {1.2, 2.5};
```

2.5 Operators in C

Operators are applied to variables and other objects in expressions and they cause some conditions or some computations to occur.

mikroC Pro for PIC language supports the following operators:

*   Arithmetic operators
*   Logical operators
*   Bitwise operators
*   Conditional operators
*   Assignment operators
*   Relational operators
*   Preprocessor operators

2.6 Modifying the Flow of Control

Statements are normally executed sequentially from the beginning to the end of a program. We can use control statements to modify the normal sequential flow of control in a C program. The following control statements are available in mikroC Pro for PIC programs:

*   Selection statements
*   Unconditional modification of flow
*   Iteration statements

There are two selection statements: **If** and **switch**.

If Statement

The general format of the **if** statement is

```
if(expression)
  Statement1;
else
  Statement2;
```

or,

```
if(expression)Statement1; else Statement2;
```

In the following example, if the value of **x** is greater than **MAX** then variable **P** is incremented by 1, otherwise it is decremented by 1:

```
if(x > MAX)
  P++;
else
  P--;
```

We can have more than one statement by enclosing the statements within curly brackets. For example,

```
if(x > MAX)
{
  P++;
  Cnt = P;
  Sum = Sum + Cnt;
}
else
  P--;
```

In the above example, if **x** is greater than **MAX** then the three statements within the curly brackets are executed, otherwise the statement **P−−** is executed.

Another example using the **if** statement is given below:

```
if(x > 0 && x < 10)
{
  Total + = Sum;
  Sum++;
}
else
{
  Total = 0;
  Sum = 0;
}
```

The **switch** statement is used when there are a number of conditions and different operations are performed when a condition is true. The syntax of the **switch** statement is

```
switch (condition)
{
 case condition1:
  Statements;
  break;
 case condition2:
  Statements;
```

```
   break;
.....................
.....................
 case condition:
  Statements;
  break;
 default:
  Statements;
 }
```

In mikroC Pro for PIC there are four ways that iteration can be performed and we will look at each one with examples:

- Using **for** statement
- Using **while** statement
- Using **do** statement
- Using **goto** statement

for Statement

The syntax of the **for** statement is

```
for(initial expression; condition expression; increment expression)
{
   Statements;
}
```

The following example shows how a loop can be set up to execute 10 times. In this example, variable **i** starts from 0 and increments by 1 at the end of each iteration. The loop terminates when i = 10, in which case the condition i < 10 becomes false. On exit from the loop, the value of **i** is 10:

```
for(i = 0; i < 10; i++)
{
   statements;
}
```

The parameters of a **for** loop are all optional and can be omitted. If the **condition expression** is left out, it is assumed to be true. In the following example, an endless loop is formed where the **condition expression** is always true and the value of **i** starts with 0 and is incremented after each iteration:

```
/* Endless loop with incrementing i */
for(i = 0; ; i++)
{
   Statements;
}
```

Another example of an endless loop is given below where all the parameters are omitted:

```
/* Example of endless loop */
for(; ;)
{
   Statements;
}
```

while *Statement*

This is another statement that can be used to create iteration in programs. The syntax of the **while** statement is

```
while (condition)
{
   Statements;
}
```

The following code shows how to set up a loop to execute 10 times using the **while** statement:

```
/* A loop that executes 10 times */
k = 0;
while (k < 10)
{
   Statements;
   k++;
}
```

At the beginning of the code, variable **k** is 0. Since **k** is less than 10, the **while** loop starts. Inside the loop, the value of **k** is incremented by 1 after each iteration. The loop repeats as long as k < 10 and is terminated when k = 10. At the end of the loop, the value of **k** is 10.

Notice that an endless loop will be formed if **k** is not incremented inside the loop:

```
/* An endless loop */
k = 0;
while (k < 10)
{
   Statements;
}
```

An endless loop can also be formed by setting the **condition** to be always true:

```
/* An endless loop */
while (k = k)
{
   Statements;
}
```

do *Statement*

The **do** statement is similar to the **while** statement but here the loop executes until the **condition** becomes false, or the loop executes as long as the **condition** is true. The **condition** is tested at the end of the loop. The syntax of the **do** statement is

```
do
{
  Statements;
} while (condition);
```

The following code shows how to set up a loop to execute 10 times using the **do** statement:

```
/* Execute 10 times */
k = 0;
do
{
  Statements;
  k++;
} while (k < 10);
```

An endless loop will be formed if the condition is not modified inside the loop as shown in the following example. Here **k** is always less than 10:

```
/* An endless loop */
k = 0;
do
{
  Statements;
} while (k < 10);
```

An endless loop can also be created if the condition is set to be true all the time.

goto *Statement*

Although not recommended, the **goto** statement can be used together with the **if** statement to create iterations in a program. The following example shows how to set up a loop to execute 10 times using the **goto** and **if** statements:

```
/* Execute 10 times */
k = 0;
```

Loop:

```
Statements;
k++;
if(k < 10)goto Loop;
```

2.7 mikroC Pro for PIC Functions

An example function definition is shown below. This function, named **Mult**, receives two integer arguments **a** and **b** and returns their product. Notice that using brackets in a return statement are optional:

```
int Mult(int a, int b)
{
  return (a*b);
}
```

When a function is called, it generally expects to be given the number of arguments expressed in the function's argument list. For example, the above function can be called as

```
z = Mult(x,y);
```

Where variable **z** has the data type **int**. In the above example, when the function is called, variable **x** is copied to **a**, and variable **y** is copied to **b** on entry to function **Mult**.

Some functions do not return any data and the data type of such functions must be declared as **void**. An example is given below:

```
void LED(unsigned char D)
{
  PORTB = D;
}
```

void functions can be called without any assignment statements, but the brackets must be used to tell the compiler that a function call is made.

2.8 mikroC Pro for PIC Library Functions

mikroC Pro for PIC provides a large set of library functions that can be used in our programs. mikroC Pro for PIC user manual gives detailed descriptions of each library function with examples. Table 2.2 gives a list of the mikroC Pro for PIC library functions, organized in functional order.

2.9 Summary

This chapter presented an introduction to the mikroC Pro for PIC language. A C program may contain a number of functions and variables and a main program. The beginning of the main program is indicated by the statement **void main()**.

A variable stores a value used during the computation. All variables in C must be declared before they are used. A variable can be an 8-bit character, a 16-bit integer, a 32-bit long,

Table 2.2: mikroC Pro for PIC Library Functions

Library	Description
ADC	Analog-to-digital conversion functions
CAN	CAN bus functions
CANSPI	SPI-based CAN bus functions
Compact flash	Compact flash memory functions
EEPROM	EEPROM memory read/write functions
Ethernet	Ethernet functions
SPI ethernet	SPI-based ethernet functions
Flash memory	Flash memory functions
Graphics LCD	Standard graphics LCD functions
T6963C graphics LCD	T6963-based graphics LCD functions
I^2C	I^2C bus functions
Keypad	Keypad functions
LCD	Standard LCD functions
Manchester code	Manchester code functions
Multimedia	Multimedia functions
One Wire	One Wire functions
PS/2	PS/2 functions
PWM	PWM functions
RS-485	RS-485 communication functions
Sound	Sound functions
SPI	SPI bus functions
USART	USART serial communication functions
Util	Utilities functions
SPI graphics LCD	SPI-based graphics LCD functions
Port expander	Port expander functions
SPI LCD	SPI-based LCD functions
ANSI C Ctype	C Ctype functions
ANSI C Math	C Math functions
ANSI C Stdlib	C Stdlib functions
ANSI C String	C String functions
Conversion	Conversion functions
Trigonometry	Trigonometry functions
Time	Time functions

or a floating point number. Constants are stored in the flash program memory of PIC microcontrollers and thus using them saves valuable and limited RAM.

Various flow control and iteration statements such as **if, switch, while, do, break,** and so on have been described in the chapter with examples.

Pointers are used to store the addresses of variables. As we shall see in the next chapter, pointers can be used to pass information back and forth between a function and its calling point. For example, pointers can be used to pass variables between a main program and a function.

Library functions simplify programmers' tasks by providing ready and tested routines that can be called and used in our programs. Examples are also given on how to use various library functions in our main programs.

2.10 Exercises

1. Write a C program to set bits 0 and 7 of PORTC to logic 1.

2. Write a C program to count down continuously and send the count to PORTB.

3. Write a C program to multiply each element of a 10 element array with number 2.

4. It is required to write a C program to add two matrices **P** and **Q**. Assume that the dimension of each matrix is 3×3 and store the result in another matrix called **W**.

5. What is meant by program repetition? Describe the operation of the **while**, **do-while**, and **for** loops in C.

6. What is an array? Write example statements to define the following arrays:
   a. An array of 10 integers.
   b. An array of 30 floats.
   c. A two-dimensional array having 6 rows and 10 columns.

7. How many times do each of the following loops iterate and what is the final value of the variable **j** in each case?
   a. `for(j = 0; j < 5; j++)`
   b. `for(j = 1; j < 10; j++)`
   c. `for(j = 0; j <= 10; j++)`
   d. `for(j = 0; j <= 10; j += 2)`
   e. `for(j = 10; j > 0; j -= 2)`

8. Write a program to calculate the average value of the numbers stored in an array. Assume that the array is called **M** and it has 20 elements.

9. Derive equivalent **if-else** statements for the following tests:
   a. `(a > b) ? 0 : 1`
   b. `(x < y) ? (a > b) : (c > d)`

10. What can you say about the following **for** loop:
```
Cnt = 0;
for(;;)
{
  Cnt++;
}
```

11. Write a function to calculate the circumference of a rectangle. The function should receive the two sides of the rectangle as floating point numbers and then return the circumference as a floating point number.

12. Write a function to convert inches to centimeters. The function should receive inches as a floating point number and then calculate the equivalent centimeters.

13. An LED is connected to port pin RB7 of a PIC18F45K22 microcontroller. Write a program to flash the LED such that the ON time is 5 s and the OFF time is 3 s.

14. Write a function to perform the following operations on two-dimensional matrices:
     a. Add matrices
     b. Subtract matrices
     c. Multiply matrices

CHAPTER 3

MPLAB X IDE and MPLAB XC8 C Programming Language

Chapter Outline

In this chapter, we shall be looking at details of the other popular PIC C programming language/compiler, the MPLAB XC8, and the MPLAB X IDE (Integrated Development Environment), both developed by Microchip Inc.

This chapter is organized as a tutorial so that the user can quickly become familiar and start using the compiler and the IDE. The popular PICDEM PIC18 Explorer Development Board is used in the examples given in this chapter.

Before going into the details of the MPLAB XC8 programming language and the MPLAB X IDE, it is worthwhile to look at the features of the PICDEM Explorer board.

It is recommended that if you are not familiar with programming using the C language, then you should read Chapter 2 before continuing with this chapter.

3.1 The PICDEM PIC18 Explorer Development Board

The PICDEM PIC18 Explorer development board (called the Explorer board from now on) is a low-cost development board for PIC18 family of microcontrollers. The board is fitted with a PIC18F8722-type microcontroller.

Figure 3.1 shows the Explorer board where each feature is indicated with a number. The board provides the following features:

1. PIC18F8722 microcontroller,
2. PIM header to connect an alternate PIC18 microcontroller,
3. In-circuit Debugger connector,
4. PICkit 3 programmer/debugger connector,

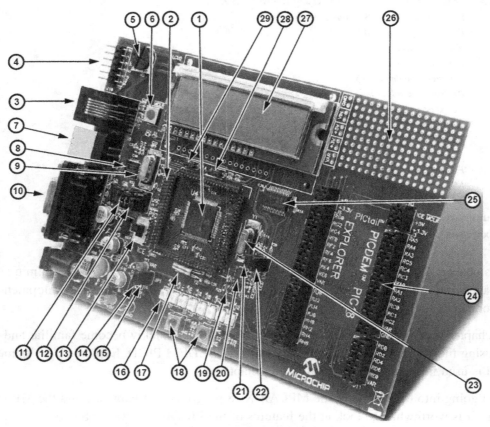

Figure 3.1: PICDEM PIC18 Explorer Board.

5. Potentiometer for analog input,
6. Reset switch,
7. RS232-universal serial bus (USB) connector,
8. PIC18LF2450 microcontroller (for converting an RS232 to a USB),
9. Crystal for the PIC18LF2450 (12 MHz),
10. Nine-pin RS232 connector,
11. Jumper J13 for routing RS232 serial communication to the USB port or the RS232 socket,
12. Jumper J4 for programming the main microcontroller or the PIC18LF2450,
13. Switch S4 for selecting the main microcontroller as either the mounted PIC18F8722 or a PIM module,
14. Power indication light emitting diode (LED),
15. JP1 for disconnecting the eight LEDs,
16. LEDs,
17. A 32,768-kHz crystal (for Timer 1),
18. Push-button switches S1 and S2,
19. MPC9701A analog temperature sensor,
20. 25LC256 electrically erasable programmable read-only memory (EEPROM),
21. Jumper JP2 to enable/disable EEPROM,
22. Jumper JP3 to enable/disable an LCD,
23. Crystal for the main microcontroller (10 MHz),
24. PICtail daughter board connector,
25. I/O expander for the LCD,
26. User prototype area,
27. LCD,
28. Jumper J2 for selecting between 3.3 and 5 V (by default the V_{DD} voltage is +5 V),
29. Jumper J14 for use with a PIM.

To use the Explorer board with the on-board PIC18F8722 microcontroller, the following switches and jumpers should be configured:

- Set Switch S4 to PIC to select the on-board microcontroller,
- Enable LEDs by connecting a jumper at JP1,
- Enable the LCD by connecting a jumper at JP3,
- Connect Jumper J4 to MAIN to program the main microcontroller (PIC18F8722).

The Explorer can be programmed by using several hardware tools, such as the PICkit 2/3, ICD 2/3, and Real In Circuit Emulator (ICE). In this chapter, we shall be seeing how to program and debug a program using the PICkit 3 and the ICD 3 programmer/debugger tools.

3.1.1 Module Connections on the Explorer Board

The various modules on the Explorer board are connected as follows:

- Eight LEDs are connected to PORTD of the microcontroller (the LEDs can be disconnected by removing Jumper JP1).
- The LCD is controlled by port pins RC3, RC4, and RC5.
- Switch S1 is connected to port pin RB0 (active LOW).
- Switch S2 is connected to port pin RA5 (active LOW).
- Switch S3 is connected to the Master Clear (MCLR) reset input.
- Switch S4 is an MCLR select switch.
- The potentiometer is connected to the AN0 input through a resistor, and it can be adjusted from 0 V to V_{DD}.
- The analog temperature sensor MCP9701A is connected to port pin RA1.
- The RS232-USB port is connected to pins RC6 and RC7.

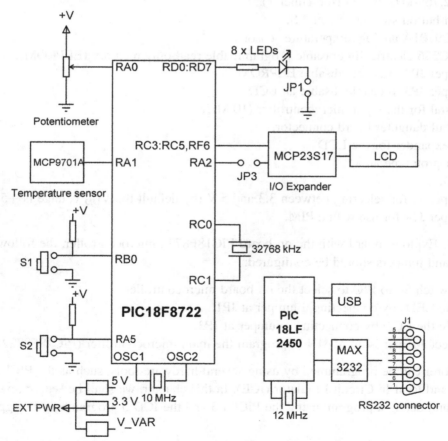

Figure 3.2: The Simplified Block Diagram of the Explorer Board.

A simplified block diagram showing the module connections on the Explorer board is shown in Figure 3.2. Note that the various chips on the board are fed with the adjustable voltage source V_VAR, which is set to +5 V by default. The board is powered from an external 9 V DC mains adapter capable of providing up to 1.3 A.

3.1.2 Using the PICkit 3 to Program/Debug

A PICkit 3 programmer/debugger can be attached to the 6-pin connector mounted at the top-left corner of the Explorer board for programming and debugging user programs.

3.2 MPLAB X IDE

The MPLAB X IDE is the integrated development environment (just like the mikroC Pro for PIC IDE) that enables the user to create a source file, edit, compile, simulate, debug, and send the generated code to the target microcontroller with the help of a compatible programmer/debugger device. The Microchip Inc. web site (http://www.microchip.com/pagehandler/en-us/family/mplabx/) contains detailed information on the features and the use of the MPLAB X IDE. In this chapter, we will be looking at the program development steps using this IDE.

MPLAB X IDE is available at the Microchip Inc. web site (http://www.microchip.com/pagehandler/en-us/family/mplabx/#downloads), and it must be installed on your PC before it can be used. At the time of writing this book, the latest version of the MPLAB X IDE was v1.85.

3.3 MPLAB XC8 Compiler

The MPLAB XC8 compiler is a powerful C compiler developed for the PIC10/12/16/18 family of microcontrollers (there are also versions for the 24- and 32-bit PIC microcontrollers). The MPLAB XC8 compiler has three versions: Pro, Standard, and Free. In this book, we will be using the Free version. The main difference between the different versions is the level of optimization applied during the compilation.

The XC8 compiler must be installed after installing the MPLAB X IDE. The compiler can be installed during the last stage of installation of the MPLAB X IDE. Alternatively, it can be installed from the Microchip Inc. web site (http://www.microchip.com/pagehandler/en_us/devtools/mplabxc/). At the time of writing this book, the latest version of the compiler was v1.20.

The XC8 language has many similarities to the mikroC Pro for PIC language. In this chapter, we will be looking at the steps of developing a simple XC8-based project. The similarities and differences between the two languages will also be explained.

Example 3.1—A Simple Project

A simple project is given in this section to show the steps in creating a source file using the MPLAB X IDE, compiling the file, and downloading the generated hex file to the PIC18F8722 microcontroller on the Explorer board using the PICkit 3.

In this project, we will be using a push-button switch S1 and the LED connected to port pin RD0. The program will turn the LED on whenever the button is pressed.

Solution 3.1

The steps are given below.

Step 1. Double click the icon to start the MPLAB X IDE. You should see the opening window as in Figure 3.3.

Step 2. Move the right-hand cursor down and click on icon *Create New Project*. Select the default (Category: *Microchip Embedded*, Projects: *Standalone Project*) as in Figure 3.4 since we are creating a new standalone project.

Step 3. Click Next. Select the target microcontroller. Family: *Advanced 8-bit MCUs (PIC18)*, and Device: *PIC18F8722* as in Figure 3.5.

Step 4. Click Next and select Hardware Tools: *PICkit 3* as in Figure 3.6.

Step 5. Click Next. Select compiler XC8 as in Figure 3.7.

Step 6. Click Next. Give your project a name. In this example, the project is given the name *BUTTON-LED* and is stored in the folder *C:\Users\Dogan\MPLABXProjects*. Click *Set as main project* option as shown in Figure 3.8.

Step 7. Click Finish to create the required project files.

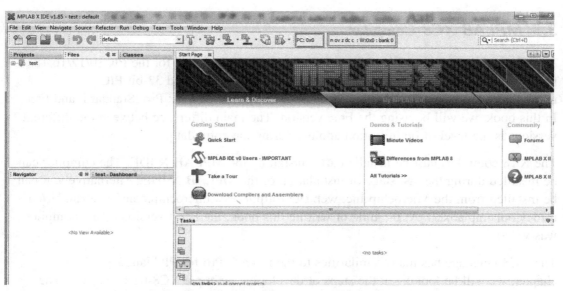

Figure 3.3: Opening Window of MPLAB X IDE.

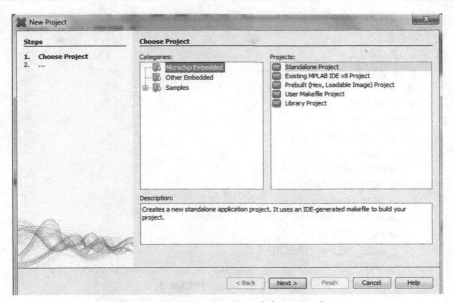

Figure 3.4: Create a Standalone Project.

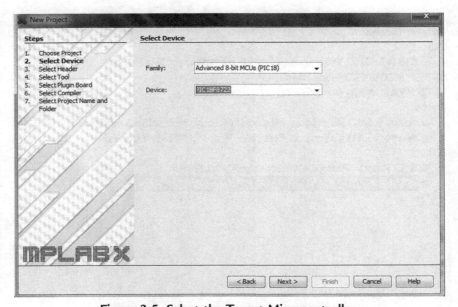

Figure 3.5: Select the Target Microcontroller.

Step 8. Right click *Source Files* on the left-hand window and select *New → C Main File*. Name the new source file as *NEWMAIN* (extension .C) as in Figure 3.9.
Step 9. Click Finish, and you should get an empty template C file as shown in Figure 3.10.

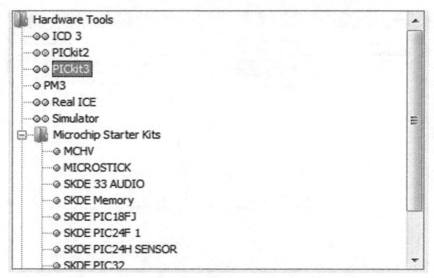

Figure 3.6: Select PICkit 3.

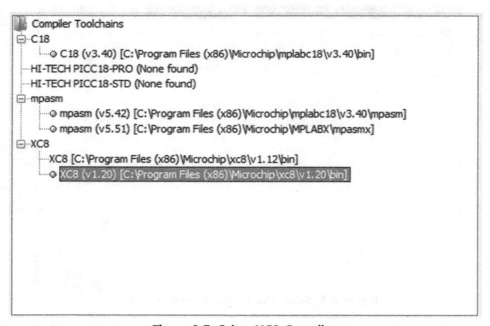

Figure 3.7: Select XC8 Compiler.

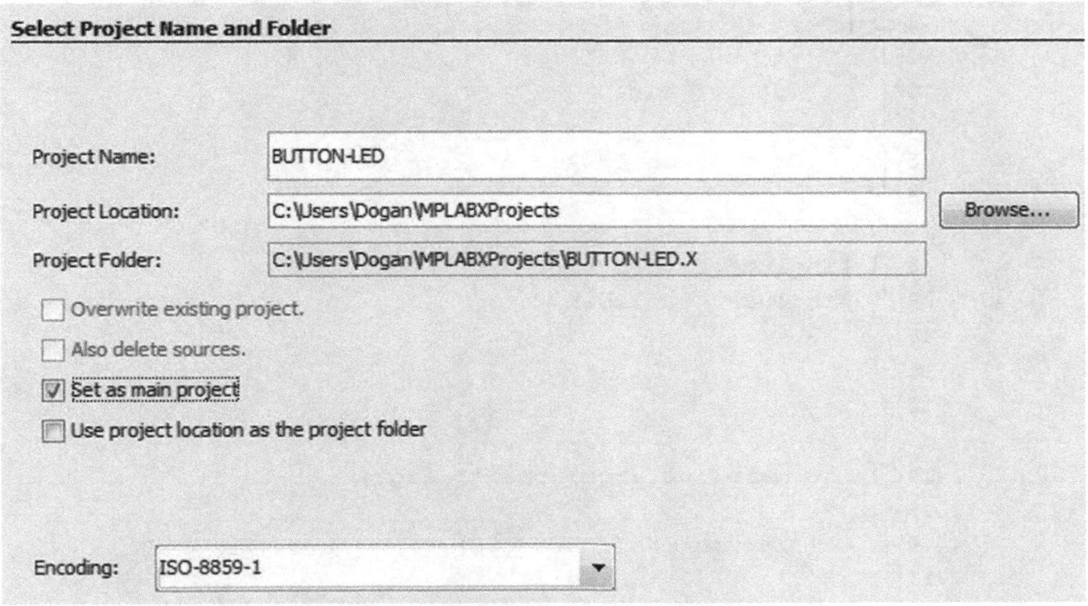

Select Project Name and Folder

Project Name: BUTTON-LED

Project Location: C:\Users\Dogan\MPLABXProjects   Browse...

Project Folder: C:\Users\Dogan\MPLABXProjects\BUTTON-LED.X

☐ Overwrite existing project.

☐ Also delete sources.

☑ Set as main project

☐ Use project location as the project folder

Encoding: ISO-8859-1 ▼

Figure 3.8: Give the Project a Name.

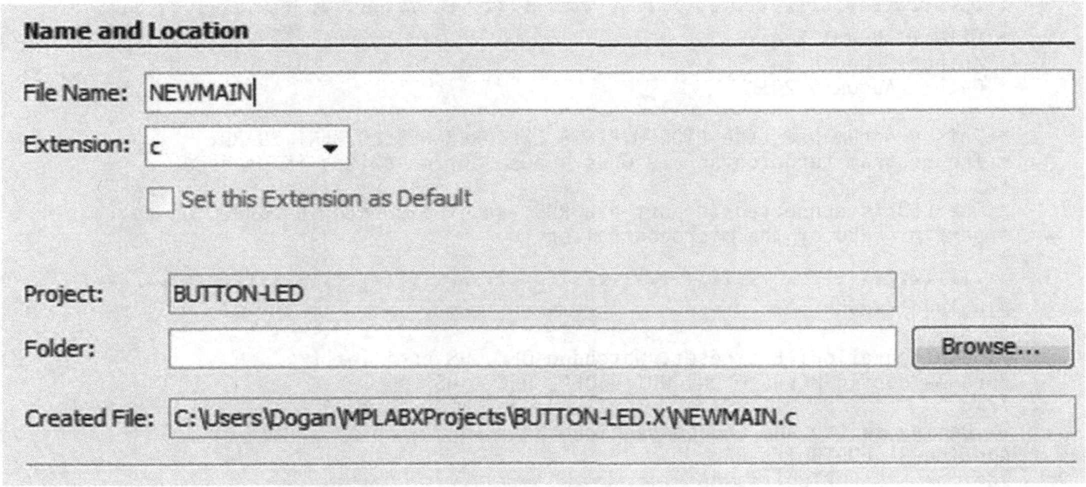

Name and Location

File Name: NEWMAIN

Extension: c ▼

☐ Set this Extension as Default

Project: BUTTON-LED

Folder: Browse...

Created File: C:\Users\Dogan\MPLABXProjects\BUTTON-LED.X\NEWMAIN.c

Figure 3.9: Create the New Source File.

```
 1 ┌ /*
 2 │    * File:    NEWMAIN.c
 3 │    * Author: Dogan
 4 │    *
 5 │    * Created on 03 August 2013, 13:54
 6 └    */
 7
 8 ┌ #include <stdio.h>
 9 └ #include <stdlib.h>
10
11 ┌ /*
12 │    *
13 └    */
14 ┌ int main(int argc, char** argv) {
15 │
16 │        return (EXIT_SUCCESS);
17 └ }
18
```

Figure 3.10: Template C File.

Step 10. Modify the file by inserting the following lines for our program. The program turns on the LED connected to port pin RD0 whenever push-button switch S1 (connected to port pin RB0) is pressed. See Figure 3.11 for part of the program listing in MPLAB X IDE:

```
*********************************************************************
 * File:   NEWMAIN.c
 * Author: Dogan
 * Date:   August, 2013
 *
 * This program uses the PICDEM PIC18 EXPLORER DEVELOPMENT BOARD.
 * The program turns on an LED when a push-button switch is pressed.
 *
 * The LED is connected to port pin RD0, and the switch is connected to
 * port pin RB0 of the microcontroller.
 *
 *********************************************************************/
#include <xc.h>

// Configuration: Ext reset, Watchdog OFF, HS oscillator
#pragma config MCLRE = ON, WDT = OFF, OSC = HS
//
// Define switch and LED connections
#define S1 PORTBbits.RB0
#define LED PORTDbits.RD0
//
// Define clock frequency
#define _XTAL_FREQ 10000000

//
```

```
// Start of main program
//
int main()
{
 TRISBbits.TRISB0 = 1;           // Configure RB0 as input
 TRISDbits.TRISD0 = 0;           // Configure PORTD as outputs
 MEMCONbits.EBDIS = 1;           // Enable PORTD I/O functions

 for(;;)                         // Do FOREVER
 {
  if(S1 == 0)LED = 1; else LED = 0;
 }
}
```

The description of the program is as follows:
- The *#include <xc.h>* statement at the beginning of the program identifies the microcontroller in use and calls the appropriate header files to include the processor specific definitions at the beginning of the program (notice that the mikroC Pro for the PIC compiler does not require a header file).
- The configuration statement *#pragma config MCLRE = ON, WDT = OFF, OSC = HS* defines the processor configuration. Here, master clear (reset) is enabled, watchdog timer is turned off, and the external high-speed crystal is selected as the clock source. The file *pic18_chipinfo.html* in the docs directory of the XC8 compiler installation (usually the folder: *C:\Program Files (x86)\Microchip\xc8\v1.20\docs\pic18_chipinfo.html*) contains a list of all the processors and a list of all possible configuration options for each processor.
- The statement *#define S1 PORTBbits.RB0* defines symbol S1 as port pin RB0. Similarly, the statement *#define LED PORTDbits.RD0* defines symbol LED as port pin RD0 of the microcontroller.
- The microcontroller clock frequency is then defined as 10 MHz.
- At the beginning of the main program, port pin RB0 is configured as an input port. Similarly, RD0 is configured as an output port.

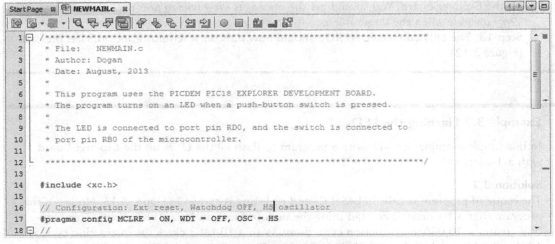

Figure 3.11: The Program Listing.

Figure 3.12: Turning the LED on.

- PORTD I/O functions are enabled by setting MEMCON bit EBDIS (see the PIC18F8722 data sheet).
- The program then enters an infinite loop where switch S1 is checked. Whenever the switch is pressed (i.e. when S1 becomes 0), the LED is turned on.

Step 11. Compile the program by clicking the *Build Main Project* button (shown as a hammer). The program should compile successfully and *Loading completed* message should be displayed.

Step 12. Connect the PICkit 3 programmer/debugger to the Explorer board. Click *Make and Program Device Main Project* button to load the program to the target microcontroller on the Explorer board. You should get the messages *Programming* and then *Programming/ Verify complete* when the target microcontroller is programmed.

Step 13. The LED connected to RD0 should now turn on when push-button S1 is pressed (Figure 3.12).

Example 3.2 Flashing the LEDs

In this simple example, we will write a program to flash all the LEDs on the Explorer board with a 1-s interval.

Solution 3.2

The required program is called FLASH.C, and its listing is shown in Figure 3.13. Notice in this program that a 1-s delay is created using the built-in function Delay10KTCYx(*n*). This function creates a 10,000*n instruction cycle delay. With a 10-MHz clock the instruction cycle is $10/4 = 2.5$ MHz, which has the period of 0.4 μs. Thus, each unit of Delay10KTCYx

```
/**************************************************************************
 * File:   FLASH.c
 * Author: Dogan
 * Date:   August, 2013
 *
 * This program uses the PICDEM PIC18 EXPLORER DEVELOPMENT BOARD.
 * The program flashes all the LEDs on the board with 1 s interval
 *
 * The LEDs are connected to PORTD of the microcontroller
 *
 **************************************************************************/

#include <xc.h>

// Configuration: Ext reset, Watchdog OFF, HS oscillator
#pragma config MCLRE = ON, WDT = OFF, OSC = HS
//
// Define LED connections
#define LEDS PORTD
//
// Define clock frequency
#define _XTAL_FREQ 10000000

//
// Start of main program
//
int main()
{
    TRISD = 0;                      // Configure PORTD as outputs
    MEMCONbits.EBDIS = 1;           // Enable PORTD I/O functions

    for(;;)                         // Do FOREVER
    {
      LEDS = 0;                     // LEDs OFF
      Delay10KTCYx(250);            // 1 s delay
      LEDS = 0xFF;                  // LEDs ON
      Delay10KTCYx(250);            // 1 s delay
    }
}
```

Figure 3.13: Flashing all the LEDs.

corresponds to 0.4 μs × 10,000 = 4 ms. To generate a 1-s delay, the argument should be 1000/4 = 250.

Example 3.3—Running in the Debug Mode

In this section, we will see how to debug the program developed in Example 3.2. The steps for debugging the program are given below:

- Compile the program for debugging by clicking *Build for Debugging Main Project* (Figure 3.14).

Figure 3.14: Compile for Debugging.

Figure 3.15: Load the Target Microcontroller.

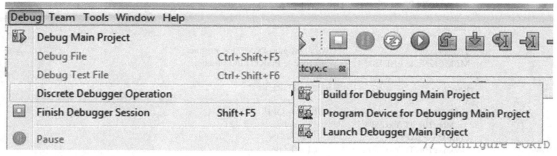

Figure 3.16: Launch the Debugger.

- Load the target microcontroller by clicking *Program Device for Debugging Main Project* (Figure 3.15).
- Start the debugger by clicking *Debug → Discrete Debugger Operation → Launch Debugger Main Project* (Figure 3.16). You should see the message *Target reset* displayed.
- Single step through the program by pressing F7. Step over the delay functions by pressing F8. As you single step through the program, you should see the LEDs turning on and off.

You can set breakpoints in the program by clicking the mouse on the numbers at the left-hand column of the program. Alternatively, breakpoints can be set by clicking Debug → New Breakpoint.

The program memory, Special Function Register (SFR), configuration bits and EE data can be observed by clicking *Window → PIC Memory Views* and then selecting the required display. Figure 3.17 shows a list of the SFR.

Watches	Trace ...	Variables	Call Stack	Breakp...	Output	Ta:

Address /	Name	Hex	Decimal	Binary	..
	TMR0_Internal	0x0000	0	00000000 00000000	..
	TMR0_Prescale	0x00	0	00000000	..
	TMR1_Internal	0x0000	0	00000000 00000000	..
	TMR1_Prescale	0x00	0	00000000	..
	TMR2_Prescale	0x00	0	00000000	..
	TMR3_Internal	0x0000	0	00000000 00000000	..

Figure 3.17: Displaying the SFR.

The program variables, breakpoints, call stack, etc. can be watched by clicking *Window - → Debugging* and selecting the required feature.

3.3.1 Programming Other Boards Using the MPLAB X

In some applications, a program may be developed using the MPLAB XC8 compiler, but the development board we are using may not be a Microchip board. Under these circumstances, we can compile the program using the MPLAB XC8 and generate the hex code. An external programming device or a development board with an on-board programmer can then be used to load the program (hex code) to the target microcontroller.

An example is given here where the program developed in Example 3.2 is loaded to the microcontroller on the popular EasyPIC V7 development board (www.mikroe.com). This board includes an ICD 3 compatible socket for programming/debugging using Microchip programming/debugging hardware tools. In this example, the ICD 3 debugger/programmer device is used to program the EasyPIC V7. We shall be using the EasyPIC V7 development board together with mikroC Pro for PIC and MPLAB XC8 compilers in the project sections of this book.

Example 3.4

In this example, we will modify the program in Example 3.2 to flash the PORTD LEDs on the EasyPIC V7 development board. This board is equipped with a PIC18F45K22 microcontroller running with an 8-MHz crystal.

Solution 3.4

Create a project as described in Example 3.1. Select the microcontroller device as PIC18F45K22 and the hardware tool as ICD 3. The required program listing is shown in Figure 3.18. Note that there are some differences in the code since a different microcontroller is used here.

```
/****************************************************************************
* File:   FLASH.c
* Author: Dogan
* Date:   August, 2013
*
* This program uses the EasyPIC V7 development board.
* The program flashes all the LEDs on the board with 1 s interval
*
* The LEDs are connected to PORTD of the microcontroller
*
****************************************************************************/

#include <xc.h>

// Configuration: Ext reset, Watchdog OFF, HS oscillator
#pragma config MCLRE = EXTMCLR, WDTEN = OFF, FOSC = HSHP
//
// Define LED connections
#define LEDS PORTD
//
// Define clock frequency
#define _XTAL_FREQ 8000000

//
// Start of main program
//
int main()
{
    TRISD = 0;                    // Configure PORTD as outputs

    for(;;)                       // Do FOREVER
    {
       LEDS = 0;                  // LEDs OFF
       Delay10KTCYx(250);         // 1 s delay
       LEDS = 0xFF;               // LEDs ON
       Delay10KTCYx(250);         // 1 s delay
    }
}
```

Figure 3.18: The Program Listing.

Compile the program as described previously. Now, we will transfer the program to the microcontroller on the EasyPIC V7 board. The steps are given below:

- Connect the USB port of ICD 3 to the PC.
- Connect the ICD 3 plug into the ICD socket on the EasyPIC V7 board.
- Turn on power to the development board.
- Load the program to the target microcontroller by clicking *Make and Program Device Main Project* in MPLAB X IDE.
- Enable the PORTD LED on the EasyPIC V7 board by setting switch SW3 PORTD to ON. You should see the LEDs flashing.

If you already have the older MPLAB IDE installed on your computer, then you may need to select the correct ICD 3 driver before loading the target microcontroller. The steps for this are as follows:

- Select All Programs → Microchip → MPLAB X IDE → MPLAB driver switcher. You should see a window as in Figure 3.19.
- Select ICD3 and MPLAB X, and click *Apply Changes* as in Figure 3.20.

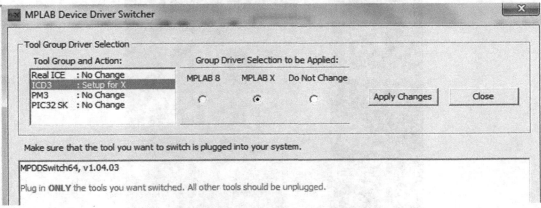

Figure 3.19: Driver Switcher Window.

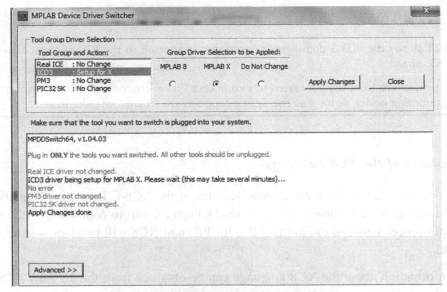

Figure 3.20: Select the ICD 3 Driver.

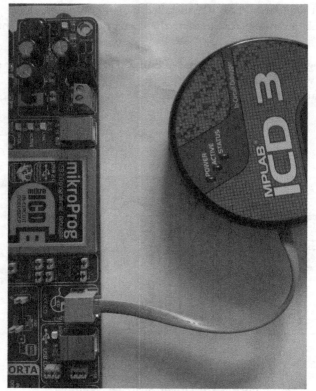

Figure 3.21: Connecting ICD 3 to the EasyPIC V7 Board.

Figure 3.21 shows the ICD 3 debugger/programmer connected to the ICD socket of the EasyPIC V7 board.

It is also possible to load the generated hex code to a PIC microcontroller using any type of PIC programming device as long as the device supports the microcontroller in use.

3.3.2 Features of the XC8 Language

In this section, we shall be looking at some features of the XC8 C language. Readers who are not familiar with the C language should read Chapter 4 before continuing with this chapter. Differences between the mikroC Pro for PIC and XC8 will be given where appropriate.

Detailed information about the XC8 language can be obtained from the *MPLAB XC8 Compiler User's Guide*, available from the Microchip Inc. web site.

Program Template

When a new XC8 program is created, the compiler generates the template shown in Figure 3.22. In our programs, we shall be modifying this template and use the one given in Figure 3.23 instead.

```
/*
 * File: newmain.c
 * Author: Dogan Ibrahim
 *
 * Created on November 6, 2013, 4:21 PM
 */
#include <stdio.h>
#include <stdlib.h>
/*
 *
 */
int main(int argc, char** argv) {
return (EXIT_SUCCESS);
}
```

Figure 3.22: XC8 Program Template Created by the Compiler.

```
/*************************************************************************
 * File:   Filename.c
 * Author: Author name
 * Date:   Date program created
 *
 * Write a brief description of the program here, including the type of
 * microcontroller used, I/O connections etc.
 *
 *
 *************************************************************************/

#include <xc.h>

// Define Configuration fuse settings
#pragma config = .........
//
// Define clock frequency
#define _XTAL_FREQ ........


//
// Start of main program
//
int main()
{
   Program body......
}
```

Figure 3.23: Modified Program Template.

An XC8 program has the following structure:

```
Program description
#include <xc.h>
Configuration bits
Oscillator frequency
Global variables
Functions
Main program
```

A single header <xc.h> must be declared at the beginning of a program to declare all compiler and device-specific types and SFRs.

Variables Types

XC8 supports the variable types shown in Table 3.1. Notice that variable type *char* on its own is same as *unsigned char*.

In addition, XC8 compiler supports 24- or 32-bit floating point variables, declared using keywords *double* and *float*.

Constants

Constant objects are read only, and they are stored in the program memory of the microcontroller. Objects that do not change their values in a program should be stored as constants to save the limited random access memory space.

Table 3.1: Variable Types Supported by XC8

Type	Size (Bits)
Bit	1
Unsigned char	8
Signed char	8
Unsigned short	16
Signed short	16
Unsigned int	16
Signed int	16
Unsigned short long	24
Signed short long	24
Unsigned long	32
Signed long	32
Unsigned long long	32
Signed long long	32

Examples of constant declarations are as follows:

```
const int Max = 10;                    // constant integer
const int Tbl[] = {0, 2, 4,6, 8};      // constant table
const Tbl[] @ 0x100 = {2, 4, 6, 8};    // constant table
```

In the last example, the table constants are stored starting from program memory location 0x100.

Persistent Qualifier

The persistent qualifier can be used to indicate that variables should not be cleared by the runtime startup code. An example is given below:

```
static persistent int x;
```

Accessing Individual I/O Pins

An individual I/O pin can be accessed by specifying the port name, followed by the word bits, then a dot symbol, and the name of the port pin. An example is given below:

```
PORTBbits.RB0 = 1;  // Set RB0 of PORTB to 1
```

Accessing Individual Bits of a Variable

Individual bits of a variable can be set or reset using the following macro definitions:

```
#define setbit(var, bit) ((var)| =(1 << (bit)))
#define clrbit(var, bit) ((var) = ~(1 << (bit)))
```

The following example sets bit 2 of variable Count:

```
setbit(Count, 2);
```

Specifying Configuration Bits

The #pragma config directive should be used to program the configuration bits for a microcontroller. An example is given below:

```
#pragma config = MCLR = ON, WDT = OFF
```

Assembly Language Instructions in C Programs

Assembly language instructions can be inserted into C programs using the asm statement. An example is given below:

```
asm("MOVLW 12");
```

Interrupt Service Routines

An interrupt service routine is recognized by the keyword *interrupt*, followed by the name of the routine. An example is given below:

```
void interrupt Myint(void)
{
     Body of the interrupt service routine
}
```

The interrupt priority can be specified after the keyword interrupt. For example,

```
void interrupt low_priority Myint(void)
{
     Body of the interrupt service routine
}
```

If variables are to be accessed between the main program and the interrupt service routine, then it is recommended to declare such variables as *volatile*.

The statements *ei()* and *di()* enable and disable global interrupts, respectively.

Program Startup

The function main() is the first function executed after Reset. However, after Reset additional code provided by the compiler, known as the startup code, is executed first. The startup code transfers control to function main(). During the startup code, the global variables with assigned values are loaded with these values, and global variables with no assigned values are cleared to zeros. A jump to address 0 (Reset) is included at the end of function main(). Thus, if a return statement is included after the last instruction in main() or if the code execution reaches the final terminating bracket at the end of main(), then the program performs a software reset. It is recommended that a loop should be added to the end of a program so that the program never performs a soft reset at the end.

MPLAB XC8 Software Library Functions

The MPLAB XC8 compiler includes a large number of software libraries that can be very useful during program development. In this section, we shall be looking at some of the commonly used library functions.

__delay_ms, __delay_us _delay, _delay3

Functions __delay_ms and __delay_us can be used to create small millisecond and microsecond delays in our programs. Before using these functions, the clock frequency should be declared using the definition _XTAL_FREQ. Assuming the clock frequency is 8 MHz, the following code generates a 20-ms delay:

```
#define _XTAL_FREQ 8000000
__delay_ms(20);
```

Function _delay is used to create delays based on the instruction cycle specified in the argument. In the following example, the delay is 20 instruction cycles:

```
_delay(20);
```

Function _delay3 is used to create delays based on 3 times the instruction cycle. In the following example, the delay is 60 instruction cycles:

```
_delay3(20);
```

__EEPROM_DATA

This function stores data in the EEPROM memory. The data must be specified in blocks of 8 bytes. An example is given below:

```
__EEPROM_DATA(0x01,0x03,0x20,0x3A,0x00,0x78,0xAA,0x02);
```

ab, labs

Returns the absolute value of an integer (abs) or a long (labs). Header file <stdlib.h> must be declared at the beginning of the program. An example is given below:

```
#define <xc.h>
#define <stdlib.h>

signed int x, y;
x = -3;              // x = -3
y = abs(x);          // y = 3
```

cos, sin, tan

These functions return the results of trigonometric functions. The argument must be in radians. The header file <math.h> must be included at the beginning of the program. An example is given below to calculate the sin of 30° and store the result in variable s:

```
#include <xc.h>
#include <math.h>

#define conv 3.14159/180.0
float s;
s = sin(30 * conv);
```

cosh, sinh, tanh

These functions implement the hyperbolic functions cosh, sinh, and tanh. The header file <math.h> must be included at the beginning of the program. An example is given below to calculate the sinh of 3.2:

```
#include <xc.h>
#include <math.h>

float s;
s = sinh(3.2);
```

acos, asin, atan, atan2

These functions return the inverses of trigonometric functions in radians. The header file <math.h> must be included at the beginning of the program.

itoa

This function converts a number into a string with the specified number base. The header file <math.h> must be included at the beginning of the program. In the following example, number 25 is converted into a string in variable *bufr* with a hexadecimal base:

```
#include <xc.h>
#include <math.h>

char bufr[5];
itoa(bufr, 25, 16);
```

log, log10

Function log returns the natural logarithm of a floating point number. The function log10 returns the logarithm to base 10. The header file <math.h> must be included at the beginning of the program.

memcmp

This function fills *n* bytes of memory with the specified byte. The header file <string.h> must be included at the beginning of the program. In the following example, *bufr* is filled with 10 character 'x' s:

```
#include <xc.h>
#include <string.h>

char bufr[10];
memset(bufr, 'x', 10);
```

rand

This is a random number generator function. It returns an integer between 0 and 32,767 that changes on each call to the function. The header file <stdlib.h> must be included at the beginning of the program. The starting point is set using function *srand*. An example is given below:

```
#include <xc.h>
#include <stdlib.h>

srand(5);
j = rand();
```

round

This function rounds the argument to the nearest integer value in floating point format. The header file <math.h> must be included at the beginning of the program. An example is given below:

```
#include <xc.h>
#include <math.h>

double rnd;
rnd = round(23.456);
```

SLEEP

Used to put the microcontroller into the sleep mode.

sqrt

This function calculates the square root of a floating point number. The header file <math.h> must be included at the beginning of the program.

String Functions

Some of the string functions provided are as follows:

```
Strcat, strncat:     string concatenate
Strchr, strrchr:     string search
Strcmp, strncmp:     string compare
Strcpy, strncpy:     string copy
Strlen:              string length
Strstr, Strpbrk:     occurrence of a character in a string
```

tolower, toupper, toascii

Convert a lower case character to the upper case character, upper case character to the lower case character, and to ASCII.

trunc

This function rounds the argument to the nearest integer. The header file <math.h> must be included at the beginning of the program.

MPLAB XC8 Peripheral Libraries

In addition to the useful functions XC8 compiler offers a number of peripheral libraries that can be useful while developing complex projects using peripheral devices. Some of these libraries are for an LCD, SD card, USB port, CAN bus, I^2C bus, SPI bus, and so on.

3.4 Summary

This chapter has described the basic features of the MPLAB XC8 C compiler. Step-by-step examples are given to show how to use the MPLAB X IDE to create a project and how to load the executable code to the target microcontroller.

Examples are given to show how to load the target microcontrollers on the two popular development boards: the PICDEM 18 Explorer board and the EasyPIC V7 development board.

Finally, a list of some commonly used MPLAB XC8 functions are given. Interested readers can obtain further details from the *MPLAB XC8 Compiler User's Guide*.

3.5 Exercises

1. Write an XC8 C program to set bits 0 and 7 of PORT C to logic 1.

2. Write an XC8 C program to count down continuously and send the count to PORTB.

3. Write a C program to multiply each element of a 10-element array with number 2.

4. Explain how the individual bits of a port can be accessed. Write the code to clear bit 3 of PORTB.

5. Write an XC8 C program to calculate the average value of the numbers stored in an array. Assume that the array is called **M** and that it has 20 elements.

6. Write a function to convert inches to centimeters. The function should receive inches as a floating point number and then calculate the equivalent centimeters.

7. An LED is connected to port pin RB7 of a PIC18F8722 microcontroller. Write a program to flash the LED such that the ON time is 5 s, and the OFF time is 3 s.

8. Write a program to calculate the trigonometric sine of angles from 0 to 90° in steps of 10°. Store the results in a floating point array.

9. Explain how a compiled XC8 C program can be downloaded to the target microcontroller on the PICDEM 18 Explorer board.

10. Explain how a program can be debugged using the PICkit 3 programmer/debugger and the PICDEM 18 Explorer board.

11. Explain how a compiled XC8 program can be downloaded to the target microcontroller on the EasyPIC V7 development board.

Microcontroller Program Development

Before writing a program, it is always helpful first to think and plan the structure of the program. Although simple programs can easily be developed by writing the code directly without any preparation, the development of complex programs almost always become easier if an algorithm is first derived. Once the algorithm is ready, coding the actual program is not a difficult task.

A program's algorithm can be described in a variety of graphical and text-based methods, such as flow charts, structure charts, data flow diagrams, and program description languages (PDLs). The problem with graphical techniques is that it can be very time consuming to draw shapes with text inside them. Also, it is a tedious task to modify an algorithm described using graphical techniques.

Flow charts can be very useful to describe the flow of control and data in small programs where there are only a handful of diagrams, usually not extending beyond a page or two. The PDL can be useful to describe the flow of control and data in small-to-medium size programs. The main advantage of the PDL description is that it is very easy to modify a given PDL since it only consists of text.

In this book, we will mainly be using the PDL, but flow charts will also be given where it is felt to be useful. The next few sections briefly describe the basic building blocks of the

PDL and its equivalent flow charts. It is left to the readers to decide which method to use during the development of their programs.

4.1 Using the PDL and Flow Charts

PDL is a free-format English-like text that describes the flow of control and data in a program. PDL is not a programming language. It is a collection of some keywords that enable a programmer to describe the operation of a program in a stepwise and logical manner. In this section, we will look at the basic PDL statements and their flow chart equivalents. The superiority of the PDL over flow charts will become obvious when we have to develop medium-to-large programs.

4.1.1 BEGIN–END

Every PDL program description should start with a BEGIN and end with an END statement. The keywords in a PDL description should be highlighted to make the reading easier. The program statements should be indented and described between the PDL keywords. An example is shown in Figure 4.1 together with the equivalent flow diagram.

4.1.2 Sequencing

For normal sequencing, the program statements should be written in English text to describe the operations to be performed. An example is shown in Figure 4.2 together with the equivalent flow chart.

4.1.3 IF–THEN–ELSE–ENDIF

IF, THEN, ELSE, and ENDIF should be used to conditionally change the flow of control in a program. Every IF keyword should be terminated with a THEN, and every IF block should be terminated with an ENDIF keyword. The use of the ELSE statement is optional and depends on the application. Figure 4.3 shows an example of using IF–THEN–ENDIF, while Figure 4.4 shows the use of IF–THEN–ELSE–ENDIF statements in a program and their equivalent flow charts.

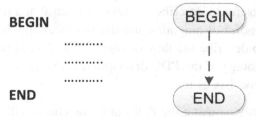

Figure 4.1: BEGIN–END Statement and Equivalent Flow Chart.

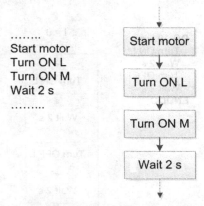

```
........
Start motor
Turn ON L
Turn ON M
Wait 2 s
........
```

Figure 4.2: Sequencing and Equivalent Flow Chart.

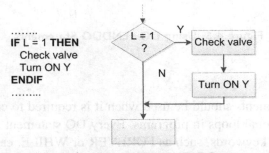

```
........
IF L = 1 THEN
    Check valve
    Turn ON Y
ENDIF
........
```

Figure 4.3: Using IF—THEN—ENDIF Statements.

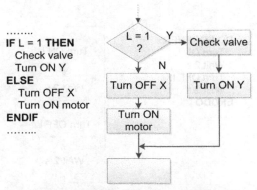

```
........
IF L = 1 THEN
    Check valve
    Turn ON Y
ELSE
    Turn OFF X
    Turn ON motor
ENDIF
........
```

Figure 4.4: Using IF—THEN—ELSE—ENDIF Statements.

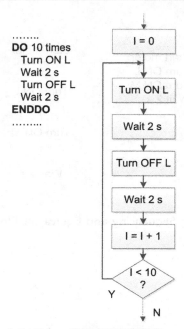

```
........
DO 10 times
  Turn ON L
  Wait 2 s
  Turn OFF L
  Wait 2 s
ENDDO
........
```

Figure 4.5: Using DO—ENDDO Statements.

4.1.4 DO—ENDDO

The DO—ENDDO statements should be used when it is required to create iterations, or conditional or unconditional loops in programs. Every DO statement should be terminated with an ENDDO. Other keywords, such as FOREVER or WHILE, can be used after the DO statement to indicate an endless loop or a conditional loop, respectively. Figure 4.5 shows an example of a DO—ENDDO loop executed 10 times. Figure 4.6 shows an endless loop created using the FOREVER statement. The flow chart equivalents are also shown in the figures.

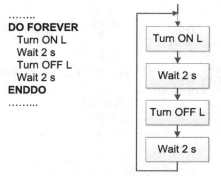

```
........
DO FOREVER
  Turn ON L
  Wait 2 s
  Turn OFF L
  Wait 2 s
ENDDO
........
```

Figure 4.6: Using DO—FOREVER Statements.

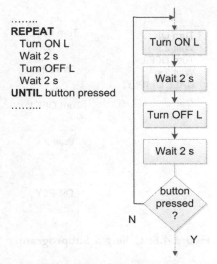

```
........
REPEAT
   Turn ON L
   Wait 2 s
   Turn OFF L
   Wait 2 s
UNTIL button pressed
........
```

Figure 4.7: Using REPEAT—UNTIL Statements.

4.1.5 REPEAT—UNTIL

REPEAT—UNTIL is similar to DO—WHILE, but here the statements enclosed by the REPEAT—UNTIL block are executed at least once, while the statements enclosed by DO—WHILE may not execute at all if the condition is not satisfied just before entering the DO statement. An example is shown in Figure 4.7, with the equivalent flow chart.

4.1.6 Calling Subprograms

In some applications, a program consists of a main program and a number of subprograms (or functions). A subprogram activation in PDL should be shown by adding the CALL statement before the name of the subprogram. In flow charts, a rectangle with vertical lines at each side should be used to indicate the invocation of a subprogram. An example call to a subprogram is shown in Figure 4.8 for both a PDL description and a flow chart. Optionally, the input—output data to a function can be listed if desired. The following example shows how the temperature can be passed to function DISPLY as an input:

```
CALL DISPLY(I: temperature)
```

In the following function call the temperature is passed to the function called CONV. The function formats the temperature for display and returns it to the calling program:

```
CALL CONV(I: temperature, O: formatted temperature)
```

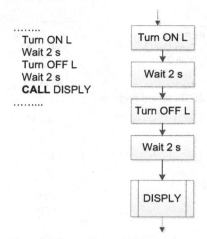

```
........
    Turn ON L
    Wait 2 s
    Turn OFF L
    Wait 2 s
    CALL DISPLY
........
```

Figure 4.8: Calling a Subprogram.

4.1.7 Subprogram Structure

A subprogram should begin and end with the keywords BEGIN/name and END/name, respectively, where *name* is the name of the subprogram. In flow chart representation, a horizontal line should be drawn inside the BEGIN box, and the name of the subprogram should be written at the lower half of the box. An example subprogram structure is shown in Figure 4.9 for both a PDL description and a flow chart.

Interrupt service routines can be shown using the same method, but the keyword ISR can be inserted in front of the function name to identify that the function is actually an

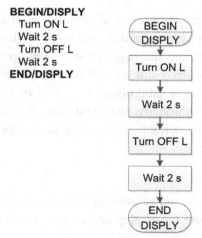

```
BEGIN/DISPLY
    Turn ON L
    Wait 2 s
    Turn OFF L
    Wait 2 s
END/DISPLY
```

Figure 4.9: Subprogram Structure.

interrupt service routine. For example, in Figure 4.9, assuming that function DISPLY is an interrupt service routine, the function body can be written as

```
BEGIN/ISR:DISPLY
    Turn ON L
    Wait 2 s
    Turn OFF L
    Wait 2 s
END/ISR:DISPLY
```

4.2 Examples

Some examples are given in this section to show how the PDL and flow charts can be used in program development.

Example 4.1

It is required to write a program to convert hexadecimal numbers "A" to "F" into the decimal format. Show the algorithm using a PDL and also draw the flow chart. Assume that the number to be converted is called HEX_NUM, and the output number is called DEC_NUM.

Solution 4.1
The required PDL is

```
BEGIN
    IF HEX_NUM = "A" THEN
      DEC_NUM = 10
    ELSE IF HEX_NUM = "B" THEN
      DEC_NUM = 11
    ELSE IF HEX_NUM = "C" THEN
      DEC_NUM = 12
    ELSE IF HEX_NUM = "D" THEN
      DEC_NUM = 13
    ELSE IF HEX_NUM = "E" THEN
      DEC_NUM = 14
    ELSE IF HEX_NUM = "F" THEN
      DEC_NUM = 15
    ENDIF
END
```

The required flow chart is shown in Figure 4.10. Note that it is much easier to write PDL statements than it is to draw flow chart shapes and write text inside them.

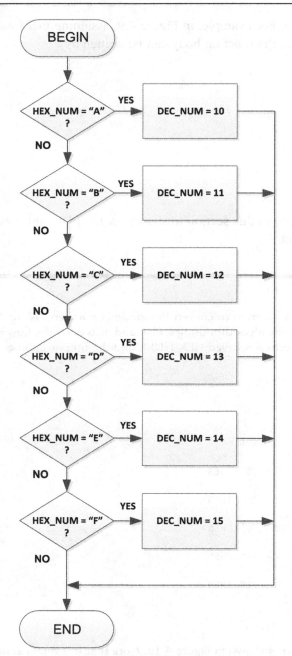

Figure 4.10: Flow Chart Solution.

Example 4.2

The PDL of part of a program is given as follows:

```
J = 0
M = 0
DO WHILE J < 10
    DO WHILE M < 20
        Flash the LED
        Increment M
    ENDDO
    Increment J
ENDDO
```

Show how this PDL can be implemented by a flow chart.

Solution 4.2

The required flow chart is shown in Figure 4.11. Here again, note how complicated the flow chart can be even for a simple nested DO WHILE loop.

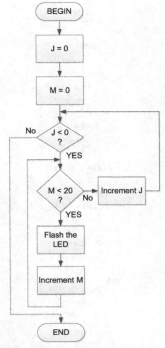

Figure 4.11: Flow Chart Solution.

Example 4.3

It is required to write a program to calculate the sum of integer numbers between 1 and 100. Show the algorithm using a PDL and also draw the flow chart. Assume that the sum will be stored in a variable called SUM.

Solution 4.3

The required PDL is

```
BEGIN
    SUM = 0
    I = 1
    DO 100 TIMES
        SUM = SUM + I
        Increment I
    ENDDO
END
```

The required flow chart is shown in Figure 4.12.

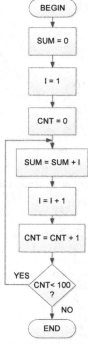

Figure 4.12: Flow Chart Solution.

Example 4.4

It is required to write a program to calculate the sum of all the even numbers between 1 and 10 inclusive. Show the algorithm using a PDL and also draw the flow chart. Assume that the sum will be stored in a variable called SUM.

Solution 4.4
The required PDL is

```
BEGIN
    SUM = 0;
    CNT = 1
    REPEAT
        IF CNT is even number THEN
            SUM = SUM + CNT
        ENDIF
        INCREMENT CNT
        UNTIL CNT > 10
END
```

The required flow chart is shown in Figure 4.13. Note how complicated the flow chart can be for a very simple problem such as this.

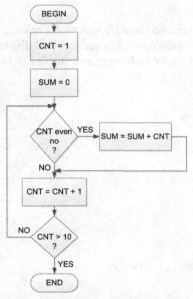

Figure 4.13: Flow Chart Solution.

Example 4.5

It is required to write a program to turn ON a light emitting diode (LED) when a button is pressed and to turn it OFF when the button is released. Assuming that initially the LED is to be OFF, write the PDL statements for this example.

Solution 4.5

The required PDL statements are as follows:

```
BEGIN
    Turn OFF LED
    DO FOREVER
        IF Button is pressed THEN
            Turn ON LED
        ELSE
            Turn OFF LED
        ENDIF
    ENDDO
END
```

Example 4.6

A temperature sensor is connected to the A/D input of a microcontroller. It is required to write a program to read the temperature every second and display it on a liquid crystal display (LCD). Use function DISPLAY to format and display the temperature. Show the PDL statements for this example.

Solution 4.6

The required PDL statements are

```
BEGIN
    Configure the A/D port
    DO FOREVER
        Read Temperature from A/D port
        CALL DISPLAY
        Wait 1 s
    ENDDO
END
BEGIN/DISPLAY
    Format temperature for LCD display
    Display temperature on LCD
END/DISPLAY
```

4.3 Representing for Loops in Flow Charts

Most programs include some form of iteration or looping. One of the easiest ways to create a loop in a C program is by using the *for* statement. This section shows how a *for* loop can be represented in a flow chart. As shown below, there are several methods of representing a *for* loop in a flow chart.

Suppose that we have a *for* loop as shown below and we wish to draw an equivalent flow chart.

```
for(m = 0; m < 10; m++)
{
    Cnt = Cnt + 2* m;
}
```

4.3.1 Method 1

Figure 4.14 shows one of the methods for representing the above *for* loop as with a flow chart. Here, the flow chart is drawn using the basic primitive components.

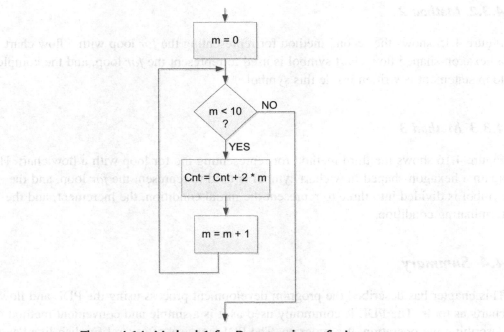

Figure 4.14: Method 1 for Representing a *for* Loop.

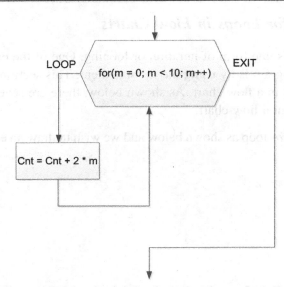

Figure 4.15: Method 2 for Representing a *for* Loop.

4.3.2 Method 2

Figure 4.15 shows the second method for representing the *for* loop with a flow chart. Here, a hexagon-shaped flow chart symbol is used to represent the *for* loop, and the complete *for* loop statement is written inside this symbol.

4.3.3 Method 3

Figure 4.16 shows the third method for representing the for loop with a flow chart. Here, again a hexagon-shaped flow chart symbol is used to represent the *for* loop, and the symbol is divided into three to represent the initial condition, the increment, and the terminating condition.

4.4 Summary

This chapter has described the program development process using the PDL and flow charts as tools. The PDL is commonly used as it is a simple and convenient method of describing the operation of a program. The PDL consists of several English-like keywords. Although the flow chart is also a useful tool, it can be very tedious in large programs to draw shapes and write text inside them.

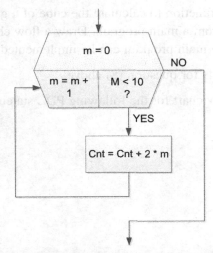

Figure 4.16: Method 3 for Representing a *for* Loop.

4.5 Exercises

1. Describe the various shapes used in drawing flow charts.

2. Describe how the various keywords used in PDL can be used to describe the operation of a program.

3. What are the advantages and disadvantages of flow charts?

4. It is required to write a program to calculate the sum of numbers from 1 to 10. Draw a flow chart to show the algorithm for this program.

5. Write the PDL statements for question (4) above.

6. It is required to write a program to calculate the roots of a quadratic equation, given the coefficients. Draw a flow chart to show the algorithm for this program.

7. Write the PDL statements for question (6) above.

8. Draw the equivalent flow chart for the following PDL statements:

   ```
   DO WHILE count < 10
     Increment J
     Increment count
   ENDDO
   ```

9. It is required to write a function to calculate the sum of numbers from 1 to 10. Draw a flow chart to show how the function subprogram and the main program can be implemented.

10. Write the PDL statements for question (9) above.

11. It is required to write a function to calculate the cube of a given integer number and then call this function from a main program. Draw a flow chart to show how the function subprogram and the main program can be implemented.

12. Write the PDL statements for question (8) above.

13. Draw the equivalent flow chart for the following PDL statements:
```
BEGIN
  J = 0
  K = 0
  REPEAT
    Flash LED A
    Increment J
    REPEAT
      Flash LED B
      Increment K
    UNTIL K = 10
  UNTIL J > 15
END
```

14. It is required to write a function to convert meters into inches and then call this function from a main program. Draw a flow chart to show how the function subprogram and the main program can be implemented.

15. Write the PDL statements for question (14) above.

16. Draw the equivalent flow chart for the following PDL statements:
```
BEGIN
  Configure I/O ports
  Turn OFF motor
  Turn OFF buzzer
  DO FOREVER
    IF button 1 is pressed THEN
      Turn ON motor
      IF button 2 is pressed THEN
        Turn ON buzzer
        Wait 3 s
        Turn OFF buzzer
      ENDIF
    ELSE
      Wait for 10 s
      Turn OFF motor
    ENDIF
  ENDDO
END
```

Simple PIC18 Projects

In this chapter, we will be looking at the design of simple PIC18 microcontroller-based projects, with the idea of becoming familiar with basic interfacing techniques and learning how to use the various microcontroller peripheral registers. We will look at the design of projects using light emitting diodes (LEDs), push-button switches, keyboards, LED arrays, liquid crystal displays (LCDs), sound devices, and so on. We will be developing programs in C language using both the *mikroC Pro for PIC* and the *MPLAB XC8* compilers. The fully tested and working code will be given for both compilers. All the projects given in this chapter can easily be built on a simple breadboard, but we will be using the low-cost and highly popular development board EasyPIC V7. The required board jumper settings will be given where necessary. We will start with very simple projects and then proceed to more complex ones. It is recommended that the reader moves through the projects in the given order to benefit the most.

The following are provided for each project:

- Project name,
- Description of the project,
- Block diagram of the project (where necessary),
- Circuit diagram of the project,
- Description of the hardware,
- Algorithm description (in PDL),
- Program listing (mikroC pro for PIC and MPLAB XC8),
- Suggestions for further development.

Project 5.1—Chasing LEDs

Project Description

In this project, eight LEDs are connected to PORTC of a PIC18F45K22-type microcontroller, and the microcontroller is operated from an 8-MHz crystal. When the power is applied to the microcontroller (or when the microcontroller is reset), the LEDs turn ON alternately in an anticlockwise manner where only one LED is ON at any time. A 1-s delay is used between each output so that the LEDs can be seen turning ON and OFF.

An LED can be connected to a microcontroller output port in two different modes: the *current-sinking* mode and the *current-sourcing* mode.

Current Sinking

As shown in Figure 5.1, in the current-sinking mode, the anode leg of the LED is connected to the +5-V supply, and the cathode leg is connected to the microcontroller output port through a current limiting resistor.

The voltage drop across an LED varies between 1.4 and 2.5 V, with a typical value of 2 V. The brightness of the LED depends on the current through the LED, and this current can vary between 8 and 16 mA, with a typical value of 10 mA.

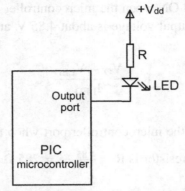

Figure 5.1: LED Connected in the Current-Sinking Mode.

The LED is turned ON when the output of the microcontroller is at logic 0 so that current flows through the LED. Assuming that the microcontroller output voltage is about 0.4 V when the output is low, we can calculate the value of the required resistor as follows:

$$R = \frac{V_S - V_{LED} - V_L}{I_{LED}}, \tag{5.1}$$

where

V_S is the supply voltage (5 V),
V_{LED} is the voltage drop across the LED (2 V),
V_L is the maximum output voltage when the output port is low (0.4 V),
I_{LED} is the current through the LED (10 mA).

By substituting the values into Eqn (5.1), we get $R = \frac{5-2-0.4}{10} = 260\ \Omega$. The nearest physical resistor is 270 Ω.

Current Sourcing

As shown in Figure 5.2, in the current-sourcing mode, the anode leg of the LED is connected to the microcontroller output port, and the cathode leg is connected to the ground through a current limiting resistor.

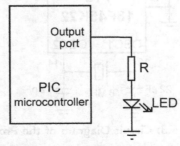

Figure 5.2: LED Connected in the Current-Sourcing Mode.

In this mode, the LED is turned ON when the microcontroller output port is at logic 1, that is, $+5$ V. In practice, the output voltage is about 4.85 V, and the value of the resistor can be determined as follows:

$$R = \frac{V_O - V_{LED}}{I_{LED}}, \tag{5.2}$$

where

V_O is the output voltage of the microcontroller port when at logic 1 ($+4.85$ V).

Thus, the value of the required resistor is $R = \frac{4.85 - 2}{10} = 285\ \Omega$. The nearest physical resistor is 290 Ω.

Project Hardware

The circuit diagram of the project is shown in Figure 5.3. LEDs are connected to PORTC in the current-sourcing mode with eight 290-Ω resistors. An 8-MHz crystal is

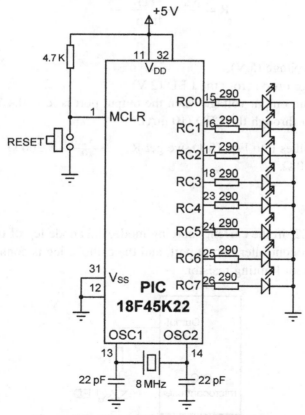

Figure 5.3: Circuit Diagram of the Project.

```
BEGIN
      Configure PORTC pins as digital output
      Initialize J = 1
      DO FOREVER
            Set PORTC = J
            Shift left J by 1 digit
            IF J = 0 THEN
                  J = 1
            ENDIF
            Wait 1 s
      ENDDO
END
```

Figure 5.4: PDL of the Project.

connected between the OSC1 and OSC2 pins of the microcontroller. Also, an external reset push button is connected to the Master Clear (MCLR) input to reset the microcontroller when required.

If you are using the EasyPIC V7 development board, then make sure that the following jumper is configured:

DIP switch SW3: PORTC ON

Project PDL

The PDL of this project is very simple and is given in Figure 5.4.

Project Program

mikroC Pro for PIC

The mikroC Pro for PIC program is named MIKROC-LED1.C, and the program listing is given in Figure 5.5. At the beginning of the program, PORTC pins are configured as outputs by setting TRISC = 0. Then, an endless **for** loop is formed, and the LEDs are turned ON alternately in an anticlockwise manner to give the chasing effect. The program checks continuously so that when LED 7 is turned ON the next LED to be turned ON is LED 0.

The program is compiled using the mikroC compiler. Project settings should be configured to an 8-MHz clock, XT crystal mode, and WDT OFF. The HEX file (MIKROC-LED1.HEX) should be loaded to the PIC18F45K22 microcontroller using an in-circuit debugger, a programming device, or the EasyPIC V7 development board.

When using the mikroC Pro for PIC compiler, the configuration fuses can be modified from the Edit Project window that is entered by clicking Project → Edit Project.

```
/***************************************************************************
                        CHASING LEDS
                        ============

In this project 8 LEDs are connected to PORTC of a PIC18F45K22 microcontroller and the
microcontroller is operated from an 8 MHz crystal.

The program turns ON the LEDs in an anticlockwise manner with 1 s delay between
each output. The net result is that LEDs seem to be chasing each other.


Author:         Dogan Ibrahim
Date:           August 2013
File:           MIKROC-LED1.C
***************************************************************************/

void main()
{
    unsigned char J = 1;

    ANSELC = 0;                         // Configure PORTC as digital
    TRISC = 0;                          // Configure PORTC as output
    for(;;)                             // Endless loop
    {
        PORTC = J;                      // Send J to PORTC
        Delay_ms(1000);                 // Delay 1 s
        J = J << 1;                     // Shift left J
        if(J == 0) J = 1;               // If last LED, move to first one
    }
}
```

Figure 5.5: mikroC Pro for PIC Program Listing.

MPLAB XC8

The MPLAB XC8 program is named XC8-LED1.C, and the program listing is given in
Figure 5.6. The program is basically the same as in Figure 5.5, except that here a 1-s delay is
created using the basic XC8 __delay_ms function in a loop as it is not possible to create large
delays using the __delay_ms function. Function Delay_Seconds creates delay in seconds
where the amount of delay is specified by the argument of the function. Note also that the
header file <xc.h> must be included at the beginning of the program. Also, the MPLAB IDE
must be configured for the PIC18F45K22 type microcontroller and In-Circuit Debugger
(ICD) 3 device (hardware tool). The ICD 3 device should be connected to the ICD socket on
the EasyPIC V7 development board (top middle part, labeled as EXT ICD). The generated
code can then be loaded to the target microcontroller using the MPLAB IDE (see Chapter 5
for more details).

When using the MPLAB XC8 compiler, the configuration fuses can be modified by
specifying the "#pragma config" statements at the beginning of the program. It is important to
note that different PIC microcontrollers have different sets of configuration fuses. Appendix
A gives a list of the valid configuration fuses for the PIC18F45K22 microcontroller.

```
/****************************************************************************
                            CHASING LEDS
                            ============

In this project 8 LEDs are connected to PORTC of a PIC18F45K22 microcontroller and the
microcontroller is operated from an 8 MHz crystal.

The program turns ON the LEDs in an anticlockwise manner with 2 s delay between
each output. The net result is that LEDs seem to be chasing each other.


Author:         Dogan Ibrahim
Date:           August 2013
File:           XC8-LED1.C
****************************************************************************/
#include <xc.h>
#pragma config MCLRE = EXTMCLR, WDTEN = OFF, FOSC = HSHP
#define _XTAL_FREQ 8000000

//
// This function creates seconds delay. The argument specifies the delay time in seconds.
//
void Delay_Seconds(unsigned char s)
{
   unsigned char i,j;

   for(j = 0; j < s; j++)
   {
      for(i = 0; I <  100; i++)__delay_ms(10);
   }
}

void main()
{
   unsigned char J = 1;

   ANSELC = 0;                          // Configure PORTC as digital
   TRISC = 0;                           // Configure PORTC as output

   for(;;)                              // Endless loop
   {
      PORTC = J;                        // Send J to PORTC
      Delay_Seconds(1);                 // Delay 1 s
      J = J << 1;                       // Shift left J
      if(J == 0) J = 1;                 // If last LED, move to first one
   }
}
```

Figure 5.6: MPLAB XC8 Program Listing.

Further Development

The project can be modified such that the LEDs chase each other in both directions. For example, while moving in an anticlockwise direction, when LED RB7 is ON, the direction can be changed so that the next LED to turn ON is RB6, RB5, and so on.

Project 5.2—Complex Flashing LED
Project Description

In this project, one LED is connected to port pin RC0 (PORTC bit 0) of a PIC18F45K22-type microcontroller, and the microcontroller is operated from an 8-MHz crystal. The LED flashes continuously with the following pattern:

3 flashes with 200 ms delay between each flash
2 s delay

Project Hardware

The circuit diagram of the project is shown in Figure 5.7. An LED is connected to port pin RC0 in the current-sourcing mode with eight 290-Ω resistors. An 8-MHz crystal is connected between the OSC1 and OSC2 pins of the microcontroller. Also, an external reset push button is connected to the MCLR input to reset the microcontroller when required.

If you are using the EasyPIC V7 development board, then make sure that the following jumper is configured:

DIP switch SW3: PORTC ON

Project PDL

The PDL of this project is very simple and is given in Figure 5.8.

Project Program
mikroC Pro for PIC

The mikroC Pro for PIC program is named MIKROC-LED2.C, and the program listing is given in Figure 5.9. Using a *for* loop, the LED is flashed three times with a 200-ms delay between each flash. Then, this process is repeated after 2 s of delay.

MPLAB XC8

The MPLAB XC8 program is named XC8-LED2.C, and the program listing is given in Figure 5.10. The program uses a function called DelayMs to create milliseconds of delay

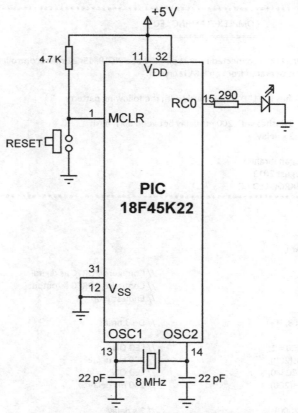

Figure 5.7: Circuit Diagram of the Project.

```
BEGIN
        Configure PORTC pins as digital output
        DO FOREVER
                DO 3 times
                        Turn ON LED
                        Wait 200 ms
                        Turn OFF LED
                        Wait 200 ms
                ENDDO
                Wait 2 s
        ENDDO
END
```

Figure 5.8: PDL of the Project.

```
/*******************************************************************************
                    COMPLEX FLASHING LED
                    ====================

In this project an LEDs is connected to port pin RC0 of a PIC18F45K22 microcontroller and the
microcontroller is operated from an 8 MHz crystal.

The program flashes the LED continuously with the following pattern:

                    3 flashes with 200 ms delay between each flash
                    2 s delay

Author:     Dogan Ibrahim
Date:       August 2013
File:       MIKROC-LED2.C
*******************************************************************************/

void main()
{
   unsigned char i;

   ANSELC = 0;                          // Configure PORTC as digital
   TRISC = 0;                           // Configure PORTC as output
   for(;;)                              // Endless loop
   {
     for(i = 0; i < 3; i++)             // Do 3 times
     {
       PORTC.RC0 = 1;                   // LED ON
       Delay_ms(200);                   // 200 ms delay
       PORTC.RC0 = 0;                   // LED OFF
       Delay_ms(200);                   // 200 ms delay
     }
     Delay_ms(2000);                    // 2 s delay
   }
}
```

Figure 5.9: mikroC Pro for PIC Program Listing.

from 1 to 65535 ms (maximum value of an unsigned integer) where the required delay is passed in the function argument.

Project 5.3—Random Flashing LEDs
Project Description

In this project, eight LEDs are connected to PORTC of a PIC18F45K22-type microcontroller, and the microcontroller is operated from an 8-MHz crystal. An integer random number is generated between 1 and 255 every second, and the LEDs are turned ON to indicate this number in binary. The net result is that the LEDs flash in a random fashion, and it is interesting to watch them flashing.

```
/******************************************************************************
                    COMPLEX FLASHING LED
                    ====================

In this project an LEDs is connected to port pin RC0 of a PIC18F45K22 microcontroller and the
the microcontroller is operated from an 8 MHz crystal.

The program flashes the LED continuously with the following pattern:

    3 flashes with 200 ms delay between each flash
    2 s delay

Author:    Dogan Ibrahim
Date:      August 2013
File:      XC8-LED2.C
******************************************************************************/
#include <xc.h>
#pragma config MCLRE = EXTMCLR, WDTEN = OFF, FOSC = HSHP
#define _XTAL_FREQ 8000000

//
// This function creates milliseconds delay. The argument specifies the delay time.
// The delay can be 1 to 65535 ms
//
void DelayMs(unsigned int ms)
{
    unsigned int j;

    for(j = 0; j < ms; j++)__delay_ms(1);
}

void main()
{
    unsigned char i;

    ANSELC = 0;                         // Configure PORTC as digital
    TRISC = 0;                          // Configure PORTC as output
    for(;;)                             // Endless loop
    {
        for(i = 0; i < 3; i++)          // Do 3 times
        {
            PORTCbits.RC0 = 1;          // LED ON
            DelayMs(200);               // 200 ms delay
            PORTCbits.RC0 = 0;          // LED OFF
            DelayMs(200);               // 200 ms delay
        }
        DelayMs(2000);
    }
}
```

Figure 5.10: MPLAB XC8 Program Listing.

Project Hardware

The circuit diagram of the project is as shown in Figure 5.3.

If you are using the EasyPIC V7 development board, then make sure that the following jumper is configured:

DIP switch SW3: PORTC ON

Project PDL

The PDL of this project is very simple and is given in Figure 5.11.

Project Program

mikroC Pro for PIC

The mikroC Pro for PIC program is named MIKROC-LED3.C and the program listing is given in Figure 5.12. The random number is generated using the built-in library function rand. The function should be initialized once by calling function srand with an integer argument before it is used (any integer number can be used). Then, every time rand is called, it generates a pseudorandom number between 0 and 32,767. To generate a number between 1 and 255, we can divide the generated number by 128. Although the generated number is not exactly random, it is good enough for our application.

The random number seed is initialized by calling function srand with integer 10. Then, an endless *for* loop is established, a random number is generated in variable p, divided by 128, and sent to PORTC. This process is repeated with a 1-s delay between each output.

MPLAB XC8

The MPLAB XC8 program is named XC8-LED3.C, and the program listing is given in Figure 5.13. As in Figure 5.6, the required 1-s delay is created using a function called Delay_Seconds. Random numbers are generated as in the mikroC Pro for PIC version of the program. Note that the header file <stdlib.h> must be included at the beginning of the program.

```
BEGIN
        Configure PORTC pins as digital output
        Initialize random number seed
        DO FOREVER
                Get a random number between 1 and 255
                Send the random number to PORTC
                Wait 1 s
        ENDDO
END
```

Figure 5.11: PDL of the Project.

```
/*******************************************************************************
                    RANDOM FLASHING LEDS
                    ====================

In this project 8 LEDs are connected to PORTC of a PIC18F45K22 microcontroller and the
microcontroller is operated from an 8 MHz crystal.

The program uses a pseudorandom number generator to generate a number between 0 and
32767. This number is then divided by 128 to limit it between 1 and 255. The resultant number
is sent to PORTC of the microcontroller. This process is repeated every second.


Author:       Dogan Ibrahim
Date:         August 2013
File:         MIKROC-LED3.C
*******************************************************************************/

void main()
{
    unsigned int p;

    ANSELC = 0;                             // Configure PORTC as digital
    TRISC = 0;                              // Configure PORTC as output
    srand(10);                              // Initialize random number seed

    for(;;)                                 // Endless loop
    {
        p = rand();                         // Generate a random number
        p = p / 128;                        // Number between 1 and 255
        PORTC = p;                          // Send to PORTC
        Delay_ms(1000);                     // 1 s delay
    }
}
```

Figure 5.12: mikroC Pro for PIC Program Listing.

Project 5.4—Logic Probe

Project Description

This project is a simple logic probe. A logic probe is used to indicate the logic status of an
unknown digital signal. In a typical application, a test lead (probe) is used to detect the
unknown signal, and two different color LEDs are used to indicate the logic status. If, for
example, the signal is logic 0, then the RED color LED is turned ON. If on the other hand
the signal is logic 1, then the GREEN LED is turned ON.

Project Hardware

The circuit diagram of the project is as shown in Figure 5.14. Port pin RC0 is used as the
probe input. Port pins RB6 and RB7 are connected to RED LED and GREEN LED,
respectively.

```
/********************************************************************************
                        RANDOM FLASHING LEDS
                        ====================
```

In this project 8 LEDs are connected to PORTC of a PIC18F45K22 microcontroller and the
microcontroller is operated from an 8 MHz crystal.

The program uses a pseudorandom number generator to generate a number between 0 and
32767. This number is then divided by 128 to limit it between 1 and 255. The resultant number
is sent to PORTC of the microcontroller. This process is repeated every second.


```
Author:      Dogan Ibrahim
Date:        August 2013
File:        XC8-LED3.C
********************************************************************************/
#include <xc.h>
#include <stdlib.h>
#pragma config MCLRE = EXTMCLR, WDTEN = OFF, FOSC = HSHP
#define _XTAL_FREQ 8000000

//
// This function creates seconds delay. The argument specifies the delay time in seconds.
//
void Delay_Seconds(unsigned char s)
{
  unsigned char i,j;

  for(j = 0; j < s; j++)
  {
    for(i = 0; i <  100; i++)__delay_ms(10);
  }
}


void main()
{
  unsigned int p;

  ANSELC = 0;                          // Configure PORTC as digital
  TRISC = 0;                           // Configure PORTC as output
  srand(10);                           // Initialize random number seed

  for(;;)                              // Endless loop
  {
    p = rand();                        // Generate a random number
    p = p/128;                         // Number between 1 and 255
    PORTC = p;                         // Send to PORTC
    Delay_Seconds(1);                  // 1 s delay
  }
}
```

Figure 5.13: MPLAB XC8 Program Listing.

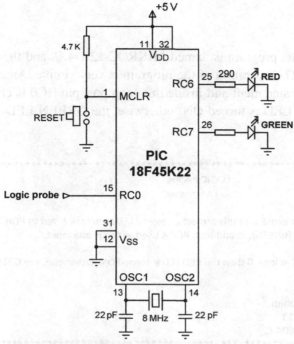

Figure 5.14: Circuit Diagram of the Project.

If you are using the EasyPIC V7 development board, then make sure that the following jumper is configured:

DIP switch SW3: PORTC ON

Project PDL

The PDL of this project is very simple and is given in Figure 5.15.

BEGIN
    Configure RC0 as digital input
    Configure RC6, RC7 as digital outputs
    DO FOREVER
        IF RC0 is logic 0 **THEN**
            Turn OFF GREEN LED
            Turn on RED LED
        ELSE
            Turn OFF RED LED
            Turn ON GREEN LED
        ENDIF
    ENDDO
END

Figure 5.15: PDL of the Project.

Project Program

mikroC Pro for PIC

The mikroC Pro for PIC program is named MIKROC-LED4.C, and the program listing is given in Figure 5.16. The operation of the program is very simple. An endless loop is established using a *for* statement and inside this loop port pin RC0 is checked. If RC0 is at logic 0, then the RED LED is turned ON; otherwise, the GREEN LED is turned ON. Note

```
/*********************************************************************
                        LOGIC PROBE
                        ===========

This is a logic probe project. In this project 2 colored LEDs are connected to PORTC pins
RC6 (RED) and RC7 (GREEN). In addition, RC0 is used as the probe input.

If the logic probe is at logic 0 then the RED LED is turned ON. Otherwise, the GREEN LED is
turned ON.

Author:  Dogan Ibrahim
Date:    August 2013
File:    MIKROC-LED4.C
*********************************************************************/
#define PROBE PORTC.RC0
#define RED_LED PORTC.RC6
#define GREEN_LED PORTC.RC7

void main()
{
    ANSELC = 0;                         // Configure PORTC as digital
    TRISC0_bit = 1;                     // Configure RC0 as input
    TRISC6_bit = 0;                     // Configure RC6 as output
    TRISC7_bit = 0;                     // Configure RC7 as output

    for(;;)                             // Endless loop
    {
      if(PROBE == 0)                    // If the signal is LOW
      {
        GREEN_LED = 0;                  // Turn OFF GREEN LED
        RED_LED = 1;                    // Torn ON RED LED
      }
      else
      {
        RED_LED = 0;                    // Turn OFF RED LED
        GREEN_LED = 1;                  // Turn ON GREEN LED
      }
    }
}
```

Figure 5.16: mikroC Pro for PIC Program Listing.

in the program that the individual PORTC pins are configured as input or outputs. We could instead set TRISC = 1 to configure RC0 as the input and others (RC1:RC7) as outputs.

MPLAB XC8

The MPLAB XC8 program is named XC8-LED4.C, and the program listing is given in Figure 5.17.

Further Development

The problem with this logic probe is that one of the LEDs is always ON even if the probe is not connected to any digital signal, or if the signal is at high-impedance state (i.e. tristate). We can develop the project further so that the high-impedance state can also be detected, and none of the LEDs are turned ON.

Figure 5.18 shows the modified circuit diagram. Note here that a transistor (BC108) is used at the front end of the circuit. The operation of the circuit is as follows.

The transistor is configured as an emitter−follower stage with the base controlled from output pin RC4 of the microcontroller through a 100-K resistor. The external signal is also applied to the base of the transistor through a 10-K resistor. The emitter of the transistor is connected to input pin RC3 of the microcontroller through a 100-K resistor. Pin RC4 applies logic levels to the base of the transistor, and then pin RC3 determines the state of the external signal (probe) as shown in Table 5.1. For example, if after setting RC4 = 1, we detect RC3 = 1 and also after setting RC4 = 0 we again detect RC3 = 1 then the probe must be at logic 1.

Figure 5.19 shows the mikroC Pro for the PIC program listing of the modified project. The program is named MIKROC-LED4-1.C. At the beginning of the program, symbols are given to the used I/O pins. Then, these I/O pins are configured as either inputs or outputs. The program executes in an endless *for* loop where the logic in Table 5.1 is implemented. A small delay is used (1 ms, although a few microseconds should be enough) after the transistor base signal is changed to allow the transistor to settle down.

Figure 5.20 shows the equivalent MPLAB XC8 program named XC8-LED4-1.C.

Project 5.5—LED Dice
Project Description

This is a simple dice project based on LEDs, a push-button switch, and a PIC18F45K22 microcontroller operating with an 8-MHz crystal. The block diagram of the project is shown in Figure 5.21.

```
/*******************************************************************************
                              LOGIC PROBE
                              ===========

This is a logic probe project. In this project 2 colored LEDs are connected to PORTC pins RC6
(RED) and RC7 (GREEN). In addition, RC0 is used as the probe input.

If the logic probe is at logic 0 then the RED LED is turned ON. Otherwise, the GREEN LED is
turned ON.

Author:    Dogan Ibrahim
Date:      August 2013
File:      XC8-LED4.C
*******************************************************************************/
#include <xc.h>
#pragma config MCLRE = EXTMCLR, WDTEN = OFF, FOSC = HSHP
#define _XTAL_FREQ 8000000

#define PROBE PORTCbits.RC0
#define RED_LED PORTCbits.RC6
#define GREEN_LED PORTCbits.RC7


void main()
{
    ANSELC = 0;                              // Configure PORTC as digital
    TRISCbits.RC0 = 1;                       // Configure RC0 as input
    TRISCbits.RC6 = 0;                       // Configure RC6 as output
    TRISCbits.RC7 = 0;                       // Configure RC7 as output

    for(;;)                                  // Endless loop
    {
      if(PROBE == 0)                         // If the signal is LOW
      {
        GREEN_LED = 0;                       // Turn OFF GREEN LED
        RED_LED = 1;                         // Turn ON RED LED
      }
      else
      {
        RED_LED = 0;                         // Turn OFF RED LED
        GREEN_LED = 1;                       // Turn ON GREEN LED
      }
    }
}
```

Figure 5.17: MPLAB XC8 Program Listing.

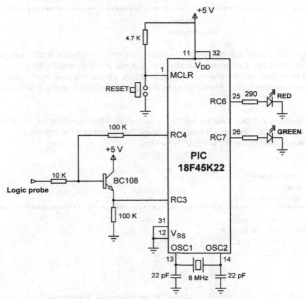

Figure 5.18: Modified Circuit Diagram.

As shown in Figure 5.22, the LEDs are organized such that when they turn ON, they indicate the numbers as in a real dice. Operation of the project is as follows: Normally, the LEDs are all OFF to indicate that the system is ready to generate a new dice number. Pressing the switch generates a random dice number between 1 and 6 and displays on the LEDs for 3 s. After 3 s, the LEDs turn OFF again.

Project Hardware

The circuit diagram of the project is shown in Figure 5.23. Seven LEDs representing the faces of a dice are connected to PORTC of a PIC18F45K22 microcontroller in current-sourcing

Table 5.1: Applied and Detected Logic Levels for Figure 5.18

Probe State	Applied to RC4	Detected at RC3
Probe at high impedance	1	1
	0	0
Probe at logic 1	1	1
	0	1
Probe at logic 0	1	0
	0	0

```
/*******************************************************************************
                          LOGIC PROBE
                          ===========
```

This is a logic probe project. In this project 2 colored LEDs are connected to PORTC pins RC6 (RED) and RC7 (GREEN). A transistor is used at the front end of the project. Pin RC4 is connected to the base of the transistor. The emitter of the transistor is connected to pin RC3 of the microcontroller. Table 7.1 in the text explains how the logic state of the probe signal is detected.

If the logic probe is at logic 0 then the RED LED is turned ON. If the logic level is 1 the GREEN LED is turned ON. If the probe is not connected to any logic signal or if the signal is at high-impedance state then none of the LEDs turn on.

```
Author:    Dogan Ibrahim
Date:      August 2013
File:      MIKROC-LED4-1.C
*******************************************************************************/
#define RED_LED PORTC.RC6
#define GREEN_LED PORTC.RC7
#define TSTOUT PORTC.RC4
#define TSTIN PORTC.RC3

void main()
{
    ANSELC = 0;                          // Configure PORTC as digital
    TRISC6_bit = 0;                      // Configure RC6 as output
    TRISC7_bit = 0;                      // Configure RC7 as output
    TRISC4_bit = 0;                      // Configure RC4 as output
    TRISC3_bit = 1;                      // Configure RC3 as input

    for(;;)                              // Endless loop
    {
      TSTOUT = 1;                        // Set RC4 = 1
      Delay_ms(1);                       // Small delay
      if(TSTIN == 1)                     // Check RC3. If RC3 = 1
      {
        TSTOUT = 0;                      // Set RC4 = 0
        Delay_ms(1);                     // Small delay
        if(TSTIN == 0)                   // Check RC3. If RC3 = 0
        {
          GREEN_LED = 0;                 // High-impedance state
          RED_LED = 0;                   // High-impedance state
        }
        else
        {
          RED_LED = 0;
          GREEN_LED = 1;                 // Probe at logic 1
        }
      }
      else
      {
        TSTOUT = 0;
        Delay_ms(1);                     // Small delay
        if(TSTIN == 0)
        {
          GREEN_LED = 0;
          RED_LED = 1;                   // Probe at logic 0
        }
      }
    }
}
```

Figure 5.19: mikroC Pro for PIC Program Listing.

```
/**********************************************************************
                    LOGIC PROBE
                    ===========

This is a logic probe project. In this project 2 colored LEDs are connected to PORTC pins RC6
(RED) and RC7 (GREEN). A transistor is used at the front end of the project. Pin RC4 is
connected to the base of the transistor. The emitter of the transistor is connected to pin RC3
of the microcontroller. Table 7.1 in the text explains how the logic state of the probe signal is
detected.

If the logic probe is at logic 0 then the RED LED is turned ON. If the logic level is 1 the GREEN
LED is turned ON. If the probe is not connected to any logic signal or if the signal is at
high-impedance state then none of the LEDs turn on.


Author:    Dogan Ibrahim
Date:      August 2013
File:      XC8-LED4-1.C
**********************************************************************/

#include <xc.h>
#pragma config MCLRE = EXTMCLR, WDTEN = OFF, FOSC = HSHP
#define _XTAL_FREQ 8000000

#define RED_LED PORTCbits.RC6
#define GREEN_LED PORTCbits.RC7
#define TSTOUT PORTCbits.RC4
#define TSTIN PORTCbits.RC3


void main()
{
    ANSELC = 0;                     // Configure PORTC as digital
    TRISCbits.RC6 = 0;              // Configure RC6 as output
    TRISCbits.RC7 = 0;              // Configure RC7 as output
    TRISCbits.RC4 = 0;              // Configure RC4 as output
    TRISCbits.RC3 = 1;              // Configure RC3 as input

    for(;;)                         // Endless loop
    {
      TSTOUT = 1;                   // Set RC4 = 1
      __delay_ms(1);                // Small delay
      if(TSTIN == 1)                // Check RC3. If RC3 = 1
      {
        TSTOUT = 0;                 // Set RC4 = 0
        __delay_ms(1);              // Small delay
        if(TSTIN == 0)              // Check RC3. If RC3 = 0
        {
          GREEN_LED = 0;            // High-impedance state
          RED_LED = 0;
        }
```

Figure 5.20: MPLAB XC8 Program Listing.

```
            else
            {
              RED_LED = 0;
              GREEN_LED = 1;                    // Probe at logic 1
            }
          }
          else
          {
            TSTOUT = 0;                         // Set RC4 = 0
            __delay_ms(1);                      // Small delay
            if(TSTIN == 0)                      // Check RC3. If RC3 = 0
            {
              GREEN_LED = 0;
              RED_LED = 1;                       // Probe at logic 0
            }
          }
        }
    }
```

Figure 5.20
cont'd

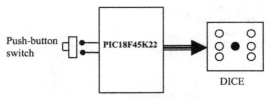

Figure 5.21: Block Diagram of the Project.

mode using 290-Ω current limiting resistors. A push-button switch is connected to bit 0 of PORTB (RB0) using a pull-up resistor. The microcontroller is operated from an 8-MHz crystal connected between pins OSC1 and OSC2.

If you are using the EasyPIC V7 development board, then make sure that the following jumpers are configured. Push-button switch RB0 on the board can be pressed to generate a dice number:

DIP switch SW3: PORTC ON
Jumper J17 (Button Press Level): GND

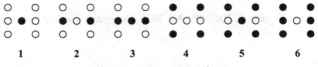

Figure 5.22: LED Dice.

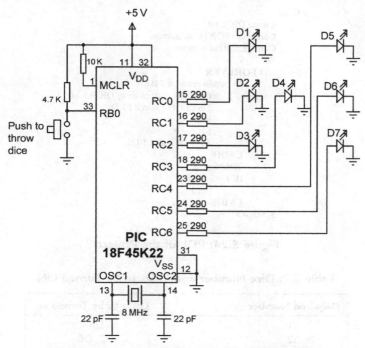

Figure 5.23: Circuit Diagram of the Project.

Project PDL

The operation of the project is described in the PDL given in Figure 5.24. At the beginning of the program, PORTC pins are configured as outputs, and bit 0 of PORTB (RB0) is configured as input. The program then executes in a loop continuously and increments a variable between 1 and 6. The state of the push-button switch is checked, and when the switch is pressed (switch output at logic 0), the current number is sent to the LEDs. A simple array is used to find out the LEDs to be turned ON corresponding to the generated dice number.

Table 5.2 gives the relationship between a dice number and the corresponding LEDs to be turned ON to imitate the faces of a real dice. For example, to display number 1 (i.e. only the middle LED is ON), we have to turn on D4. Similarly, to display number 4, we have to turn ON D1, D3, D5, and D7.

The relationship between the required number and the data to be sent to PORTC to turn on the correct LEDs is given in Table 5.3. For example, to display dice number 2, we have to send hexadecimal 0x22 to PORTC. Similarly, to display number 5, we have to send hexadecimal 0x5D to PORTC and so on.

BEGIN
    Create DICE table
    Configure PORTC as outputs
    Configure RB0 as input
    Set J = 1
    DO FOREVER
        IF button pressed **THEN**
            Get LED pattern from DICE table
            Turn ON required LEDs
            Wait 3 s
            Set J = 0
            Turn OFF all LEDs
        ENDIF
        Increment J
        IF J = 7 **THEN**
            Set J = 1
        ENDIF
    ENDDO
END

Figure 5.24: PDL of the Project.

Table 5.2: Dice Number and LEDs to be Turned ON

Required Number	LEDs to be Turned on
1	D4
2	D2, D6
3	D2, D4, D6
4	D1, D3, D5, D7
5	D1, D3, D4, D5, D7
6	D1, D2, D3, D5, D6, D7

Project Program

mikroC Pro for PIC

The mikroC Pro for PIC program is called MIKROC-LED5.C and the program listing is given in Figure 5.25. At the beginning of the program, **Switch** is defined as bit 0 of PORTB, and **Pressed** is defined as 0. The relationship between the dice numbers and the LEDs to be turned on are stored in an array called **DICE**. Variable **J** is used as the dice

Table 5.3: Required Number and PORTC Data

Required Number	PORTC Data (Hex)
1	0×08
2	0×22
3	$0 \times 2A$
4	0×55
5	$0 \times 5D$
6	0×77

```
/******************************************************************************
                                SIMPLE DICE
                                ==========

In this project 7 LEDs are connected to PORTC of a PIC18F45K22 microcontroller and the
microcontroller is operated from an 8 MHz crystal. The LEDs are organized as the faces
of a real dice. When a push-button switch connected to RB0 is pressed a dice pattern is
displayed on the LEDs. The display remains in this state for 3 s and after this period
the LEDs all turn OFF to indicate that the system is ready for the button to be pressed again.


Author:        Dogan Ibrahim
Date:          August 2013
File:          MIKROC-LED5.C
******************************************************************************/

#define Switch PORTB.RB0
#define Pressed 0

void main()
{
    unsigned char J = 1;
    unsigned char Pattern;
    unsigned char DICE[] = {0,0x08,0x22,0x2A,0x55,0x5D,0x77};

    ANSELC = 0;                          // Configure PORTC as digital
    ANSELB = 0;                          // Configure PORTB as digital
    TRISC = 0;                           // Configure PORTC as outputs
    TRISB = 1;                           // Configure RB0 as input
    PORTC = 0;                           // Turn OFF all LEDs

    for(;;)                              // Endless loop
    {
        if(Switch == Pressed)            // Is switch pressed ?
        {
            Pattern = DICE[J];           // Get LED pattern
            PORTC = Pattern;             // Turn on LEDs
            Delay_ms(3000);              // Delay 3 s
            PORTC = 0;                   // Turn OFF all LEDs
            J = 0;                       // Initialize J
        }
        J++;                             // Increment J
        if(J == 7) J = 1;                // Back to 1 if >6
    }
}
```

Figure 5.25: mikroC Pro for PIC Program Listing.

number. Variable **Pattern** is the data sent to the LEDs. The program then enters an endless
for loop where the value of variable **J** is incremented very fast between 1 and 6. When the
push-button switch is pressed, the LED pattern corresponding to the current value of **J** is
read from the array and sent to the LEDs. The LEDs remain at this state for 3 s (using
function Delay_ms with the argument set to 3000 ms), and after this time, they all turn
OFF to indicate that the system is ready to generate a new dice number.

MPLAB XC8

The MPLAB XC8 program is named XC8-LED5.C, and the program listing is given in Figure 5.26. The 3-s delay is created using function Delay_Seconds as before.

Using a Random Number Generator

In the above project, the value of variable **J** changes very fast between 1 and 6, and when the push-button switch is pressed, the current value of this variable is taken and used as the dice number. Because the values of **J** are changing very fast, we can say that the numbers generated are random, that is, new numbers do not depend on the previous numbers.

In this section, we shall see how a pseudorandom number generator function can be used to generate the dice numbers. The modified program listing is shown in Figure 5.27 (MIKROC-LED5-1.C). In this program, a function called **Number** generates the dice numbers. Here, we could have used the built-in *rand* function as in Project 5.3, but a pseudorandom number generator function has been created instead to show how such a function works. The function receives the upper limit of the numbers to be generated (6 in this example), and also a seed value that defines the number set to be generated. In this example, the seed is set to 1. Every time the function is called, a number will be generated between 1 and 6.

The operation of the program is basically the same as in Figure 5.25. When the push-button switch is pressed, function **Number** is called to generate a new dice number between 1 and 6, and this number is used as an index in array **DICE** to find the bit pattern to be sent to the LEDs.

Figure 5.28 shows the MPLAB XC8 version of the program (XC8-LED5-1.C). In this version, the random number generator function is used. The function is divided by 6553 and then 1 is added to generate a number between 1 and 6.

Project 5.6—Two-Dice Project
Project Description

This project is similar to the previous project, but here a pair of dice are used instead of one. In many dice games (e.g. backgammon), a pair of dice are thrown together and then the player takes action based on the result.

The circuit given in Figure 5.23 can be modified by adding another set of seven LEDs for the second dice. For example, the first set of LEDs can be driven from PORTC, the second set from PORTD, and the push-button switch can be connected to RB0 as before. Such a design will require the use of 14 output ports just for the LEDs. Later on, we will see how

```
/********************************************************************************
                              SIMPLE DICE
                              ==========

In this project 7 LEDs are connected to PORTC of a PIC18F45K22 microcontroller and the
microcontroller is operated from an 8 MHz crystal. The LEDs are organized as the faces
of a real dice. When a push-button switch connected to RB0 is pressed a dice pattern is
displayed on the LEDs. The display remains in this state for 3 s and after this period
the LEDs all turn OFF to indicate that the system is ready for the button to be pressed again.


Author:         Dogan Ibrahim
Date:           August 2013
File:           XC8-LED5.C
********************************************************************************/

#include <xc.h>
#pragma config MCLRE = EXTMCLR, WDTEN = OFF, FOSC = HSHP
#define _XTAL_FREQ 8000000

#define Switch PORTBbits.RB0
#define Pressed 0

//
// This function creates seconds delay. The argument specifies the delay time in seconds.
//
void Delay_Seconds(unsigned char s)
{
    unsigned char i,j;

    for(j = 0; j < s; j++)
    {
        for(i = 0; i <  100; i++)__delay_ms(10);
    }
}


void main()
{
    unsigned char J = 1;
    unsigned char Pattern;
    unsigned char DICE[] = {0,0x08,0x22,0x2A,0x55,0x5D,0x77};

    ANSELC = 0;                          // Configure PORTC as digital
    ANSELB = 0;                          // Configure PORTB as digital
    TRISC = 0;                           // Configure PORTC as outputs
    TRISB = 1;                           // Configure RB0 as input
    PORTC = 0;                           // Turn OFF all LEDs

    for(;;)                              // Endless loop
    {
```

Figure 5.26: MPLAB XC8 Program Listing.

```
          if(Switch == Pressed)                     // Is switch pressed ?
          {
            Pattern = DICE[J];                       // Get LED pattern
            PORTC = Pattern;                         // Turn on LEDs
            Delay_Seconds(3);                        // Delay 3 s
            PORTC = 0;                               // Turn OFF all LEDs
            J = 0;                                   // Initialize J
          }
          J++;                                       // Increment J
          if(J == 7) J = 1;                          // Back to 1 if >6
        }
      }
```

Figure 5.26
cont'd

the LEDs can be combined to reduce the I/O requirements. Figure 5.29 shows the block diagram of the project.

Project Hardware

The circuit diagram of the project is shown in Figure 5.30. The circuit is basically the same as in Figure 5.23, with the addition of another set of LEDs, connected to PORTD.

If you are using the EasyPIC V7 development board, then make sure that the following jumpers are configured. Push-button switch RB0 on the board can be pressed to generate a dice number:

DIP switch SW3: PORTC ON
DIP switch SW3: PORTD ON
Jumper J17 (Button Press Level): GND

Project PDL

The operation of the project is very similar to that in the previous project. Figure 5.31 shows the PDL for this project. At the beginning of the program, PORTC and PORTD pins are configured as outputs, and bit 0 of PORTB (RB0) is configured as input. The program then executes in a loop continuously and checks the state of the push-button switch. When the switch is pressed, two pseudorandom numbers are generated between 1 and 6, and these numbers are sent to PORTC and PORTD. The LEDs remain at this state for 3 s, and after this time, all the LEDs are turned OFF to indicate that the push button can be pressed again for the next pair of numbers.

```
/*******************************************************************************
                            SIMPLE DICE
                            ===========

In this project 7 LEDs are connected to PORTC of a PIC18F45K22 microcontroller and the
microcontroller is operated from an 8 MHz crystal. The LEDs are organized as the faces of a real
dice. When a push-button switch connected to RB0 is pressed a dice pattern is displayed on the
LEDs. The display remains in this state for 3 s and after this period the LEDs all turn OFF to
indicate that the system is ready for the button to be pressed again.

In this program a pseudorandom number generator function is used to generate the
dice numbers between 1 and 6.

Author:     Dogan Ibrahim
Date:       August 2013
File:       MIKROC-LED5-1.C
*******************************************************************************/

#define Switch PORTB.RB0
#define Pressed 0

//
// This function generates a pseudorandom integer number
// between 1 and Lim
//
unsigned char Number(int Lim, int Y)
{
    unsigned char Result;
    static unsigned int Y;

    Y = (Y * 32719 + 3) % 32749;
    Result = ((Y % Lim) + 1);
    return Result;
}


void main()
{
    unsigned char J, Pattern, Seed = 1;
    unsigned char DICE[] = {0,0x08,0x22,0x2A,0x55,0x5D,0x77};

    ANSELC = 0;                          // Configure PORTC as digital
    ANSELB = 0;                          // Configure PORTB as digital
    TRISC = 0;                           // Configure PORTC as outputs
    TRISB = 1;                           // Configure RB0 as input
    PORTC = 0;                           // Turn OFF all LEDs

    for(;;)                              // Endless loop
    {
        if(Switch == Pressed)            // Is switch pressed ?
        {
```

Figure 5.27: Modified mikroC Pro for the PIC Dice Program.

```
J = Number(6, seed);          // Generate a Number 1 to 6
Pattern = DICE[J];            // Get LED pattern
PORTC = Pattern;              // Turn on LEDs
Delay_ms(3000);               // Delay 3 s
PORTC = 0;                    // Turn OFF all LED
}
}
}
```

Figure 5.27
cont'd

Project Program

mikroC Pro for PIC

The program is called MIKROC-LED6.C, and the program listing is given in Figure 5.32. At the beginning of the program, **Switch** is defined as bit 0 of PORTB, and **Pressed** is defined as 0. The relationship between the dice numbers and the LEDs to be turned on are stored in an array called **DICE** as in the previous project. Variable **Pattern** is the data sent to the LEDs. The program enters an endless **for** loop where the state of the push-button switch is checked continuously. When the switch is pressed, two random numbers are generated by calling function **Numbers**. The bit pattern to be sent to the LEDs are then determined and sent to PORTC and PORTD. The program then repeats inside the endless loop checking the state of push-button switch.

MPLAB XC8

The MPLAB X8 version of the program is shown in Figure 5.33. Here again, the built-in random number generator function rand is used to generate the dice numbers.

Project 5.7—Two-Dice Project Using Fewer I/O Pins
Project Description

This project is similar to Project 5.6, but here, LEDs are shared, which uses fewer input/output pins.

The LEDs in Table 5.2 can be grouped as shown in Table 5.4. Looking at this Table we can say that

- D4 can appear on its own,
- D2 and D6 are always together,
- D1 and D3 are always together,
- D5 and D7 are always together.

Thus, we can drive D4 on its own, and then drive the D2, D6 pair together in series, D1, D3 pair together in series, and also D5, D7 pair together in series (actually, we could share

```
/*******************************************************************************
                        SIMPLE DICE
                        ==========

In this project 7 LEDs are connected to PORTC of a PIC18F45K22 microcontroller and the
microcontroller is operated from an 8 MHz crystal. The LEDs are organized as the faces of a
real dice. When a push-button switch connected to RB0 is pressed a dice pattern is displayed
on the LEDs. The display remains in this state for 3 s and after this period the LEDs all
turn OFF to indicate that the system is ready for the button to be pressed again.

 The random number generator function is used in this program.
 *
Author:        Dogan Ibrahim
Date:          August 2013
File:          XC8-LED5-1.C
*******************************************************************************/

#include <xc.h>
#include <stdlib.h>
#pragma config MCLRE = EXTMCLR, WDTEN = OFF, FOSC = HSHP
#define _XTAL_FREQ 8000000

#define Switch PORTBbits.RB0
#define Pressed 0

//
// This function creates seconds delay. The argument specifies the delay time in seconds.
//
void Delay_Seconds(unsigned char s)
{
   unsigned char i,j;

   for(j = 0; j < s; j++)
   {
      for(i = 0; i <  100; i++)__delay_ms(10);
   }
}


void main()
{
    unsigned char J, Pattern, Seed = 1;
    unsigned char DICE[] = {0,0x08,0x22,0x2A,0x55,0x5D,0x77};

    ANSELC = 0;                          // Configure PORTC as digital
    ANSELB = 0;                          // Configure PORTB as digital
    TRISC = 0;                           // Configure PORTC as outputs
    TRISB = 1;                           // Configure RB0 as input
    PORTC = 0;                           // Turn OFF all LEDs
    srand(1);                            // Initialize rand function
```

Figure 5.28: Modified MPLAB XC8 Dice Program.

```
for(;;)                              // Endless loop
{
    if(Switch == Pressed)            // Is switch pressed ?
    {
      J = rand() / 6553 + 1;         // Generate a number 1 to 6
      Pattern = DICE[J];             // Get LED pattern
      PORTC = Pattern;               // Turn on LEDs
      Delay_Seconds(3);              // Delay 3 s
      PORTC = 0;                     // Turn OFF all LEDs
    }
  }
}
```

Figure 5.28
cont'd

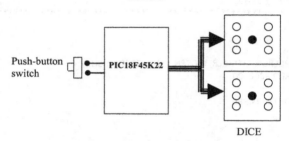

Figure 5.29: Block Diagram of the Project.

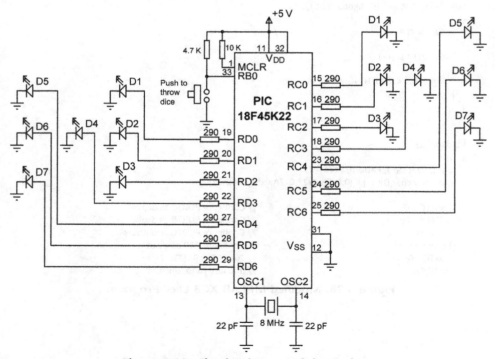

Figure 5.30: Circuit Diagram of the Project.

BEGIN
> Create DICE table
> Configure PORTC as outputs
> Configure PORTD as outputs
> Configure RB0 as input
> **DO FOREVER**
>> **IF** button pressed **THEN**
>>> Get a random number between 1 and 6
>>> Find bit pattern
>>> Turn ON LEDs on PORTC
>>> Get second random number between 1 and 6
>>> Find bit pattern
>>> Turn on LEDs on PORTD
>>> Wait 3 s
>>> Turn OFF all LEDs
>> **ENDIF**
> **ENDDO**
END

Figure 5.31: PDL of the Project.

D1, D3, D5, D7, but this would require 8 V to drive if the LEDs are connected in series. Connecting these LEDs in parallel will require excessive current and a driver IC will be required). Altogether four lines will be required to drive seven LEDs of a dice. Similarly, four lines will be required to drive the second dice. Thus, a pair of dice can easily be driven from an 8-bit output port.

Project Hardware

The circuit diagram of the project is shown in Figure 5.34. PORTC of a PIC18F45K22 microcontroller is used to drive the LEDs as follows:

- RC0 drives D2, D6 of the first dice,
- RC1 drives D1, D3 of the first dice,
- RC2 drives D5, D7 of the first dice,
- RC3 drives D4 of the first dice,
- RC4 drives D2, D6 of the second dice,
- RC5 drives D1, D3 of the second dice,
- RC6 drives D5, D7 of the second dice,
- RC7 drives D4 of the second dice.

Since we are driving two LEDs on some outputs, we can calculate the required value of the current limiting resistors. Assuming that the voltage drop across each LED is 2 V, the current through the LED is 10 mA, and the output high voltage of the microcontroller is 4.85 V, the required resistors are $R = \frac{4.85 - 2 - 2}{10} = 85\ \Omega$. We will choose 100-$\Omega$ resistors.

We now need to find the relationship between the dice numbers and the bit pattern to be sent to the LEDs for each dice. Table 5.5 shows the relationship between the first dice

```
/**************************************************************************
                              TWO DICE
                              ========

In this project 7 LEDs are connected to PORTC of a PIC18F452 microcontroller and 7 LEDs to
PORTD. The microcontroller is operated from an 8 MHz crystal. The LEDs are organized as the
faces of a real dice. When a push-button switch connected to RB0 is pressed a dice pattern is
displayed on the LEDs. The display remains in this state for 3 s and after this period the
LEDs all turn OFF to indicate that the system is ready for the button to be pressed again.

In this program a pseudorandom number generator function is used to generate the dice
numbers between 1 and 6.

Author:         Dogan Ibrahim
Date:           August 2013
File:           MIKROC-LED6.C
**************************************************************************/

#define Switch PORTB.F0
#define Pressed 0

//
// This function generates a pseudorandom integer number between 1 and Lim.
//
unsigned char Number(int Lim, int Y)
{
    unsigned char Result;
    static unsigned int Y;

    Y = (Y * 32719 + 3) % 32749;
    Result = ((Y % Lim) + 1);
    return Result;
}


//
// Start of MAIN program
//
void main()
{
    unsigned char J,Pattern,Seed = 1;
    unsigned char DICE[] = {0,0x08,0x22,0x2A,0x55,0x5D,0x77};

    ANSELC = 0;                         // Configure PORTC as digital
    ANSELD = 0;                         // Configure PORTD as digital
    ANSELB = 0;                         // Configure PORTB as digital
    TRISC = 0;                          // Configure PORT as outputs
    TRISD = 0;                          // Configure PORTD as outputs
    TRISB = 1;                          // Configure RB0 as input
    PORTC = 0;                          // Turn OFF all LEDs
    PORTD = 0;                          // Turn OFF all LEDs
```

Figure 5.32: mikroC Pro for PIC Program Listing.

```
        for(;;)                              // Endless loop
        {
            if(Switch == Pressed)            // Is switch pressed ?
            {
            J = Number(6,seed);              // Generate first dice number
            Pattern = DICE[J];               // Get LED pattern
            PORTC = Pattern;                 // Turn on LEDs for first dice
            J = Number(6,seed);              // Generate second dice number
            Pattern = DICE[J];               // Get LED pattern
            PORTD = Pattern;                 // Turn on LEDs for second dice
            Delay_ms(3000);                  // Delay 3 seconds
            PORTC = 0;                       // Turn OFF all LEDs
            PORTD = 0;                       // Turn OFF all LEDS
            }
        }
    }
```

Figure 5.32
cont'd

numbers and the bit pattern to be sent to port pins RC0—RC3. Similarly, Table 5.6 shows the relationship between the second dice numbers and the bit pattern to be sent to port pins RC4—RC7.

We can now find the 8-bit number to be sent to PORTC to display both dice numbers as follows:

- Get the first number from the number generator, call this P.
- Index DICE table to find the bit pattern for a low nibble, that is, $L = DICE[P]$.
- Get the second number from the number generator, call this P.
- Index DICE table to find the bit pattern for a high nibble, that is, $U = DICE[P]$.
- Multiply the high nibble with 16 and add a low nibble to find the number to be sent to PORTC, that is, $R = 16 * U + L$ where R is the 8-bit number to be sent to PORTC to display both dice values.

If you are using the EasyPIC V7 development board, then make sure that the following jumpers are configured. Push-button switch RB0 on the board can be pressed to generate a dice number:

DIP switch SW3: PORTC ON
DIP switch SW3: PORTD ON
Jumper J17 (Button Press Level): GND

Project PDL

Figure 5.35 shows the PDL for this project. At the beginning of the program, PORTC pins are configured as outputs, and bit 0 of PORTB (RB0) is configured as the input. The

```
/*******************************************************************************
                              TWO DICE
                              ========

In this project 7 LEDs are connected to PORTC of a PIC18F452 microcontroller and 7 LEDs to
PORTD. The microcontroller is operated from an 8 MHz crystal. The LEDs are organized as the
faces of a real dice. When a push-button switch connected to RB0 is pressed a dice pattern is
displayed on the LEDs. The display remains in this state for 3 s and after this period the
LEDs all turn OFF to indicate that the system is ready for the button to be pressed again.

In this program a pseudorandom number generator function is used to generate the dice numbers
between 1 and 6.

Author:        Dogan Ibrahim
Date:          August 2013
File:          XC8-LED6.C
*******************************************************************************/

#include <xc.h>
#include <stdlib.h>
#pragma config MCLRE = EXTMCLR, WDTEN = OFF, FOSC = HSHP
#define _XTAL_FREQ 8000000

#define Switch PORTBbits.RB0
#define Pressed 0

//
// This function creates seconds delay. The argument specifies the delay time in seconds.
//
void Delay_Seconds(unsigned char s)
{
  unsigned char i,j;

  for(j = 0; j < s; j++)
  {
    for(i = 0; i <  100; i++)__delay_ms(10);
  }
}


void main()
{
    unsigned char J, Pattern, Seed = 1;
    unsigned char DICE[] = {0,0x08,0x22,0x2A,0x55,0x5D,0x77};

    ANSELC = 0;                          // Configure PORTC as digital
    ANSELD = 0;                          // Configure PORTD as digital
    ANSELB = 0;                          // Configure PORTB as digital
    TRISC = 0;                           // Configure PORT as outputs
    TRISD = 0;                           // Configure PORTD as outputs
    TRISB = 1;                           // Configure RB0 as input
```

Figure 5.33: MPLAB XC8 Program Listing.

```
        PORTC = 0;                    // Turn OFF all LEDs
        PORTD = 0;                    // Turn OFF all LEDs
        srand(1);                     // Initialize rand function

        for(;;)                       // Endless loop
        {
          if(Switch == Pressed)       // Is switch pressed ?
          {
            J = rand() / 6553 + 1;    // Generate a number 1 to 6
            Pattern = DICE[J];        // Get LED pattern
            PORTC = Pattern;          // Turn on PORTC LEDs
            J = rand() / 6553 + 1;    // Generate a number 1 to 6
            Pattern = DICE[J];        // Get LED pattern
            PORTD = Pattern;          // Turn on PORTD LEDs
            Delay_Seconds(3);         // Delay 3 s
            PORTC = 0;                // Turn OFF PORTC LEDs
            PORTD = 0;                // Turn OFF PORTD LEDs
          }
        }
      }
```

Figure 5.33
cont'd

program then executes in a loop continuously and checks the state of the push-button switch. When the switch is pressed, two pseudorandom numbers are generated between 1 and 6, and the bit pattern to be sent to PORTC is found using the method described above. This bit pattern is then sent to PORTC to display both dice numbers at the same time. The display shows the dice numbers for 3 s, and then, all the LEDs turn OFF to indicate that the system is waiting for the push button to be pressed again to display next set of numbers.

Project Program

mikroC Pro for PIC

The mikroC Pro for the PIC program is called MIKROC-LED7.C, and the program listing is given in Figure 5.36. At the beginning of the program, **Switch** is defined as bit 0 of

Table 5.4: Grouping the LEDs

Required Number	LEDs to be Turned on
1	D4
2	D2, D6
3	D2, D6, D4
4	D1, D3, D5, D7
5	D1, D3, D5, D7, D4
6	D2, D6, D1, D3, D5, D7

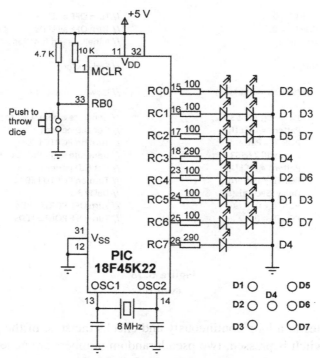

Figure 5.34: Circuit Diagram of the Project.

PORTB, and **Pressed** is defined as 0. The relationship between the dice numbers and the LEDs to be turned on are stored in an array called **DICE** as in the previous project. Variable **Pattern** is the data sent to the LEDs. The program enters an endless **for** loop where the state of the push-button switch is checked continuously. When the switch is pressed, two random numbers are generated by calling function **Numbers**. Variables **L** and **U** store the lower and the higher nibbles of the bit pattern to be sent to PORTC. The bit pattern to be sent to PORTC is then determined using the method described in the Project hardware section and stored in variable **R**. This bit pattern is then sent to PORTC to display both dice numbers at the same time. The dice is displayed for 3 s, and after this period, the LEDs are turned OFF to indicate that the system is ready.

Table 5.5: First Dice Bit Patterns

Dice Number	RC3	RC2	RC1	RC0	Hex Value
1	1	0	0	0	8
2	0	0	0	1	1
3	1	0	0	1	9
4	0	1	1	0	6
5	1	1	1	0	E
6	0	1	1	1	7

Table 5.6: Second Dice Bit Patterns

Dice Number	RC7	RC6	RC5	RC4	Hex Value
1	1	0	0	0	8
2	0	0	0	1	1
3	1	0	0	1	9
4	0	1	1	0	6
5	1	1	1	0	E
6	0	1	1	1	7

MPLAB XC8

The MPLAB X8 version of the program is shown in Figure 5.37 (XC8-LED7.C). Here, again the built-in random number generator function rand is used to generate the dice numbers.

Modifying the Program

The program given in Figure 5.36 can be modified and made more efficient by combining the two dice nibbles into a single table value. The new program is described in this section.

There are 36 possible combinations of two dice values. Referring to Tables 5.5 and 5.6 and Figure 5.34, we can create Table 5.7 to show all the possible two dice values and the corresponding numbers to be sent to PORTC.

The modified program (program name MIKROC-LED7-1.C) is given in Figure 5.38. In this program, array **DICE** contains the 36 possible dice values. Program enters an endless **for** loop, and inside this loop, the state of the push-button switch is checked. Also, a variable is incremented from 1 to 36, and when the button is pressed, the value of this variable is used as an index to array **DICE** to determine the bit pattern to be sent to

```
BEGIN
        Create DICE table
        Configure PORTC as outputs
        Configure RB0 as input
        DO FOREVER
                IF button pressed THEN
                        Get a random number between 1 and 6
                        Find low nibble bit pattern
                        Get second random number between 1 and 6
                        High high nibble bit pattern
                        Calculate data to be sent to PORTC
                        Wait 3 s
                        Turn OFF all LEDs
                ENDIF
        ENDDO
END
```

Figure 5.35: PDL of the Project.

```
/************************************************************************
                    TWO DICE - FEWER I/O COUNT
                    ===========================
```

In this project LEDs are connected to PORTC of a PIC18F45K22 microcontroller and
the microcontroller is operated from an 8 MHz crystal. The LEDs are organized as the
faces of a real dice. When a push-button switch connected to RB0 is pressed a dice
pattern is displayed on the LEDs. The display remains in this state for 3 s and
after this period the LEDs all turn OFF to indicate that the system is ready for the
button to be pressed again.

In this program a pseudorandom number generator function is used to generate
the dice numbers between 1 and 6.

```
Author:        Dogan Ibrahim
Date:          August 2013
File:          MIKROC-LED7.C
************************************************************************/

#define Switch PORTB.RB0
#define Pressed 0

//
// This function generates a pseudorandom integer Number between 1 and Lim.
//
unsigned char Number(int Lim, int Y)
{
        unsigned char Result;
        static unsigned int Y;

        Y = (Y * 32719 + 3) % 32749;
        Result = ((Y % Lim) + 1);
        return Result;
}

//
// Start of MAIN program
//
void main()
{
    unsigned char J,L,U,R,Seed = 1;
    unsigned char DICE[] = {0,0x08,0x01,0x09,0x06,0x0E,0x07};

    ANSELC = 0;                              // Configure PORTC as digital
    ANSELB = 0;                              // Configure PORTB as digital
    TRISC = 0;                               // Configure PORTC as outputs
    TRISB = 1;                               // Configure RB0 as input
    PORTC = 0;                               // Turn OFF all LEDs

    for(;;)                                  // Endless loop
```

Figure 5.36: mikroC Pro for PIC Program Listing.

```
        {
          if(Switch == Pressed)                // Is switch pressed ?
          {
            J = Number(6,seed);                // Generate first dice number
            L = DICE[J];                       // Get LED pattern
            J = Number(6,seed);                // Generate second dice number
            U = DICE[J];                       // Get LED pattern
            R = 16 * U + L;                    // Bit pattern to send to PORTC
                    PORTC = R;                 // Turn on LEDs for both dice
            Delay_ms(3000);                    // Delay 3 s
            PORTC = 0;                         // Turn OFF all LEDs
          }
        }
      }
```

Figure 5.36
cont'd

PORTC. As before, the program displays the dice numbers for 3 s and then turns OFF all LEDs to indicate that it is ready.

The MPLAB XC8 version of the modified program is shown in Figure 5.39 (XC8-LED7-1.C).

Project 5.8—7-Segment LED Counter
Project Description

This project describes the design of a 7-segment LED based counter that counts from 0 to 9 continuously with a 1-s delay between each count. The project shows how a 7-segment LED can be interfaced and used in a PIC microcontroller project.

Seven-segment displays are used frequently in electronic circuits to show numeric or alphanumeric values. As shown in Figure 5.40, a 7-segment display basically consists of seven LEDs connected such that numbers from 0 to 9 and some letters can be displayed. Segments are identified by letters from **a** to **g**, and Figure 5.41 shows the segment names of a typical 7-segment display.

Figure 5.42 shows how numbers from 0 to 9 can be obtained by turning ON different segments of the display.

Seven-segment displays are available in two different configurations: **common cathode** and **common anode**. As shown in Figure 5.43, in common-cathode configuration, all the cathodes of all segment LEDs are connected together to the ground. The segments are turned ON by applying a logic 1 to the required segment LED via current limiting resistors. In the common-cathode configuration, the 7-segment LED is connected to the microcontroller in the current-sourcing mode.

```
/*******************************************************************************
                      TWO DICE - FEWER I/O COUNT
                      ===========================
```

In this project LEDs are connected to PORTC of a PIC18F45K22 microcontroller and the
microcontroller is operated from an 8 MHz crystal. The LEDs are organized as the faces of a
real dice. When a push-button switch connected to RB0 is pressed a dice pattern is displayed
on the LEDs. The display remains in this state for 3 s and after this period the LEDs all
turn OFF to indicate that the system is ready for the button to be pressed again.

In this program a pseudorandom number generator function is used to generate the dice
numbers between 1 and 6.

```
Author:        Dogan Ibrahim
Date:          August 2013
File:          XC8-LED7.C
*******************************************************************************/

#include <xc.h>
#include <stdlib.h>
#pragma config MCLRE = EXTMCLR, WDTEN = OFF, FOSC = HSHP
#define _XTAL_FREQ 8000000

#define Switch PORTBbits.RB0
#define Pressed 0

//
// This function creates seconds delay. The argument specifies the delay time in seconds.
//
void Delay_Seconds(unsigned char s)
{
  unsigned char i,j;

  for(j = 0; j < s; j++)
  {
    for(i = 0; i < 100; i++)__delay_ms(10);
  }
}

void main()
{
    unsigned char J,L,U,R;
    unsigned char DICE[] = {0,0x08,0x22,0x2A,0x55,0x5D,0x77};

    ANSELC = 0;                          // Configure PORTC as digital
    ANSELB = 0;                          // Configure PORTB as digital
    TRISC = 0;                           // Configure PORTC as outputs
    TRISB = 1;                           // Configure RB0 as input
    PORTC = 0;                           // Turn OFF all LEDs
    srand(1);                            // Initialize rand function
```

Figure 5.37: MPLAB XC8 Program Listing.

```
for(;;)                              // Endless loop
{
    if(Switch == Pressed)            // Is switch pressed ?
    {
        J = rand() / 6553 + 1;       // Generate a number 1 to 6
        L = DICE[J];                 // Get dice pattern
        J = rand() / 6553 + 1;       // Generate another number
        U = DICE[J];                 // Get dice pattern
        R = 16 * U + L;              // Bit pattern to send to PORTC
        PORTC = R;                   // Send bit pattern to PORTC
        Delay_Seconds(3);            // Delay 3 s
        PORTC = 0;                   // Turn OFF PORTC LEDs
    }
}
}
```

Figure 5.37

cont'd

In a common anode configuration, the anode terminals of all the LEDs are connected together as shown in Figure 5.44. This common point is then normally connected to the supply voltage. A segment is turned ON by connecting its cathode terminal to logic 0 via a current limiting resistor. In the common anode configuration, the 7-segment LED is connected to the microcontroller in the current-sinking mode.

In this project, a *Kingbright SA52-11* model red common anode 7-segment display is used. This is a 13-mm (0.52-in) display with 10 pins, and it also has a segment LED for the decimal point. Table 5.8 shows the pin configuration of this display.

Table 5.7: Two Dice Combinations and Number to be Sent to PORTC

Dice Numbers	PORTC Value	Dice Numbers	PORTC Value
1,1	0 x 88	4,1	0 x 86
1,2	0 x 18	4,2	0 x 16
1,3	0 x 98	4,3	0 x 96
1,4	0 x 68	4,4	0 x 66
1,5	0 x E8	4,5	0 x E6
1,6	0 x 78	4,6	0 x 76
2,1	0 x 81	5,1	0 x 8E
2,2	0 x 11	5,2	0 x 1E
2,3	0 x 91	5,3	0 x 9E
2,4	0 x 61	5,4	0 x 6E
2,5	0 x E1	5,5	0 x EE
2,6	0 x 71	5,6	0 x 7E
3,1	0 x 89	6,1	0 x 87
3,2	0 x 19	6,2	0 x 17
3,3	0 x 99	6,3	0 x 97
3,4	0 x 69	6,4	0 x 67
3,5	0 x E9	6,5	0 x E7
3,6	0 x 79	6,6	0 x 77

```
/******************************************************************************
                        TWO DICE - LESS I/O COUNT
                        ===========================
```

In this project LEDs are connected to PORTC of a PIC18F452 microcontroller and the microcontroller is operated from an 8 MHz crystal. The LEDs are organized as the faces of a real dice. When a push-button switch connected to RB0 is pressed a dice pattern is displayed on the LEDs. The display remains in this state for 3 s and after this period the LEDs all turn OFF to indicate that the system is ready for the button to be pressed again.

In this program a pseudorandom number generator function is used to generate the dice numbers between 1 and 6.

```
Author:         Dogan Ibrahim
Date:           August 2013
File:           MIKROC-LED7-1.C
******************************************************************/

#define Switch PORTB.F0
#define Pressed 0

//
// Start of MAIN program
//
void main()
{
    unsigned char Pattern, J = 1;
    unsigned char DICE[] = {0,0x88,0x18,0x98,0x68,0xE8,0x78,
                0x81,0x11,0x91,0x61,0xE1,0x71,
                0x89,0x19,0x99,0x69,0xE9,0x79,
                0x86,0x16,0x96,0x66,0xE6,0x76,
                0x8E,0x1E,0x9E,0x6E,0xEE,0x7E,
                0x87,0x17,0x97,0x67,0xE7,0x77};

    ANSELC = 0;                    // Configure PORTC as digital
    ANSELB = 0;                    // Configure PORTB as digital
    TRISC = 0;                     // Configure PORTC as outputs
    TRISB = 1;                     // Configure RB0 as input
    PORTC = 0;                     // Turn OFF all LEDs

    for(;;)                        // Endless loop
    {
        if(Switch == Pressed)      // Is switch pressed ?
        {
            Pattern = DICE[J];     // Number to send to PORTC
            PORTC = Pattern;       // Send to PORTC
            Delay_ms(3000);        // 3 seconds delay
            PORTC = 0;             // Clear PORTC
        }
        J++;                       // Increment J

        if(J == 37) J = 1;         // If J = 37, reset to 1
    }
}
```

Figure 5.38: Modified mikroC pro for PIC Program.

```
/*****************************************************************************
                      TWO DICE - LESS I/O COUNT
                      ==========================

In this project LEDs are connected to PORTC of a PIC18F452 microcontroller and the
microcontroller is operated from an 8 MHz crystal. The LEDs are organized as the faces
of a real dice. When a push-button switch connected to RB0 is pressed a dice pattern
is displayed on the LEDs. The display remains in this state for 3 s and after this
period the LEDs all turn OFF to indicate that the system is ready for the button to be
pressed again.

In this program a pseudorandom number generator function is used to generate
the dice numbers between 1 and 6.

Author:        Dogan Ibrahim
Date:          August 2013
File:          XC8-LED7-1.C
*****************************************************************************/

#include <xc.h>
#pragma config MCLRE = EXTMCLR, WDTEN = OFF, FOSC = HSHP
#define _XTAL_FREQ 8000000

#define Switch PORTBbits.RB0
#define Pressed 0

//
// This function creates seconds delay. The argument specifies the delay time
// in seconds.
//
void Delay_Seconds(unsigned char s)
{
    unsigned char i,j;

    for(j = 0; j < s; j++)
    {
        for(i = 0; i <  100; i++)__delay_ms(10);
    }
}


void main()
{
    unsigned char Pattern, J = 1;
    unsigned char DICE[] = {0,0x88,0x18,0x98,0x68,0xE8,0x78,
                0x81,0x11,0x91,0x61,0xE1,0x71,
                0x89,0x19,0x99,0x69,0xE9,0x79,
                0x86,0x16,0x96,0x66,0xE6,0x76,
                0x8E,0x1E,0x9E,0x6E,0xEE,0x7E,
                0x87,0x17,0x97,0x67,0xE7,0x77};
```

Figure 5.39: Modified MPLAB XC8 Program.

```
    ANSELC = 0;                          // Configure PORTC as digital
    ANSELB = 0;                          // Configure PORTB as digital
    TRISC = 0;                           // Configure PORTC as outputs
    TRISB = 1;                           // Configure RB0 as input
    PORTC = 0;                           // Turn OFF all LEDs

    for(;;)                              // Endless loop
    {
       if(Switch == Pressed)            // Is switch pressed ?
       {
         Pattern = DICE[J];             // Number to send to PORTC
         PORTC = Pattern;               // Send to PORTC
         Delay_Seconds(3);              // 3 s delay
         PORTC = 0;                     // Clear PORTC
       }
        J++;                            // Increment J
        if(J == 37) J = 1;             // If J = 37, reset to 1
    }

}
```

Figure 5.39
cont'd

Project Hardware

The circuit diagram of the project is shown in Figure 5.44. A PIC18F45K22-type microcontroller is used with an 8-MHz crystal. Segments **a**–**g** of the display are connected to PORTC of the microcontroller through 290-Ω current limiting resistors. Before driving the display, we have to know the relationship between the numbers to be displayed and the corresponding segments to be turned ON, and this is shown in Table 5.9. For example, to

Figure 5.40: Some 7-Segment Displays. (For color version of this figure, the reader is referred to the online version of this book.)

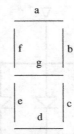

Figure 5.41: Segment Names of a 7-Segment Display.

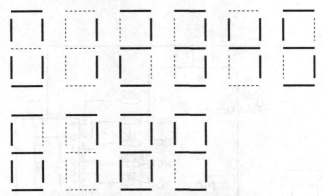

Figure 5.42: Displaying Numbers 0–9.

display number 3, we have to send the hexadecimal number 0x4F to PORTC which turns ON segments **a**, **b**, **c**, **d**, and **g**. Similarly, to display number 9, we have to send the hexadecimal number 0x6F to PORTC, which turns ON segments **a**, **b**, **c**, **d**, **f**, and **g**.

Project PDL

The operation of the project is shown in Figure 5.45 with a PDL. At the beginning of the program, an array called SEGMENT is declared and filled with the relationship between the

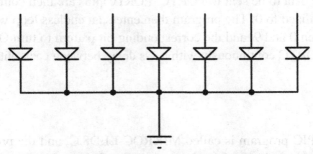

Figure 5.43: Common-Cathode 7-Segment Display.

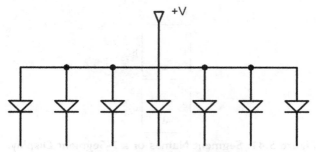

Figure 5.44: Circuit Diagram of the Project.

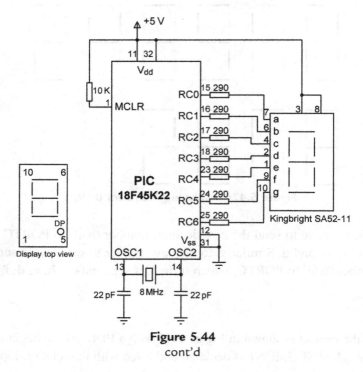

Figure 5.44
cont'd

numbers 0–9 and the data to be sent to PORTC. PORTC pins are then configured as outputs, and a variable is initialized to 0. The program then enters an endless loop where the variable is incremented between 0 and 9, and the corresponding bit pattern to turn ON the appropriate segments is sent to PORTC continuously with a 1-s delay between each output.

Project Program

mikroC Pro for PIC

The mikroC Pro for PIC program is called MIKROC-LED8.C, and the program listing is given in Figure 5.46. At the beginning of the program, character variables **Pattern** and

Table 5.8: The SA52-11 Pin Configuration

Pin Number	Segment
1	e
2	d
3	Common anode
4	c
5	Decimal point
6	b
7	a
8	Common anode
9	f
10	g

Table 5.9: Displayed Number and Data Sent to PORTC

Number	x	g	f	e	d	c	b	a	PORTC Data
0	0	0	1	1	1	1	1	1	0 x 3F
1	0	0	0	0	0	1	1	0	0 x 06
2	0	1	0	1	1	0	1	1	0 x 5B
3	0	1	0	0	1	1	1	1	0 x 4F
4	0	1	1	0	0	1	1	0	0 x 66
5	0	1	1	0	1	1	0	1	0 x 6D
6	0	1	1	1	1	1	0	1	0 x 7D
7	0	0	0	0	0	1	1	1	0 x 07
8	0	1	1	1	1	1	1	1	0 x 7F
9	0	1	1	0	1	1	1	1	0 x 6F

x is not used, taken as 0.

Cnt are declared, and **Cnt** is cleared to 0. Then, Table 5.9 is implemented using array **SEGMENT**. After configuring PORTC pins as outputs, the program enters an endless loop using the **for** statement. Inside the loop, the bit pattern corresponding to the contents of **Cnt** is found and stored in variable **Pattern**. Because we are using a common anode

```
BEGIN
        Create SEGMENT table
        Configure PORTC as digital outputs
        Initialize CNT to 0
        DO FOREVER
                Get bit pattern from SEGMENT corresponding to CNT
                Send this bit pattern to PORTC
                Increment CNT between 0 and 9
                Wait 1 s
        ENDDO
END
```

Figure 5.45: PDL of the Project.

```
/*******************************************************************************
                          7-SEGMENT DISPLAY
                          =================

In this project a common anode 7-segment LED display is connected to PORTC of a PIC18F45K22
microcontroller and the microcontroller is operated from an 8 MHz crystal. The program displays
numbers 0 to 9 on the display with a 1 s delay between each output.

Author:        Dogan Ibrahim
Date:          August 2013
File:          MIKROC-LED8.C
*******************************************************************************/

void main()
{
   unsigned char Pattern, Cnt = 0;
   unsigned char SEGMENT[] = {0x3F,0x06,0x5B,0x4F,0x66,0x6D,
     0x7D,0x07,0x7F,0x6F};

   ANSELC = 0;                          // Configure PORTC as digital
   TRISC = 0;                           // Configure PORTC as outputs

   for(;;)                              // Endless loop
   {
     Pattern = SEGMENT[Cnt];            // Number to send to PORTC
     Pattern = ~Pattern;                // Invert bit pattern
     PORTC = Pattern;                   // Send to PORTC
     Cnt++;                             // Increment Cnt
     if(Cnt == 10) Cnt = 0;             // Cnt is between 0 and 9
     Delay_ms(1000);                    // 1 s delay
   }
}
```

Figure 5.46: mikroC Pro for PIC Program Listing.

display, a segment is turned ON when it is at logic 0, and thus, the bit pattern is inverted before sending to PORTC. The value of **Cnt** is then incremented between 0 and 9, after which the program waits for a second before repeating the above sequence.

MPLAB XC8

The MPLAB X8 version of the program is shown in Figure 5.47 (XC8-LED8.C). The operation of the program is as in mikroC Pro for PIC.

Modified Program

Note that the program can be made more readable if we create a function to display the required number and then call this function from the main program.

```
/********************************************************************************
                          7-SEGMENT DISPLAY
                          =================
```

In this project a common anode 7-segment LED display is connected to PORTC of a PIC18F45K22 microcontroller and the microcontroller is operated from an 8 MHz crystal. The program displays numbers 0 to 9 on the display with a 1 s delay between each output.

```
Author:        Dogan Ibrahim
Date:          August 2013
File:          XC8-LED8.C
********************************************************************************/

#include <xc.h>
#pragma config MCLRE = EXTMCLR, WDTEN = OFF, FOSC = HSHP
#define _XTAL_FREQ 8000000

//
// This function creates seconds delay. The argument specifies the delay time
// in seconds.
//
void Delay_Seconds(unsigned char s)
{
    unsigned char i,j;

    for(j = 0; j < s; j++)
    {
        for(i = 0; i < 100; i++)__delay_ms(10);
    }
}


void main()
{
    unsigned char Pattern, Cnt = 0;
    unsigned char SEGMENT[] = {0x3F,0x06,0x5B,0x4F,0x66,0x6D,0x7D,0x07,0x7F,0x6F};

    ANSELC = 0;                              // Configure PORTC as digital
    TRISC = 0;                               // Configure PORTC as outputs

    for(;;)                                  // Endless loop
    {
        Pattern = SEGMENT[Cnt];              // Number to send to PORTC
        Pattern = ~Pattern;                  // Invert bit pattern
        PORTC = Pattern;                     // Send to PORTC
        Cnt++;                               // Increment Cnt
        if(Cnt == 10) Cnt = 0;               // Cnt is between 0 and 9
        Delay_Seconds(1);                    // 1 s delay
    }
}
```

Figure 5.47: MPLAB XC8 Program Listing.

Project 5.9—Two-Digit Multiplexed 7-Segment LED
Project Description

This project is similar to the previous project, but here, multiplexed two digits are used instead of just one digit, and a fixed number is displayed. In this project, number 25 is displayed as an example. In multiplexed LED applications (Figure 5.48), the LED segments of all the digits are tied together, and the common pins of each digit is turned ON separately by the microcontroller. By displaying each digit for several milliseconds, the eye cannot differentiate that the digits are not ON all the time. In this way we can multiplex any number of 7-segment displays together. For example, to display number 53, we have to send 5 to the first digit and enable its common pin. After a few milliseconds, number 3 is sent to the second digit, and the common point of the second digit is enabled. When this process is repeated continuously, the user sees as if both displays are ON continuously.

Some manufacturers provide multiplexed multidigit displays in single packages. For example, we can purchase two-, four-, or eight-digit multiplexed displays in a single package. The display used in this project is the DC56-11EWA which is a red 0.56-in height common-cathode two digit display having 18 pins and the pin configuration as shown in Table 5.10. This display can be controlled from the microcontroller as follows:

- Send the segment bit pattern for digit 1 to segments **a—g**.
- Enable digit 1.
- Wait for a few milliseconds.
- Disable digit 1.
- Send the segment bit patter for digit 2 to segments **a—g**.
- Enable digit 2.
- Wait for a few milliseconds.
- Disable digit 2.
- Repeat the above process continuously.

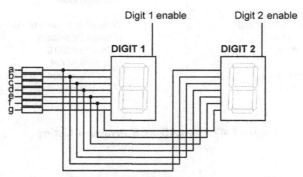

Figure 5.48: Two Multiplexed 7-Segment Displays.

Table 5.10: Pin Configuration of the
DC56-11EWA Dual Display

Pin Number	Segment
1,5	e
2,6	d
3,8	c
14	Digit 1 enable
17,7	g
15,10	b
16,11	a
18,12	f
13	Digit 2 enable
4	Decimal point 1
9	Decimal point 2

The segment configuration of the DC56-11EWA display is shown in Figure 5.49. In a multiplexed display application, the segment pins of the corresponding segments are connected together. For example, pins 11 and 16 are connected as the common **a** segment. Similarly, pins 15 and 10 are connected as the common **b** segment, and so on.

Project Hardware

The block diagram of this project is shown in Figure 5.50. The circuit diagram is given in Figure 5.51. The segments of the display are connected to PORTD of a PIC18F45K22-type microcontroller, operated with an 8-MHz crystal. Current limiting resistors are used on each segment of the display. Each digit is enabled using a BC108-type transistor switch connected to RA0 and RA1 port pins of the microcontroller. A segment is turned on when a logic 1 is applied to the base of the corresponding segment transistor.

If you are using the EasyPIC V7 development board, then make sure that the following jumpers are configured to enable two digits of the 7-segment display:

DIP switch SW4: DIS0 and DIS1 enabled to RA0 and RA1, respectively

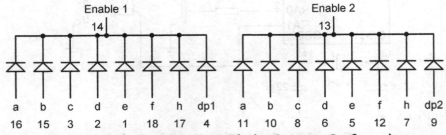

Figure 5.49: The DC56-11EWA Display Segment Configuration.

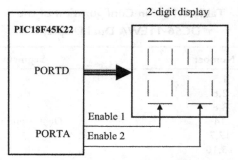

Figure 5.50: Block Diagram of the Project.

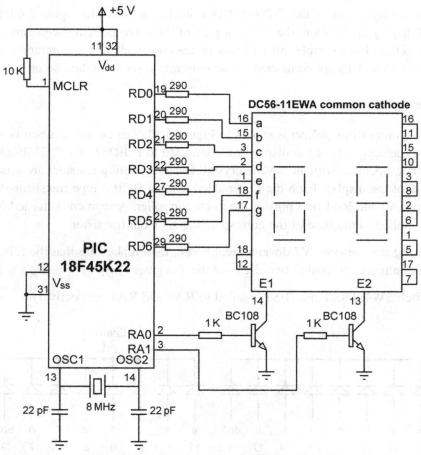

Figure 5.51: Circuit Diagram of the Project.

```
        BEGIN
            Create SEGMENT table
            Configure PORTA as digital outputs
            Configure PORTD as digital outputs
            Initialize CNT to 25
            DO FOREVER
                Find MSD digit
                Get bit pattern from SEGMENT
                Enable digit 2
                Wait for a while
                Disable digit 2
                Find LSD digit
                Get bit pattern from SEGMENT
                Enable digit 1
                Wait for a while
                Disable digit 1
            ENDDO
        END
```

Figure 5.52: PDL of the Project.

Project PDL

At the beginning of the program, PORTA and PORTD pins are configured as outputs. The program then enters an endless loop where first of all the most significant digit (MSD) of the number is calculated, function **Display** is called to find the bit pattern, and then sent to the display and digit 1 is enabled. Then, after a small delay digit 1 is disabled, the LSD of the number is calculated, function **Display** is called to find the bit pattern and then sent to the display, and digit 2 is enabled. Then, again after a small delay, digit 2 is disabled and the above process repeats indefinitely. Figure 5.52 shows the PDL of the project.

Project Program

mikroC Pro for PIC

The mikroC Pro for the PIC program is called MIKROC-LED9.C, and the program listing is given in Figure 5.53. **DIGIT1** and **DIGIT2** are defined to be equal to bit 0 and bit 1 of PORTA, respectively. The value to be displayed (number 25) is stored in variable **Cnt**. An endless loop is formed using a **for** statement. Inside the loop, the MSD of the number is calculated by dividing the number by 10. Function **Display** is then called to find the bit pattern to send to PORTD. Then, digit 2 is enabled by setting DIGIT2 = 1, and the program waits for 10 ms. After this, digit 2 is disabled, and the LSD of the number is calculated using the mod operator ("%") and sent to PORTD. At the same time, digit 1 is enabled by setting DIGIT1 = 1, and the program waits for 10 ms. After this time, digit 1 is disabled, and the program repeats forever.

```
/*******************************************************************************
                            Dual 7-SEGMENT DISPLAY
                            ======================

In this project two common cathode 7-segment LED displays are connected to PORTD of a
PIC18F45K22 microcontroller and the microcontroller is operated from an 8 MHz crystal.
Digit 1 (right digit) enable pin is connected to port pin RA0 and digit 2 (left digit) enable pin is
connected to port pin RA1 of the microcontroller. The program displays number 25 on the displays.


Author:         Dogan Ibrahim
Date:           August 2013
File:           MIKROC-LED9.C
*******************************************************************************/
#define DIGIT1 PORTA.RA0
#define DIGIT2 PORTA.RA1


//
// This function finds the bit pattern to be sent to the port to display a number
// on the 7-segment LED. The number is passed in the argument list of the function.
//
unsigned char Display(unsigned char no)
{
  unsigned char Pattern;
  unsigned char SEGMENT[] = {0x3F,0x06,0x5B,0x4F,0x66,0x6D,
                  0x7D,0x07,0x7F,0x6F};

  Pattern = SEGMENT[no];               // Pattern to return
  return (Pattern);
}


//
// Start of MAIN Program
//
void main()
{
  unsigned char Msd, Lsd, Cnt = 25;

    ANSELA = 0;                        // Configure PORTA as digital
    ANSELD = 0;                        // Configure PORTD as digital
    TRISA = 0;                         // Configure PORTA as outputs
    TRISD = 0;                         // Configure PORTD as outputs

    DIGIT1 = 0;                        // Disable digit 1
    DIGIT2 = 0;                        // Disable digit 2

    for(;;)                            // Endless loop
    {
      Msd = Cnt/10;                    // MSD digit
      PORTD = Display(Msd);            // Send to PORTD
      DIGIT2 = 1;                      // Enable digit 2
```

Figure 5.53: mikroC Pro for PIC Program Listing.

```
        Delay_Ms(10);                    // Wait a while

        DIGIT2 = 0;                      // Disable digit 2
        Lsd = Cnt % 10;                  // LSD digit
        PORTD = Display(Lsd);            // Send to PORTD
        DIGIT1 = 1;                      // Enable digit 1
        Delay_Ms(10);                    // Wait a while
        DIGIT1 = 0;                      // Disable digit 1
    }
}
```

Figure 5.53
cont'd

MPLAB XC8

The MPLAB X8 version of the program is shown in Figure 5.54 (XC8-LED9.C). The operation of the program is as in mikroC Pro for PIC.

Project 5.10—Four-Digit Multiplexed 7-Segment LED
Project Description

This project is similar to that in the previous project, but here, multiplexed four digits are used instead of two. The display digits are enabled and disabled as in the two-digit example.

Project Hardware

The block diagram of this project is shown in Figure 5.55. The circuit diagram is given in Figure 5.56. The segments of the display are connected to PORTD of a PIC18F45K22 type microcontroller, operated with an 8-MHz crystal. Current limiting resistors are used on each segment of the display. Each digit is enabled using a BC108 type transistor switch connected to RA0:RA3 port pins of the microcontroller. A segment is turned on when logic 1 is applied to the base of the corresponding segment transistor.

If you are using the EasyPIC V7 development board, then make sure that the following jumpers are configured to enable four digits of the 7-segment display:

DIP switch SW4: DIS0:DIS3 enabled to RA0:RA3 respectively

Project PDL

Figure 5.57 shows the PDL of the project. At the beginning of the program, DIGIT1—DIGIT4 connections are defined. Then, PORTA and PORTD are configured as digital outputs, and all the digits are disabled. The program executes in an endless *for*

```
/******************************************************************************
                        Dual 7-SEGMENT DISPLAY
                        =======================

In this project two common cathode 7-segment LED displays are connected to PORTD of a
PIC18F45K22 microcontroller and the microcontroller is operated from an 8 MHz crystal
Digit 1 (right digit) enable pin is connected to port pin RA0 and digit 2 (left digit) enable pin
is connected to port pin RA1 of the microcontroller. The program displays number 25 on the
displays.

Author:         Dogan Ibrahim
Date:           August 2013
File:           XC8-LED9.C
******************************************************************************/

#include <xc.h>
#pragma config MCLRE = EXTMCLR, WDTEN = OFF, FOSC = HSHP
#define _XTAL_FREQ 8000000

#define DIGIT1 PORTAbits.RA0
#define DIGIT2 PORTAbits.RA1


//
// This function finds the bit pattern to be sent to the port to display a number on the
// 7-segment LED. The number is passed in the argument list of the function.
//
unsigned char Display(unsigned char no)
{
  unsigned char Pattern;
  unsigned char SEGMENT[] = {0x3F,0x06,0x5B,0x4F,0x66,0x6D,
              0x7D,0x07,0x7F,0x6F};

  Pattern = SEGMENT[no];                      // Pattern to return
  return (Pattern);
}



void main()
{
  unsigned char Msd, Lsd, Cnt = 25;

  ANSELA = 0;                                 // Configure PORTA as digital
  ANSELD = 0;                                 // Configure PORTD as digital
  TRISA = 0;                                  // Configure PORTA as outputs
  TRISD = 0;                                  // Configure PORTD as outputs

  DIGIT1 = 0;                                 // Disable digit 1
  DIGIT2 = 0;                                 // Disable digit 2

  for(;;)                                     // Endless loop
  {
```

Figure 5.54: MPLAB XC8 Program Listing.

```
Msd = Cnt/10;                    // MSD digit
PORTD = Display(Msd);            // Send to PORTD
DIGIT2 = 1;                      // Enable digit 2
__delay_ms(10);                  // Wait a while

DIGIT2 = 0;                      // Disable digit 2
Lsd = Cnt % 10;                  // LSD digit
PORTD = Display(Lsd);            // Send to PORTD
DIGIT1 = 1;                      // Enable digit 1
__delay_ms(10);                  // Wait a while
DIGIT1 = 0;                      // Disable digit 1
    }
}
```

Figure 5.54
cont'd

loop. Inside this loop, the digits of the number to be displayed (1234) are extracted, sent to PORTD, and the corresponding digits are enabled.

Function *Display* extracts the bit pattern corresponding to a number. Each digit is displayed for 5 ms.

Project Program

mikroC Pro for PIC

The mikroC Pro for the PIC program is called MIKROC-LED10.C and the program listing is given in Figure 5.58. **DIGIT1–DIGIT4** are defined to be equal to bit 0–bit 3 of PORTA, respectively. The value to be displayed (number 1234 in this example) is stored in variable **Cnt**. An endless loop is formed using a **for** statement. Inside this loop, a digit is

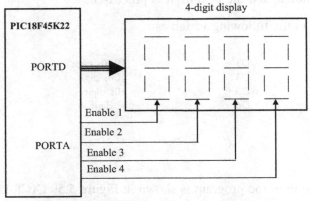

Figure 5.55: Block Diagram of the Project.

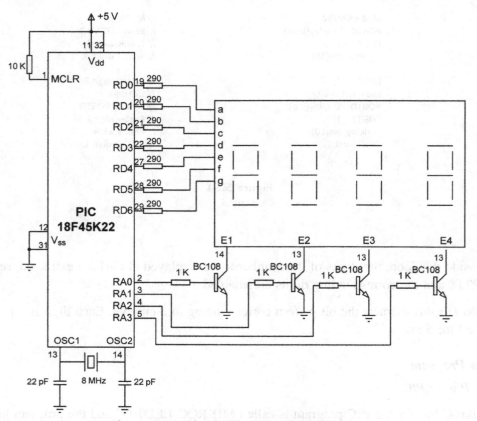

Figure 5.56: Circuit Diagram of the Project.

extracted, sent to the display, and then the corresponding digit is enabled. After a 5-ms delay, the digit is disabled, and the next digit is processed.

The digits are stored in the following variables:

D1	1000s digit
D3	100s digit
D5	10s digit
D6	1s digit

MPLAB XC8

The MPLAB X8 version of the program is shown in Figure 5.59 (XC8-LED10.C). The operation of the program is as in mikroC Pro for PIC.

BEGIN
    Define digits
    Configure PORTA as digital outputs
    Configure PORTD as digital outputs
    Initialize CNT to 1234
    DO FOREVER
        Find 1000s digit
        CALL Display to get bit pattern
        Enable digit 4
        Wait for a while
        Disable digit 4
        Find 100s digit
        CALL Display to get bit pattern
        Enable digit 3
        Wait for a while
        Disable digit 3
        Find 10s digit
        CALL Display to get bit pattern
        Enable digit 2
        Wait for a while
        Disable digit 2
        Find 1s digit
        CALL Display to get bit pattern
        Enable bit 1
        Wait for a while
        Disable digit 1
    ENDDO
END

BEGIN/Display
    Create Segment table
    Find the bit pattern corresponding to the number
    Return the bit pattern
END/Display

Figure 5.57: PDL of the Project.

Project 5.11—LED Voltmeter

Project Description

In this project, a voltmeter with an LED display is designed. The voltmeter can be used to measure voltages 0—5 V. The voltage to be measured is applied to one of the analog inputs of a PIC18F45K22 microcontroller. The microcontroller reads the analog voltage, converts into digital, formats it, and then turns ON one of the five LEDs to indicate the applied voltage range.

Figure 5.60 shows the block diagram of the project.

Project Hardware

The circuit diagram is given in Figure 5.61. The LEDs are connected to PORTD, and analog input RA0 (AN0, channel 0) is used to read the input voltage.

```
/*****************************************************************************
                        4-DIGIT 7-SEGMENT DISPLAY
                        =========================

In this project four common cathode 7-segment LED displays are connected to PORTD of a
PIC18F45K22 microcontroller and the microcontroller is operated from an 8 MHz crystal.
Four PORTA pins are used to enable/disable the LEDs. The program displays number 1234.


Author:        Dogan Ibrahim
Date:          August 2013
File:          MIKROC-LED10.C
*****************************************************************************/
#define DIGIT1 PORTA.RA0
#define DIGIT2 PORTA.RA1
#define DIGIT3 PORTA.RA2
#define DIGIT4 PORTA.RA3

//
// This function finds the bit pattern to be sent to the port to display a number
// on the 7-segment LED. The number is passed in the argument list of the function.
//
unsigned char Display(unsigned char no)
{
  unsigned char Pattern;
  unsigned char SEGMENT[] = {0x3F,0x06,0x5B,0x4F,0x66,0x6D,0x7D,0x07,0x7F,0x6F};

  Pattern = SEGMENT[no];                           // Pattern to return
  return (Pattern);
}



//
// Start of MAIN Program
//
void main()
{
    unsigned int Cnt = 1234;
    unsigned int D1,D2,D3,D4,D5,D6;
    ANSELA = 0;                          // Configure PORTA as digital
    ANSELD = 0;                          // Configure PORTD as digital
    TRISA = 0;                           // Configure PORTA as outputs
    TRISD = 0;                           // Configure PORTD as outputs

    DIGIT1 = 0;                          // Disable digit 1
    DIGIT2 = 0;                          // Disable digit 2
    DIGIT3 = 0;                          // Disable digit 3
    DIGIT4 = 0;                          // Disable digit 4

    for(;;)                              // Endless loop
    {
```

Figure 5.58: mikroC Pro for PIC Program Listing.

```
          D1 = Cnt/1000;               // 1000s  digit
          PORTD = Display(D1);         // Send to PORTD
          DIGIT4 = 1;                  // Enable digit 4
          Delay_Ms(5);                 // Wait a while
          DIGIT4 = 0;                  // Disable digit 4

          D2 = Cnt % 1000;
          D3 = D2/100;                 // 100s digit
          PORTD = Display(D3);         // Send to PORTD
          DIGIT3 = 1;                  // Enable digit 3
          Delay_Ms(5);                 // Wait a while
          DIGIT3 = 0;                  // Disable digit 3

          D4 = D2 % 100;
          D5 = D4/10;                  // 10s digit
          PORTD = Display(D5);         // Send to PORTD
          DIGIT2 = 1;                  // Enable digit 2
          Delay_Ms(5);                 // Wait a while
          DIGIT2 = 0;                  // Disable digit 2

          D6 = D4 % 10;
          PORTD = Display(D6);         // Send to PORTD
          DIGIT1 = 1;                  // Enable digit 1
          Delay_Ms(5);                 // Wait a while
          DIGIT1 = 0;          /       / Disable digit 1

      }
  }
```

Figure 5.58
cont'd

If you are using the EasyPIC V7 development board, then make sure that the following jumper is connected so that the potentiometer can be enabled at the RA0 input:

Jumper J15 (ADC INPUT): Connect RA0 inputs

Project PDL

Figure 5.62 shows the PDL of the project. At the beginning of the program, PORTD is configured as a digital output, and RA0 is configured as analog input. Then, an endless loop is formed, and the applied voltage is read and converted into analog millivolts. The corresponding LED is then turned ON. This process repeats continuously with a 10-ms delay between each iteration.

Project Program

mikroC Pro for PIC

The mikroC Pro for PIC program is called MIKROC-LED11.C, and the program listing is given in Figure 5.63. At the beginning of the program, the LEDs are given symbols to

```
/*******************************************************************************
                        4-DIGIT 7-SEGMENT DISPLAY
                        ==========================

In this project two common cathode 7-segment LED displays are connected to PORTD
of a PIC18F45K22 microcontroller and the microcontroller is operated from an 8 MHz crystal.
Four PORTA pins are used to enable/disable the LEDs.

The program displays number 1234.


Author:     Dogan Ibrahim
Date:       August 2013
File:       XC8-LED10.C
*******************************************************************************/

#include <xc.h>
#pragma config MCLRE = EXTMCLR, WDTEN = OFF, FOSC = HSHP
#define _XTAL_FREQ 8000000

#define DIGIT1 PORTAbits.RA0
#define DIGIT2 PORTAbits.RA1
#define DIGIT3 PORTAbits.RA2
#define DIGIT4 PORTAbits.RA3

//
// This function finds the bit pattern to be sent to the port to display a number
// on the 7-segment LED. The number is passed in the argument list of the function.
//
unsigned char Display(unsigned char no)
{
    unsigned char Pattern;
    unsigned char SEGMENT[] = {0x3F,0x06,0x5B,0x4F,0x66,0x6D,0x7D,0x07,0x7F,0x6F};

    Pattern = SEGMENT[no];                          // Pattern to return
    return (Pattern);
}


void main()
{
    unsigned int Cnt = 1234;
    unsigned int D1,D2,D3,D4,D5,D6;
    ANSELA = 0;                                     // Configure PORTA as digital
    ANSELD = 0;                                     // Configure PORTD as digital
    TRISA = 0;                                      // Configure PORTA as outputs
    TRISD = 0;                                      // Configure PORTD as outputs

    DIGIT1 = 0;                                     // Disable digit 1
    DIGIT2 = 0;                                     // Disable digit 2
    DIGIT3 = 0;                                     // Disable digit 3
```

Figure 5.59: MPLAB XC8 Program Listing.

```
         DIGIT4 = 0;                              // Disable digit 4

      for(;;)                                     // Endless loop
      {
         D1 = Cnt/1000;                           // 1000s  digit
         PORTD = Display(D1);                     // Send to PORTD
         DIGIT4 = 1;                              // Enable digit 4
         __delay_ms(5);                           // Wait a while
         DIGIT4 = 0;                              // Disable digit 4

         D2 = Cnt % 1000;
         D3 = D2/100;                             // 100s digit
         PORTD = Display(D3);                     // Send to PORTD
         DIGIT3 = 1;                              // Enable digit 3
         __delay_ms(5);                           // Wait a while
         DIGIT3 = 0;                              // Disable digit 3

         D4 = D2 % 100;
         D5 = D4/10;                              // 10s digit
         PORTD = Display(D5);                     // Send to PORTD
         DIGIT2 = 1;                              // Enable digit 2
         __delay_ms(5);                           // Wait a while
         DIGIT2 = 0;                              // Disable digit 2

         D6 = D4 % 10;
         PORTD = Display(D6);                     // Send to PORTD
         DIGIT1 = 1;                              // Enable digit 1
         __delay_ms(5);                           // Wait a while
         DIGIT1 = 0;                              // Disable digit 1

      }
   }
```

Figure 5.59
cont'd

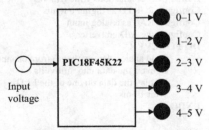

Figure 5.60: Block Diagram of the Project.

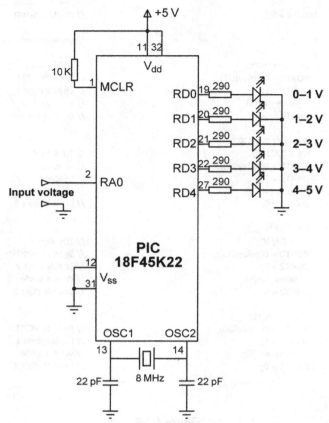

Figure 5.61: Circuit Diagram of the Project.

```
BEGIN
        Define LED connections with symbols
        Configure PORTD as digital outputs
        Configure RA0 as analog input
        Configure the A/D converter
        DO FOREVER
                Read analog data (voltage) from channel 0
                Convert the data into millivolts
                Display the data on one of the LEDs
                Wait 10 ms
        ENDO
END
```

Figure 5.62: PDL of the Project.

```
/*******************************************************************************
                        VOLTMETER WITH LED DISPLAYS
                        ============================
```

In this project 5 LEDs are connected to PORTD. Also, input port RA0 (AN0) is used as analog input. Voltage to be measured is applied to this pin. The microcontroller reads the analog voltage, converts into digital, and then turns ON one of the LEDs to indicate the voltage range.

Analog input range is 0 to 5 V. A PIC18F45K22 type microcontroller is used in this project, operated with an 8 MHz cystal.

Analog data is read using the mikroC Pro for PIC built-in function Adc_Read.

```
Author:      Dogan Ibrahim
Date:        August 2013
File:        MIKROC-LED11.C
*******************************************************************************/
#define LED01V PORTD.RD0
#define LED12V PORTD.RD1
#define LED23V PORTD.RD2
#define LED34V PORTD.RD3
#define LED45V PORTD.RD4

void main()
{
    unsigned long Vin, mV;

    ANSELD = 0;                               // Configure PORTD as digital
    ANSELA = 1;                               // COnfigure RA0 as analog
    TRISD = 0;                                // Configure PORTD as outputs
    TRISA = 1;                                // Configure RA0 as input
//
// Configure A/D converter. Channel 0 (RA0, or AN0) is used in this project.
//
    ADCON2 = 0x80;                            // Right justify the result

    for(;;)                                   // Endless loop
    {
        PORTD = 0;                            // Clear LEDs to start with
        Vin = Adc_Read(0);                    // Read from channel 0 (AN0)
        mV = (Vin * 5000) >> 10;              // mv = Vin x 5000/1024
//
// Find which LED to turn ON
//
        if(mV >= 0 && mV <= 1000)             // Between 1–2 V
        {
            LED01V = 1;                       // Turn ON LED0–1 V
        }
        else if(mV > 1000 && mV <= 2000)      // Between 2–3 V
```

Figure 5.63: mikroC Pro for PIC Program Listing.

```
        {
            LED12V = 1;                              // Turn ON LED1-2 V
        }
        else if(mV > 2000 && mV <= 3000)            // Between 2-3 V
        {
            LED23V = 1;                              // Turn ON LED2-3 V
        }
        else if(mv > 3000 && mV <= 4000)            // Between 3-4 V
        {
            LED34V = 1;                              // Turn ON LED3-4 V
        }
        else if(mV > 4000 && mV <= 5000)            // Between 4-5 V
        {
            LED45V = 1;                              // Turn ON LED4-5 V
        }
        Delay_Ms(10);                                // 10 ms delay
    }
}
```

Figure 5.63
cont'd

make it easy to identify them. Then, PORTD is configured as a digital output and RA0 is configured as an analog input. The A/D converter is configured to right justify the result. The program executes in an endless loop established using a for statement. Inside this loop, analog data are read from channel 0 using the **Adc_Read(0)** built-in function. The converted digital data are stored in variable **Vin**, which is declared as an **unsigned long**. The A/D converter is 10 bits wide, and thus, there are 1024 steps (0−1023) corresponding to the reference voltage of 5000 mV. Each step corresponds to 5000 mV/1024 = 4.88 mV. Inside the loop variable, **Vin** is converted into millivolts after multiplying by 5000 and dividing into 1024. The division is done by shifting right by 10 digits. At this point, variable **mV** contains the converted input voltage in millivolts. A number of conditional statements are then used to turn ON the required LED.

MPLAB XC8

The MPLAB X8 version of the program is shown in Figure 5.64 (XC8-LED11.C). The A/D converter process is slightly more complex. XC8 supports the following A/D converter functions (see *PIC18 Peripheral Library Help Document*).

OpenADC: This function is used to configure the A/D module. The number of arguments required depends on the type of microcontroller used (see *PIC18 Peripheral Library Help Document*). PIC18F45K22 microcontroller requires three arguments. The bits of an argument should be separated with the AND operator (&), and the arguments should be separated using commas.

```
/********************************************************************************
                        VOLTMETER WITH LED DISPLAYS
                        ===========================
```

In this project 5 LEDs are connected to PORTD. Also, input port RA0 (AN0) is used as analog input. Voltage to be measured is applied to this pin. The microcontroller reads the analog voltage, converts into digital, and then turns ON one of the LEDs to indicate the voltage range.

Analog input range is 0 to 5 V. A PIC18F45K22 type microcontroller is used in this project, operated with an 8 MHz cystal.

Analog data is read using the MPLAB XC8 built-in functions.

```
Author:         Dogan Ibrahim
Date:           August 2013
File:           XC8-LED11.C
********************************************************************************/
#include <xc.h>
#include <plib/adc.h>
#pragma config MCLRE = EXTMCLR, WDTEN = OFF, FOSC = HSHP
#define _XTAL_FREQ 8000000

#define LED01V PORTDbits.RD0
#define LED12V PORTDbits.RD1
#define LED23V PORTDbits.RD2
#define LED34V PORTDbits.RD3
#define LED45V PORTDbits.RD4

void main()
{
    unsigned long Vin, mV;

    ANSELD = 0;                                     // Configure PORTD as digital
    ANSELA = 1;                                     // COnfigure RA0 as analog
    TRISD = 0;                                      // Configure PORTD as outputs
    TRISA = 1;                                       // Configure RA0 as input
//
// Configure A/D converter
//
    OpenADC(ADC_FOSC_2 & ADC_RIGHT_JUST & ADC_12_TAD,
        ADC_CH0 & ADC_INT_OFF,
        ADC_TRIG_CTMU & ADC_REF_VDD_VDD & ADC_REF_VDD_VSS);

    for(;;)                                         // Endless loop
    {
      PORTD = 0;                                    // Clear LEDs to start with
      SelChanConvADC(ADC_CH0);                      // Select channel 0 and start conversion
      while(BusyADC());                             // Wait for completion
      Vin = ReadADC();                              // Rea converted data
      mV = (Vin * 5000) >> 10;                      // mv = Vin x 5000/1024
```

Figure 5.64: MPLAB XC8 Program Listing.

```
//
// Find which LED to turn ON
//
    if(mV <= 1000)                                  // Between 1–2 V
    {
    LED01V = 1;                                      // Turn ON LED0–1 V
    }
    else if(mV > 1000 && mV <= 2000)                 // Between 2–3 V
    {
      LED12V = 1;                                     // Turn ON LED1–2 V
    }
    else if(mV > 2000 && mV <= 3000)                 // Between 2–3 V
    {
      LED23V = 1;                                     // Turn ON LED2–3 V
    }
    else if(mV > 3000 && mV <= 4000)                 // Between 3–4 V
    {
      LED34V = 1;                                     // Turn ON LED3–4 V
    }
    else if(mV > 4000 && mV <= 5000)                 // Between 4–5 V
    {
      LED45V = 1;                                     // Turn ON LED4–5 V
    }
    __delay_ms(10);                                   // 10 ms delay
    }
}
```

Figure 5.64
cont'd

The following arguments are valid for the PIC18F45K22 microcontroller:

Argument 1:

A/D clock source
    * ADC_FOSC_2
    * ADC_FOSC_4
    * ADC_FOSC_8
    * ADC_FOSC_16
    * ADC_FOSC_32
    * ADC_FOSC_64
    * ADC_FOSC_RC
    * ADC_FOSC_MASK
A/D result justification
    * ADC_RIGHT_JUST
    * ADC_LEFT_JUST
    * ADC_RESULT_MASK

A/D acquisition time select
* ADC_0_TAD
* ADC_2_TAD
* ADC_4_TAD
* ADC_6_TAD
* ADC_12_TAD
* ADC_16_TAD
* ADC_20_TAD
* ADC_TAD_MASK

Argument 2:

Channel
* ADC_CH0
* ADC_CH1
* ADC_CH2
* ADC_CH3
* ADC_CH4
* ADC_CH5
* ADC_CH6
* ADC_CH7
* ADC_CH8
* ADC_CH9
* ADC_CH10
* ADC_CH11
* ADC_CH12
* ADC_CH13
ADC_CH14
* ADC_CH15
* ADC_CH16
* ADC_CH17
* ADC_CH18
* ADC_CH19
* ADC_CH20
* ADC_CH21
* ADC_CH22
* ADC_CH23
* ADC_CH24
* ADC_CH25
* ADC_CH26
* ADC_CH27

* ADC_CH_CTMU
* ADC_CH_DAC
* ADC_CH_FRV

A/D Interrupts
* ADC_INT_ON
* ADC_INT_OFF
* ADC_INT_MASK

Argument 3:

Special Trigger Select
* ADC_TRIG_CTMU
* ADC_TRIG_CCP5

A/D VREF + Configuration
* ADC_REF_VDD_VDD
* ADC_REF_VDD_VREFPLUS
* ADC_REF_FVR_BUF

A/D VREF− Configuration
* ADC_REF_VDD_VSS
* ADC_REF_VDD_VREFMINUS

SetChanADC:	This function selects the channel to be used for the A/D converter.
SetChanConvADC:	This function selects the channel to be used and at the same time starts the conversion.
ConvertADC:	This function starts the A/D conversion.
BusyADC:	This function returns the A/D conversion status.
ReadADC:	This function returns the A/D result.
CloseADC:	This function turns off the A/D converter module.

The MPLAB XC8 A/D conversion process is carried out as follows:

* Configure the A/D converter (OpenADC).
* Select channel to be used (SelChanConvADC).
* Wait until the conversion is complete (BusyADC).
* Read converted data (ReadADC).

Project 5.12—LCD Voltmeter
Project Description

In this project, the design of a voltmeter with an LCD display is described. The voltmeter can be used to measure voltages in the range 0–5 V. The voltage to be measured is applied to one of the analog inputs of a PIC18F45K22-type microcontroller. The microcontroller reads the analog voltage, converts into digital, and then displays on an LCD.

In microcontroller systems, the output of a measured variable is usually displayed using LEDs, 7-segment displays, or LCD-type displays. LCDs have the advantages that they can be used to display alphanumeric or graphical data. Some LCDs have ≥ 40 character lengths with the capability to display several lines. Some other LCD displays can be used to display graphics images. Some modules offer color displays while some others incorporate back lighting so that they can be viewed in dimly lit conditions.

There are basically two types of LCDs as far as the interface technique is concerned: parallel LCDs and serial LCDs. Parallel LCDs (e.g. Hitachi HD44780) are connected to a microcontroller using more than one data line, and the data are transferred in parallel form. It is common to use either four or eight data lines. Using a four-wire connection saves I/O pins, but it is slower since the data are transferred in two stages. Serial LCDs are connected to the microcontroller using only one data line, and data are usually sent to the LCD using the standard RS-232 asynchronous data communication protocol. Serial LCDs are much easier to use, but they cost more than the parallel ones do.

The programming of a parallel LCD is usually a complex task and requires a good understanding of the internal operation of the LCD controllers, including the timing diagrams. Fortunately, most high-level languages provide special library commands for displaying data on alphanumeric as well as on graphical LCDs. All the user has to do is connect the LCD to the microcontroller, define the LCD connection in the software, and then send special commands to display data on the LCD.

HD44780 LCD Module

HD44780 is one of the most popular alphanumeric LCD modules used in the industry and also by hobbyists. This module is monochrome and comes in different sizes. Modules with 8, 16, 20, 24, 32, and 40 columns are available. Depending on the model chosen, the number of rows varies between 1, 2, or 4. The display provides a 14-pin (or 16-pin) connector to a microcontroller. Table 5.11 gives the pin configuration and pin functions of a 14-pin LCD module. Below is a summary of the pin functions:

V_{SS} is the 0-V supply or ground. The V_{DD} pin should be connected to the positive supply. Although the manufacturers specify a 5-V dc supply, the modules will usually work with as low as 3 V or as high as 6 V.

Pin 3 is named V_{EE}, and this is the contrast control pin. This pin is used to adjust the contrast of the display, and it should be connected to a variable voltage supply. A potentiometer is normally connected between the power supply lines with its wiper arm connected to this pin so that the contrast can be adjusted.

Table 5.11: Pin Configuration of the HD44780 LCD Module

Pin Number	Name	Function
1	V_{SS}	Ground
2	V_{DD}	+ve supply
3	V_{EE}	Contrast
4	RS	Register select
5	R/W	Read/write
6	E	Enable
7	D0	Data bit 0
8	D1	Data bit 1
9	D2	Data bit 2
10	D3	Data bit 3
11	D4	Data bit 4
12	D5	Data bit 5
13	D6	Data bit 6
14	D7	Data bit 7

Pin 4 is the Register Select (RS), and when this pin is LOW, data transferred to the display are treated as commands. When RS is HIGH, character data can be transferred to and from the module.

Pin 5 is the Read/Write (R/W) line. This pin is pulled LOW to write commands or character data to the LCD module. When this pin is HIGH, character data or status information can be read from the module.

Pin 6 is the Enable (E) pin that is used to initiate the transfer of commands or data between the module and the microcontroller. When writing to the display, data are transferred only on the HIGH to LOW transition of this line. When reading from the display, data become available after the LOW to HIGH transition of the enable pin, and these data remain valid as long as the enable pin is at logic HIGH.

Pins 7—14 are the eight data bus lines (D0—D7). Data can be transferred between the microcontroller and the LCD module using either a single 8-bit byte, or as two 4-bit nibbles. In the latter case, only the upper four data lines (D4—D7) are used. Four-bit mode has the advantage that four less I/O lines are required to communicate with the LCD. In this book, we shall be using alphanumeric-based LCD only and look at the 4-bit interface only.

Connecting the LCD to the Microcontroller

The following pins are used in 4-bit mode:

D4:D7
E
R/S

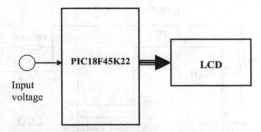

Figure 5.65: Block Diagram of the Project.

In this book, we shall be connecting the LCD to our microcontroller as in the following table:

LCD Pin	Microcontroller Pin
D4	RB0
D5	RB1
D6	RB2
D7	RB3
R/S	RB4
E	RB5

Project Hardware

Figure 5.65 shows the block diagram of the project. The microcontroller reads the analog voltage, converts into digital, formats it, and then displays on the LCD.

The circuit diagram of the project is shown in Figure 5.66. The voltage to be measured (between 0 and 5 V) is applied to port AN0 where this port is configured as an analog input in software. The LCD is connected to PORTB of the microcontroller as described earlier. A potentiometer is used to adjust the contrast of the LCD display.

Project PDL

The PDL of the project is shown in Figure 5.67. At the beginning of the program, PORTB is configured as the output, and PORTA is configured as the input. Then the LCD and the A/D converter are configured. The program then enters an endless loop where analog input voltage is converted into digital and displayed on the LCD. This process is repeated every second.

Project Program

mikroC Pro for PIC

The mikroC Pro for PIC program listing (MIKROC-LCD1.C) is given in Figure 5.68. At the beginning of the program, the connections between the LCD and the microcontroller are defined using a number of *sbit* statements. PORTB is defined as the output and PORTA as the input. Then, the LCD is configured, and the text "VOLTMETER" is displayed on

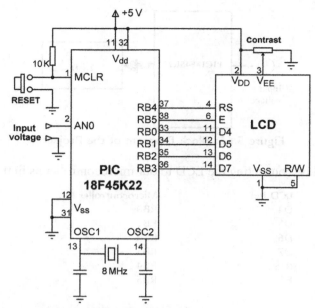

Figure 5.66: Circuit Diagram of the Project.

the LCD for 2 s. The A/D is configured by setting register ADCON1 to 0x80 so that the A/D result is right justified, V_{ref} voltage is set to V_{DD} (+5 V), and all PORTA pins are configured as analog inputs. The message "VOLTMETER" is displayed on the first row of the LCD for 2 s.

The main program loop starts with a **for** statement. Inside this loop, the LCD is cleared, analog data are read from channel 0 (pin AN0) using statement **Adc_Read(0)**. The converted digital data are stored in variable **Vin**, which is declared as an **unsigned long**. The A/D converter is 10 bits wide, and thus, there are 1024 steps (0−1023) corresponding to the reference voltage of 5000 mV. Each step corresponds to 5000 mV/1024 = 4.88 mV. Inside the loop, variable **Vin** is converted into millivolts by multiplying by 5000 and

```
BEGIN
        Configure PORTB as digital outputs
        Configure RA0 as input
        Configure the LCD
        Display heading
        Configure the A/D converter
        DO FOREVER
                Read analog data (voltage) from channel 0
                Format the data
                Display the data (voltage)
                Wait 1 s
        ENDO
END
```

Figure 5.67: PDL of the Project.

```
/*******************************************************************************
                        VOLTMETER WITH LCD DISPLAY
                        ===========================
```

In this project an LCD is connected to PORTB. Also, input port AN0 is used as analog input. Voltage to be measured is applied to AN0. The microcontroller reads the analog voltage, converts into digital, and then displays on the LCD.

Analog input range is 0 to 5 V. A PIC18F45K22 type microcontroller is used in this project, operated with an 8 MHz crystal.

Analog data is read using the Adc_Read built-in function.

The LCD is connected to the microcontroller as follows:

```
Microcontroller    LCD
============       ===

    RB0            D4
    RB1            D5
    RB2            D6
    RB3            D7
    RB4            R/S
    RB5            E

Author:        Dogan Ibrahim
Date:          August 2013
File:          MIKROC-LCD1.C
*************************************************************/
// LCD module connections
sbit LCD_RS at RB4_bit;
sbit LCD_EN at RB5_bit;
sbit LCD_D4 at RB0_bit;
sbit LCD_D5 at RB1_bit;
sbit LCD_D6 at RB2_bit;
sbit LCD_D7 at RB3_bit;

sbit LCD_RS_Direction at TRISB4_bit;
sbit LCD_EN_Direction at TRISB5_bit;
sbit LCD_D4_Direction at TRISB0_bit;
sbit LCD_D5_Direction at TRISB1_bit;
sbit LCD_D6_Direction at TRISB2_bit;
sbit LCD_D7_Direction at TRISB3_bit;
// End LCD module connections


void main()
{
    unsigned long Vin, mV;
    unsigned char op[12];
    unsigned char i,j,lcd[5];
```

Figure 5.68: mikroC Pro for PIC Program Listing.

```
        char *str;

        ANSELB = 0;                              // Configure PORTB as digital
        ANSELA = 1;                              // Configure RA0 as analog
        TRISB = 0;                               // Configure PORTB as outputs
        TRISA = 1;                               // Configure RA0 as input

//
// Configure LCD
//
    Lcd_Init();                                  // LCD is connected to PORTB
    Lcd_Cmd(_LCD_CLEAR);
    Lcd_cmd(_LCD_CURSOR_OFF);                    // Hide cursor
    Lcd_Out(1,1,"VOLTMETER");
    Delay_ms(2000);
//
// Configure A/D converter. AN0 is used in this project.
//
    ADCON1 = 0x80;                               // Use AN0 and Vref = +5 V
//
// Program loop
//
    for(;;)                                      // Endless loop
    {
    Lcd_Cmd(_LCD_CLEAR);                         // Clear LCD
    Vin = Adc_Read(0);                           // Read from channel 0 (AN0)
    Lcd_Out(1,1,"mV = ");                        // Display "mV = "
    mV = (Vin * 5000) >> 10;                     // mv = Vin x 5000/1024
    LongToStr(mV,op);                            // Convert to string in "op"
    str = Ltrim(op);                             // Remove leading spaces
//
// Display result on LCD
//
    Lcd_Out(1,6,str);                            // Output to LCD
    Delay_ms(1000);                              // Wait 1 s
    }
}
```

Figure 5.68
cont'd

dividing into 1024. The division is done by shifting right by 10 digits. At this point, variable **mV** contains the converted data in millivolts.

Function **LongToStr** is called to convert **mV** into a string in character array **op**. **LongToStr** converts a long variable into a string having a fixed width of 12 characters. If the resulting string is <12 characters, then the left of the data is filled with leading blanks. These leading blanks are removed using built-in function **Ltrim**, and the data are stored in character variable called **str**. Function **Lcd_Out** is called to display the data on the LCD starting from column 6 of row 1. For example, if the measured voltage is 1267 mV, it is displayed on the LCD as

mV = 1267

Table 5.12: MPLAB XC8 LCD Functions

Function	Description
BusyXLCD	Check if the LCD controller is busy
OpenXLCD	Configure I/O port lines for the LCD and initialize
putcXLCD	Write a byte of data to the LCD
putsXLCD	Write a string from the data memory to the LCD
putrsXLCD	Write a string from the program memory to the LCD
ReadAddrXLCD	Read the address byte from the LCD controller
ReadDataXLCD	Read a byte from the LCD controller
SetCGRamAddr	Set the character generator address
SetDDRamAddr	Set the display data address
WriteCmdXLCD	Write a command to the LCD (the LCD must not be busy before sending a command)

The above process is repeated after a 1-s delay.

MPLAB XC8

The MPLAB XC8 version of the program is slightly more complex because of the LCD functions. Table 5.12 gives a list of the MPLAB XC8 LCD functions available. Note that the header file "xlcd.h" must be included at the beginning of a program when any of these functions are used.

The LCD library requires that the following delay functions must be included in a program using the LCD library:

DelayFor18TCY	Delay for 18 cycles
DelayPORXLCD	Delay for 15 ms
DelayXLCD	Delay for 5 ms

Assuming a microcontroller clock frequency of 4 MHz, the instruction cycle time is 1 µs. With a clock frequency of 8 MHz, the instruction cycle time is 0.5 µs. Figure 5.69 shows how the above delay functions could approximately be obtained for both 4- and 8-MHz clock frequencies. The 18-cycle delay is obtained using NOP operations, where each NOP operation takes one cycle to execute. The end of a function with no "return" statement takes two cycles. When a "return" statement is used, a BRA statement branches to the end of the function where a RETURN 0 is executed to return from the function, thus adding two more cycles. For example, the following function takes four cycles to execute:

```
void test(void)
{
    nop();   ; 1 cycle
    nop();   ; 1 cycle
}            ; RETURN 0, takes 2 cycles
```

4 MHz Clock

```
#include<delays.h>

void DelayFor18TCY(void)
{
        Nop(); Nop(); Nop(); Nop();                    // 18 cycle delay
        Nop(); Nop(); Nop(); Nop();
        Nop(); Nop(); Nop(); Nop();
        Nop(); Nop();
        return;
}


void DelayPORXLCD(void)                                // 15 ms delay
{
        Delay1KTCYx(15);
}


void DelayXLCD(void)
{
        Delay1KTCYx(5);                                // 5 ms delay
}
```

8 MHz Clock

```
#include <delays.h>

void Delayfor18TCY(void)
{
        Nop(); Nop(); Nop(); Nop();                    // 18 cycle delay
        Nop(); Nop(); Nop(); Nop();
        Nop(); Nop(); Nop(); Nop();
        Nop(); Nop();
        return;
}

void DelayPORXLCD(void)                                // 15 ms delay
{
        Delay1KTCYx(30);
}

void DelayXLCD(void)
{
        Delay1KTCYx(10);                               // 5 ms delay
}
```

Figure 5.69: LCD Delay Functions for 4- and 8-MHz Clock.

and the following function takes six cycles to execute:

```
void test(void)
{
   nop();  ; 1 cycle
   nop();  ; 1 cycle
   return; ; BRA X, 2 cycles
}          ; X: RETURN 0, 2 cycles
```

Note that we could also use the *__delay_ms(15)* and *__delay_ms(5)* for the 15- and 5-ms delays, respectively, instead of the Delay1KTCYx() function.

A brief description of the commonly used C18 LCD functions is given below.

BusyXLCD

This function checks to determine whether or not the LCD controller is busy and data or commands should not be sent to the LCD if the controller is busy. The function returns a 1 if the controller is busy, and 0 otherwise. The program can be forced to wait until the LCD controller is ready by using the following statement:

   while(BusyXLCD());

Note that this function requires the RW pin of the LCD to be connected to the microcontroller (not to ground).

OpenXLCD

This function is used to configure the interface between the microcontroller I/O ports and the LCD pins. The function requires an argument to specify the interface mode (4- or 8-bit), and the LCD character mode and number of lines used. A value should be selected and logically AND ed from the following two groups:

   FOUR_BIT
   EIGHT_BIT
   LINE_5X7
   LINE_5X10
   LINES_5X7

LINES_5X7 is used for multiple row displays. For example, if we are using a four-wire connection with an LCD having a single row with 5x7 characters, then the function should be initialized as follows:

   OpenXLCD(FOUR_BIT & LINE_5X7);

For a two-row display the initialization is

   OpenXLCD(FOUR_BIT & LINES_5X7);

The actual physical connection between the LCD and microcontroller I/O ports is defined in file "xlcd.h" and the default settings use PORTB pins in the 4-bit mode where the low 4 bits of the port (RB0–RB3) are connected to the upper data lines (D4–D7) of the LCD (see the manual *MPLAB C18 C Compiler Libraries* for more information on the default connection).

The default connection between an LCD and the microcontroller is shown below. Note that E and RS are assumed to be connected to RB4 and RB5, respectively. This is the other way round if using the EasyPIC V7 development board. In addition, the RW pin is connected to the ground on the EasyPIC V7 development board.

LCD Pin	Microcontroller Pin
E	RB4
RS	RB5
RW	RB6
D4	RB0
D5	RB1
D6	RB2
D7	RB3

The connection between an LCD and a microcontroller can be modified by editing the xlcd.h file. It will then be necessary to recompile the xlcd routines and add them to the microcontroller library of the target microcontroller.

putcXLCD

This function is used to write a byte to the LCD. The byte is passed as an argument to the function. In the following example, character "A" is displayed on the LCD:

```
unsigned char x = 'A';
putcXLCD(x);
```

putsXLCD

This function writes a string of characters from the data memory to the LCD. The writing stops when a NULL character is detected. An example use of this function is given below:

```
char txt[] = "My text";
putsXLCD(txt);
```

putrsXLCD

This function writes a string of characters from the program memory to the LCD. The writing stops when a NULL character is detected. An example use of this function is given below:

```
putrsXLCD("My Computer");
```

SerDDRamAddr

This function sets the display data address. For example, using this function, we can write to a specified row and column of the LCD. The busy status of the LCD should be checked before calling this function.

Each character occupies one DDRAM address. The first row addresses are from 00 to 0x27, the second row addresses start from 0x40 to 0x67. For example, to move the cursor to the beginning of the second row, the DDRAM address will be 0x40, and the required command is

```
SetDDRamAddr(0x40);
```

WriteCmdXLCD

This function sends a command to the LCD. The following commands can be specified in the command argument:

DOFF	Turn display off
CURSOR_OFF	Enable display, hide cursor
BLINK_ON	Enable cursor blinking
BLINK_OFF	Disable cursor blinking
SHIFT_CUR_LEFT	Shift cursor left
SHIFT_CUR_RIGHT	Shift cursor right
SHIFT_DISP_LEFT	Shift display to the left
SHIFT_DISP_RIGHT	Shift display to the right

In addition, the LCD control functions given in Table 5.13 can be specified in the argument to control the LCD.

It is important that the LCD controller should not be busy (check with function BusyXLCD) when commands are sent to it. Some example commands are given below:

```
WriteCmdXLCD(BLINK_ON);  // Blink ON
WriteCmdXLCD(1);         // Clear LCD
```

Table 5.13: LCD Functions

Command	Operation
0 x 1	Clear display
0 x 2	Home cursor
0 x 0C	Cursor off
0 x 0E	Underline cursor on
0 x 0F	Blinking cursor on
0 x 10	Move the cursor left by one position
0 x 14	Move the cursor right by one position
0 x 80	Move the cursor to the beginning of the first row
0 x C0	Move the cursor to the beginning of the second row
0 x 94	Move the cursor to the beginning of the third row
0 x D4	Move the cursor to the beginning of the fourth row

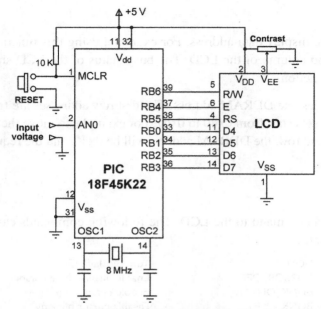

Figure 5.70: Circuit Diagram of the Project.

Figure 5.70 shows the circuit diagram of this project when used with the MPLAB XC8 compiler. Notice that this is similar to Figure 5.66 but the E and RS pins of the LCD are interchanged and also the RW pin is connected to pin RB6 of the microcontroller.

The MPLAB X8 version of the program for this project is shown in Figure 5.71 (XC8-LCD1.C). The following functions are used in the program:

Delay_Seconds	create delay by the specified number of seconds
DelayFor18TCY	18 cycles delay (required by the xlcd library)
DelayPORXLCD	15 ms delay (required by the xlcd library)
DelayXLCD	5 ms delay (required by the xlcd library)
LCD_Clear	Clear the LCD
LCD_Move	Move cursor to the specified row and column

The program configures the LCD and the A/D converter. Then, the heading "VOLTMETER" is displayed, and the program enters an endless loop. Inside this loop, the external voltage at channel 0 is read and converted into millivolts. The voltage is then converted into ASCII string using MPLAB XC8 function *itoa* and is displayed starting from column 6 of the LCD. This process is repeated after a 1-s delay.

```
/*******************************************************************************
                          VOLTMETER WITH LCD DISPLAY
                          ==========================
```

In this project an LCD is connected to PORTB. Also, input port AN0 is used as analog input. Voltage to be measured is applied to AN0. The microcontroller reads the analog voltage, converts into digital, and then displays on the LCD.

Analog input range is 0 to 5 V. A PIC18F45K22 type microcontroller is used in this project, operated with an 8 MHz crystal.

Analog data is read using the Adc_Read built-in function.

The LCD is connected to the microcontroller as follows:

Microcontroller	LCD
============	===
RB0	D4
RB1	D5
RB2	D6
RB3	D7
RB4	E
RB5	R/S
RB6	RW

```
Author:      Dogan Ibrahim
Date:        August 2013
File:        XC8-LCD1.C
*******************************************************************/

#include <xc.h>
#include <stdlib.h>
#include <plib/adc.h>
#include <plib/xlcd.h>
#include <plib/delays.h>
#pragma config MCLRE = EXTMCLR, WDTEN = OFF, FOSC = HSHP
#define _XTAL_FREQ 8000000

//
// This function creates seconds delay. The argument specifies the delay time in seconds.
//
void Delay_Seconds(unsigned char s)
{
    unsigned char i,j;

    for(j = 0; j < s; j++)
    {
        for(i = 0; i < 100; i++)__delay_ms(10);
    }
}
```

Figure 5.71: MPLAB XC8 Program Listing.

```
//
// This function creates 18 cycles delay for the xlcd library
//
void DelayFor18TCY( void )
{
Nop(); Nop(); Nop(); Nop();
Nop(); Nop(); Nop(); Nop();
Nop(); Nop(); Nop(); Nop();
Nop(); Nop();
return;
}


//
// This function creates 15 ms delay for the xlcd library
//
void DelayPORXLCD( void )
{
   __delay_ms(15);
   return;
}


//
// This function creates 5 ms delay for the xlcd library
//
void DelayXLCD( void )
{
   __delay_ms(5);
   return;
}


//
// This function clears the screen
//
void LCD_Clear()
{
   while(BusyXLCD());
   WriteCmdXLCD(0x01);
}

//
// This function moves the cursor to position row,column
//
void LCD_Move(unsigned char row, unsigned char column)
{
 char ddaddr = 40*(row – 1) + column;
 while( BusyXLCD() );
 SetDDRamAddr( ddaddr );
}
```

Figure 5.71
cont'd

```
void main()
{
    unsigned long Vin, mV;
    char op[10];

    ANSELB = 0;                                    // Configure PORTB as digital
    ANSELA = 1;                                    // COnfigure RA0 as analog
    TRISB = 0;                                     // Configure PORTB as outputs
    TRISA = 1;                                     // Configure RA0 as input
//
// Configure the LCD to use 4-bits, in multiple display mode
//
    Delay_Seconds(1);
    OpenXLCD(FOUR_BIT & LINES_5X7);
//
// Configure the A/D converter
//
    OpenADC(ADC_FOSC_2 & ADC_RIGHT_JUST & ADC_20_TAD,
            ADC_CH0 & ADC_INT_OFF,
            ADC_TRIG_CTMU & ADC_REF_VDD_VDD & ADC_REF_VDD_VSS);

    while(BusyXLCD());                             // Wait if the LCD is busy
    WriteCmdXLCD(DON);                             // Turn Display ON
    while(BusyXLCD());                             // Wait if the LCD is busy
    WriteCmdXLCD(0x06);                            // Move cursor right
    putrsXLCD("VOLTMETER");                        // Display heading
    Delay_Seconds(2);                             // 2 s delay
    LCD_Clear();                                   // Clear display

    for(;;)                                        // Endless loop
    {
        LCD_Clear();                               // Clear LCD
        SelChanConvADC(ADC_CH0);                   // Select channel 0 and start conversion
        while(BusyADC());                          // Wait for completion
        Vin = ReadADC();                           // Read converted data
        mV = (Vin * 5000) >> 10;                   // mv = Vin x 5000/1024
        LCD_Move(1,1);                             // Move to row = 1, column = 1
        putrsXLCD("mV = ");                        // Display "mV = "
        mV = (Vin * 5000) >> 10;                   // mv = Vin x 5000/1024
        itoa(op, mV, 10);                          // Convert mV into ASCII string
//
// Display result on LCD
//
        LCD_Move(1,6);                             // Move to row = 1,column = 6
        putsXLCD(op);                              // Display the measured voltage
        Delay_Seconds(1);                          // Wait 1 s
    }
}
```

Figure 5.71
cont'd

Project 5.13—Generating Sound

Project Description

This project shows how sound with different frequencies can be generated using a simple buzzer. The project shows how the simple melody "Happy Birthday" can be played using a buzzer.

Figure 5.72 shows the block diagram of the project.

Project Hardware

A buzzer is a small piezoelectric device that gives a sound output when excited. Normally, buzzers are excited using square wave signals, also called Pulse Width Modulated (PWM) signals. The frequency of the signal determines the pitch of the generated sound, and duty cycle of the signal can be used to increase or decrease the volume. Most buzzers operate in the frequency range 2—4 kHz. The PWM signal required to generate sound can be using the PWM module of the PIC microcontrollers.

In this project, a buzzer is connected to pin RC2 of a PIC18F45K22-type microcontroller through a transistor switch as shown in Figure 5.73.

If you are using the EasyPIC V7 development board, the following jumper should be selected:

J21: Select RC2

Project PDL

When playing a melody, each note is played for a certain duration and with a certain frequency. In addition, a certain gap is necessary between two successive notes. The frequencies of the musical notes starting from middle C (i.e. C4) are given below. The harmonic of a note is obtained by doubling the frequency. For example, the frequency of C5 is $2 \times 262 = 524$ Hz.

Notes	C4	C4#	D4	D4#	E4	F4	F4#	G4	G4#	A4	A4#	B4
Hz	261.63	277.18	293.66	311.13	329.63	349.23	370	392	415.3	440	466.16	493.88

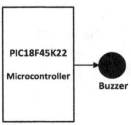

Figure 5.72: Block Diagram of the Project.

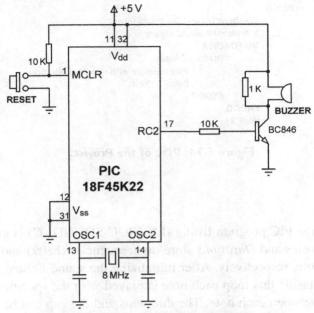

Figure 5.73: Circuit Diagram of the Project.

To play the tune of a song, we need to know its musical notes. Each note is played for a certain duration, and there is a certain time gap between two successive notes.

The next thing we want is to know how to generate a sound with a required frequency and duration. Fortunately, mikroC Pro for the PIC compiler provides a sound library with the following two functions:

Sound_Init	This function is used to specify to which port and which pin the sound device is connected to
Sound_Play	This function generates a square wave signal with the specified frequency (hertz) and duration (milliseconds) from the port pin specified by the initialization function. The frequencies given in the above table are approximated to their nearest integers since this function accepts only integer frequencies.

In this project, we will be generating the classic "Happy Birthday" melody, and thus, we need to know the notes and their durations. These are given in the table below where the durations are in milliseconds and should be multiplied by 400 to give correct values.

Note	C4	C4	D4	C4	F4	E4	C4	C4	D4	C4	G4	F4	C4	C4	C5	A4	F4	E4	D4	A4#	A4#	A4	F4	G4	F4
Duration	1	1	2	2	2	3	1	1	2	2	2	3	1	1	2	2	2	2	2	2	1	1	2	2	2	4

The PDL of the project is shown in Figure 5.74. Basically, two tables are used to store the notes and their corresponding durations. Then, the Sound_Play function is called in a loop to play all the notes. The melody repeats after a 3-s delay.

```
BEGIN
        Create Notes and Durations tables
        Initialize the Sound library
        DO FOREVER
                DO for all notes
                        Play the note with specified duration
                        Delay 100 ms
                ENDDO
        ENDDO
        Wait 3 s
END
```

Figure 5.74: PDL of the Project.

mikroC Pro for PIC

The mikroC Pro for the PIC program listing (MIKROC-SOUND1.C) is given in
Figure 5.75. Tables *Notes* and *Durations* store the frequencies (hertz) and durations
(1/400 ms) of each note, respectively. After initializing the sound library, an endless loop
is established where inside this loop each note is played with the specified duration. A
100-ms gap is used between each note. The durations and the gap can be changed to
increase or decrease the speed of the melody.

MPLAB XC8

The MPLAB XC8 compiler does not have a built-in sound library. We can however write
our own function to generate sound. There are basically two ways in which we can
generate accurate square waves for sounding the buzzer: using the processor built-in PWM
module, or writing a timer interrupt service routine. The problem with the PWM module
is that it is not possible to generate audio frequency signals using the PWM module. In
this section, we shall see how to develop a timer interrupt service routine to generate
square waves in the audio band.

The musical note frequencies we wish to generate are within the frequency range of
several hundred to several thousand Hertz. We shall be using Timer0 in the 16-bit interrupt
mode such that every time an interrupt is generated we shall toggle the output port pin
connected to the buzzer. Thus, effectively, we will be generating square waves.

When TIMER0 is operating in the 16-bit mode, the value to be loaded into registers
TMR0H and TMR0L to generate an interrupt after time T is given by

$$\text{TMR0H:TMR0L} = 65{,}536 - T/(4 \times T_{\text{osc}} \times \text{Prescaler value})$$

Our clock frequency is 8 MHz, which has the period of $T_{\text{osc}} = 0.125$ μs. If we give the
Prescaler the value 2, then, the above equation becomes

$$\text{TMR0H:TMR0L} = 65{,}536 - T$$

```
/****************************************************************************
                         PLAYING MELODY
                         =============

In this project a buzzer is connected to port pin RC2 of a PIC18F45K22 microcontroller,
operating with an 8 MHz crystal.

The program plays the classical "Happy Birthday" melody.


Author:      Dogan Ibrahim
Date:        August 2013
File:        MIKROC-SOUND1.C
*****************************************************************/
#define Max_Notes 25

void main()
{
   unsigned char i;

   unsigned int Notes[Max_Notes] =
   {
     262, 262, 294, 262, 349, 330, 262, 262, 294, 262, 392,
     349, 262, 262, 524, 440, 349, 330, 294, 466, 466, 440,
     349, 392, 349
   };

   unsigned char Durations[Max_Notes] =
   {
     1, 1, 2, 2, 2, 3, 1, 1, 2, 2, 2, 3, 1, 1, 2, 2, 2, 2, 2,
     1, 1, 2, 2, 2, 3
   };

   ANSELC = 0;                                    // Configure PORTC as digital

   Sound_Init(&PORTC, 2);                         // Initialize the Sound library

   for(;;)                                        // Endless loop
   {
     for(i = 0; i < Max_Notes; i++)               // Do for all notes
     {
       Sound_Play(Notes[i], 400*Durations[i]);    // Play the notes
       Delay_ms(100);                             // Note gap
     }
     Delay_Ms(3000);                              // Repeat after 3 s
   }
}
```

Figure 5.75: mikroC Pro for PIC Program Listing.

The time T to generate an interrupt depends upon the frequency f of the waveform we wish to generate. When an interrupt occurs, if the output signal is at logic 1, then it is cleared to 0, if it is at logic 0, then it is set to 1. Assuming that we wish to generate square waves with equal ON and OFF times, the time T in microseconds is then given by

$$T = 500,000/f$$

We can therefore calculate the value to be loaded into the timer registers as follows:

TMR0H:TMR0L = 65,536 − 500,000/f

For example, assuming that we wish to generate a square wave signal with frequency, $f = 100$ Hz, the value to be loaded into the Timer0 registers will be

TMR0H:TMR0L = 65,536 − 500,000/100 = 60,536

Here 60,536 is equivalent to 0xEC78 in hexadecimals. Thus, TMR0H = 0xEC and TMR0L = 0x78.

The MPLAB XC8 program to generate the square waves and play the melody is shown in Figure 5.76 (called XC8-SOUND1.C). The following functions are used in the program:

Interrupt isr: This is the interrupt service routine. Here, the timer registers are reloaded, the buzzer output is toggled, and the timer interrupt flag is cleared so that new timer interrupts can be accepted by the microcontroller.

Delay_Ms	This function generates millisecond delays specified by the argument.
Sound_Play	This function is similar to the mikroC Pro for PIC Sound_Play function. It generates a square wave signal with the specified frequency and duration. The values to be loaded into the timer registers are calculated as described above. After loading the timer registers, the timer counter is enabled and the program waits until the specified duration occurs. After this point, the timer counter is disabled to stop further interrupt from being generated, which effectively stops the buzzer.
Delay_Seconds	This function generates seconds delays specified by the argument.

Delay_Seconds: This function generates seconds delays specified by the argument.

It is assumed that the buzzer is connected to the RC2 pin of the microcontroller. At the beginning of the program, PORTC is configured as the digital output. Then, TIMER0 is initialized to operate in the 16-bit mode, with internal clock, and with a prescaler value of 2. Global interrupts and TIMER0 interrupts are enabled by setting the appropriate bits in register INTCON. The rest of the program is as in the mikroC Pro for the PIC where the required melody is played by generating the required notes.

Project 5.14—Generating Custom LCD Fonts
Project Description

There are some applications where we may want to create custom fonts such as special characters, symbols, or logos on the LCD. This project will show how to create the symbol of an arrow pointing right on the LCD, and then display "Right arrow <symbol of side arrow>" on the first row of the LCD.

```
/***********************************************************************
                          PLAYING MELODY
                          ==============

In this project a buzzer is connected to port pin RC2 of a PIC18F45K22 microcontroller,
operating with an 8 MHz crystal.

The program plays the classical "Happy Birthday" melody.


Author:      Dogan Ibrahim
Date:        August 2013
File:        XC8-SOUND1.C
***********************************************************/

#include <xc.h>
#pragma config MCLRE = EXTMCLR, WDTEN = OFF, FOSC = HSHP
#define _XTAL_FREQ 8000000
#define BUZZER LATCbits.LATC2
#define Max_Notes 25

unsigned char TIMER_H, TIMER_L;

unsigned int Notes[Max_Notes] =
{
  262, 262, 294, 262, 349, 330, 262, 262, 294, 262, 392,
  349, 262, 262, 524, 440, 349, 330, 294, 466, 466, 440,
  349, 392, 349
};

unsigned char Durations[Max_Notes] =
{
  1, 1, 2, 2, 2, 3, 1, 1, 2, 2, 2, 3, 1, 1, 2, 2, 2, 2, 2,
  1, 1, 2, 2, 2, 3
};

//
// Timer interrupt service routine
//
void interrupt isr(void)
{
   TMR0H = TIMER_H;
   TMR0L = TIMER_L;
   BUZZER = ~BUZZER;
   INTCONbits.TMR0IF= 0;

}

//
// This function generates millisecond delays
//
```

Figure 5.76: MPLAB XC8 Program Listing.

```
void Delay_Ms(unsigned int s)
{
  unsigned int j;
  for(j = 0; j < s; j++)__delay_ms(1);
}


//
// This function plays a note with the specified frequency and duration
//
void Sound_Play(unsigned int freq, unsigned int duration)
{
 float period;
 period = 500000.0/freq;
 period = 65536 − period;
 TIMER_H = (char)(period/256);
 TIMER_L = (char)(period − 256*TIMER_H);
 TMR0H = TIMER_H;
 TMR0L = TIMER_L;
 T0CONbits.TMR0ON = 1;
 Delay_Ms(duration);
 T0CONbits.TMR0ON = 0;
}


//
// This function creates seconds delay. The argument specifies the delay time in seconds.
//
void Delay_Seconds(unsigned char s)
{
  unsigned char i,j;

  for(j = 0; j < s; j++)
  {
    for(i = 0; i < 100; i++)__delay_ms(10);
  }
}




void main()
{
  unsigned char i;

  ANSELC = 0;                              // Configure PORTC as digital
  TRISC = 0;                               // Configure PORTC as output
  BUZZER = 0;                              // Buzzer = 0 to start with
//
// Configure TIMER0 for 16 bit, prescaler = 2
//
  T0CONbits.TMR0ON = 0;                    // Timer OFF
  T0CONbits.T08BIT = 0;                    // Timer in 16 bit mode
```

Figure 5.76
cont'd

```
    T0CONbits.T0CS = 0;              // Use internal clock
    T0CONbits.T0SE = 0;              // Low-to-high transition
    T0CONbits.PSA = 0;               // Use the prescaler
    T0CONbits.T0PS = 0;              // Prescaler = 2

    INTCON = 0xA0;                   // Enable global and TMR0 interrupts

    for(;;)                          // Endless loop
    {
      for(i = 0; i < Max_Notes; i++)     // Do for all notes
      {
        Sound_Play(Notes[i], 400 * Durations[i]);    // Play the notes
        Delay_Ms(100);               // Gap between notes
      }
      Delay_Seconds(3);              // Repeat after 3 s
    }
}
```

Figure 5.76

cont'd

mikroC Pro for the PIC compiler provides a tool that makes the creation of custom fonts very easy. The steps for creating a font of any shape are given below:

- Start mikroC Pro for PIC compiler.
- Select Tools → LCD Custom Character. You will see the LCD font editor form shown in Figure 5.77.
- Select 5 × 7 (the default).
- Click "Clear all" to clear the font editor.
- Now, draw the shape of your font by clicking on the squares in the editor window. In this project, we will be creating the symbol of a "right arrow" as shown in Figure 5.78.
- When you are happy with the font, click "mikroC Pro for PIC" tab so that the code generated will be for the mikroC Pro for PIC compiler.
- Click "Generate Code" button. You will get the code as shown in Figure 5.79.
- Click "Copy Code To Clipboard" to save the code.
- We shall see later in the project how to display this font using the generated code.

Circuit Diagram

The circuit diagram of the project is shown in Figure 5.80.

Project PDL

The PDL of this project is very simple and is given in Figure 5.81.

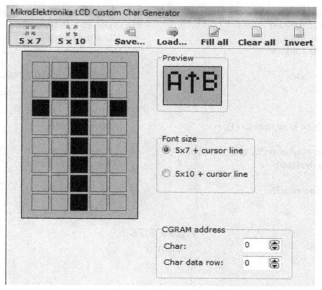

Figure 5.77: LCD Font Editor.

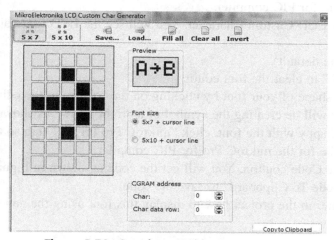

Figure 5.78: Creating a "Side Arrow" Font.

Project Program

mikroC Pro for PIC

The mikroC Pro for PIC program is named MIKROC-LCD2.C, and the program listing of the project is shown in Figure 5.82. At the beginning of the project, the connections between the microcontroller and the LCD are defined using *sbit* statements. PORTB is

| mikroC PRO | mikroPascal PRO | mikroBasic PRO |

```
const char character[] = {0,4,2,31,2,4,0,0};

void CustomChar(char pos_row, char pos_char) {
  char i;
    Lcd_Cmd(64);
    for (i = 0; i<=7; i++) Lcd_Chr_CP(character[i]);
    Lcd_Cmd(_LCD_RETURN_HOME);
    Lcd_Chr(pos_row, pos_char, 0);
}
```

Figure 5.79: Generating Code for the Font.

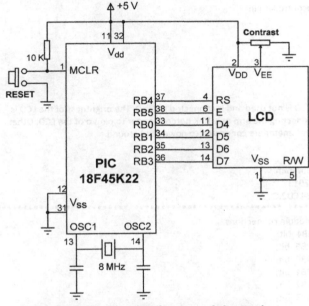

Figure 5.80: Circuit Diagram of the Project.

```
BEGIN
          Define microcontroller - LCD connections
          Define bit map of the required font
          Configure PORTB as digital and output
          Initialize LCD
          Display text on LCD
          CALL CustomChar to display the created font
END


BEGIN/CustomChar
          Display required font as character 0
END/CustomChar
```

Figure 5.81: PDL of the Project.

```
/********************************************************************
                    CREATING CUSTOM FONT ON LCD
                    =============================
```

This project displays a custom font on the LCD. A "Right arrow" is displayed with text as shown below:

           Right arrow <up arrow symbol>

The font has been created using the mikro C font editor.

In this project a HD44780 controller based LCD is connected to a PIC18F45K22 type microcontroller, operated from an 8MHz crystal.

The LCD is connected to PORTB of the microcontroller as follows:

LCD pin	Microcontroller pin
D4	RB0
D5	RB1
D6	RB2
D7	RB3
R/S	RB4
E	RB5

R/W pin of the LCD is not used and is connected to GND. The brightness of the LCD is controlled by connecting the arm of a 10 K potentiometer to pin Vo of the LCD. Other pins of the potentiometer are connected to power and ground.

Author: Dogan Ibrahim
Date: August 2013
File: MIKROC-LCD2.C
```
********************************************************************/
// Start of LCD module connections
sbit LCD_RS at RB4_bit;
sbit LCD_EN at RB5_bit;
sbit LCD_D4 at RB0_bit;
sbit LCD_D5 at RB1_bit;
sbit LCD_D6 at RB2_bit;
sbit LCD_D7 at RB3_bit;

sbit LCD_RS_Direction at TRISB4_bit;
sbit LCD_EN_Direction at TRISB5_bit;
sbit LCD_D4_Direction at TRISB0_bit;
sbit LCD_D5_Direction at TRISB1_bit;
sbit LCD_D6_Direction at TRISB2_bit;
sbit LCD_D7_Direction at TRISB3_bit;
// End of LCD module connections

//
// The following code is generated automatically by the mikroC compiler font editor
```

Figure 5.82: Program Listing of the Project.

```
                                                        //
  const char character[] = {0,4,2,31,2,4,0,0};

void CustomChar(char pos_row, char pos_char) {
 char i;
   Lcd_Cmd(64);
   for (i = 0; i <= 7; i++) Lcd_Chr_CP(character[i]);
   Lcd_Cmd(_LCD_RETURN_HOME);
   Lcd_Chr(pos_row, pos_char, 0);
}


//
// Start of Main program
//
void main()
{
     ANSELB = 0;                              // Configure PORTB as digital
     TRISB = 0;                               // Configure PORTB pins as output

     Lcd_Init();                              // Initialize LCD
     Lcd_Cmd(_LCD_CLEAR);                     // Clear display
     Lcd_Cmd(_LCD_CURSOR_OFF);                // Cursor off

     Lcd_Out(1, 1, "Right arrow");            // Display text "Right arrow"
     CustomChar(1, 13);                       // Display the "right arrow" symbol

     while(1);                                // End of program, wait here forever
}
```

Figure 5.82
cont'd

Figure 5.83: The LCD Display.

configured as a digital output port. The LCD is initialized, cleared, and the cursor is turned OFF. Then, the LCD_Out function is called to display text "Right arrow", starting row 1 and column 1 of the LCD. Function CustomChar is generated by the compiler, and this function displays the created font at the specified row and column positions.

Figure 5.83 shows a picture of the LCD display.

Project 5.15—Digital Thermometer
Project Description

In this project, the design of a digital thermometer is described. An analog temperature sensor is used to sense the temperature, and the temperature is displayed on an LCD. The block diagram of the digital thermometer is shown in Figure 5.84.

An LM35DZ type analog temperature sensor is used in this project. This is a three-pin small sensor that provides an analog output voltage directly proportional to the measured temperature. The output voltage is given by

$$V_o = 10 \text{ mV/}°C$$

Thus, for example, at 20 °C, the output voltage is 200 mV, at 25 °C it is 250 mV, and so on.

Circuit Diagram

The circuit diagram of the project is shown in Figure 5.85. The temperature sensor is connected to analog input AN0 (RA0) of the microcontroller. The LCD is connected to PORTB as in the previous LCD projects.

Project PDL

The PDL of this project is very simple and is given in Figure 5.86.

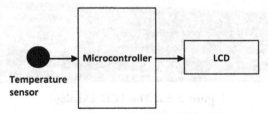

Figure 5.84: Block Diagram of the Digital Thermometer.

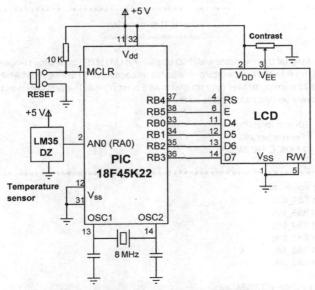

Figure 5.85: Circuit Diagram of the Project.

Project Program

mikroC Pro for PIC

The mikroC Pro for the PIC program is named MIKROC-THERMO.C, and the program listing of the project is shown in Figure 5.87. At the beginning of the project, the connections between the microcontroller and the LCD are defined using *sbit* statements.

```
BEGIN
        Define LCD connections
        Configure PORTA and PORTB as digital
        Configure RA0 (AN0) as input
        Initialize LCD
        DO FOREVER
                Read temperature from channel 0
                Convert reading into millivolts
                Divide by 10 to find the temperature in Degrees C
                Convert temperature into string
                Clear display
                Display Heading "Temperature"
                Display the temperature
                Wait 1 s
        ENDDO
END
```

Figure 5.86: PDL of the Project.

```
/*******************************************************************************
                              Digital Thermometer
                              ===================

This project is a digital thermometer with LCD display. An LM35DZ type analog temperature
sensor is used to sense the temperature. The sensor is connected to analog input AN0 (RA0)
of a PIC18F45K22 microcontroller operating with an 8 MHz crystal. The program reads the
temperature every second and displays on the LCD.

Programmer:    Dogan Ibrahim
Date:          September 2013
File:          MIKROC-THERMO.C

*******************************************************************************/
// LCD module connections
sbit LCD_RS at RB4_bit;
sbit LCD_EN at RB5_bit;
sbit LCD_D4 at RB0_bit;
sbit LCD_D5 at RB1_bit;
sbit LCD_D6 at RB2_bit;
sbit LCD_D7 at RB3_bit;

sbit LCD_RS_Direction at TRISB4_bit;
sbit LCD_EN_Direction at TRISB5_bit;
sbit LCD_D4_Direction at TRISB0_bit;
sbit LCD_D5_Direction at TRISB1_bit;
sbit LCD_D6_Direction at TRISB2_bit;
sbit LCD_D7_Direction at TRISB3_bit;
// End LCD module connections

void main()
{
    unsigned char Txt[14];
    unsigned int temp;
    float mV;

    ANSELA = 0;                          // Configure PORTA as digital
    ANSELB = 0;                          // Configure PORTB as digital
    TRISA.RA0 = 1;                       // RA0 is input
    Lcd_Init();                          // Initialize LCD

    Lcd_Cmd(_LCD_CURSOR_OFF);             // Disable cursor

    for(;;)
    {
      temp = ADc_Read(0);                // Read from channel 0
      mV = temp * 5000.0/1024.0;         // Convert to mV
      mV = mV/10.0;                      // mV is now the Degrees C
      floatToStr(mV, Txt);               // Convert to string
      Lcd_cmd(_LCD_CLEAR);               // Clear LCD
      Lcd_out(1,1,"Temperature");        // Display heading
      Lcd_Out(2,1,Txt);
      Delay_ms(1000);                    // Display temperature
    }                                    // Wait 1 s and repeat
}
```

Figure 5.87: mikroC Pro for PIC Program Listing.

PORTA is configured as a digital input. The LCD is initialized, cleared, and the cursor is turned OFF. The remainder of the program is executed in an endless loop. Here, the temperature is read from analog channel 0 (AN0 or RA0), converted into millivolts by multiplying with 5000, and dividing by the A/D converter resolution (10 bits), and divided by 10 to find the temperature. The temperature is then converted into a string and is displayed on the LCD. In this program, floating point arithmetic is used to find and display the temperature for higher accuracy.

Intermediate PIC18 Projects

Chapter Outline

In this chapter, we will be developing more complex projects using various peripheral devices. As in the previous chapter, the project description, hardware design, PDL, full program listing, and description of the program for each project will be given in detail.

Project 6.1—Four-Digit Multiplexed Seven-Segment Light Emitting Diode Event Counter Using an External Interrupt

Project Description

This project is similar to Project 5.10, but here, the timer interrupt of the microcontroller is used to refresh the displays. In Project 5.10, the microcontroller was busy updating the displays continuously, and thus, it could not perform any other tasks. For example, if we wish to make a counter with a 1 s delay between each count, the program given in Project 5.10 could not be used as the displays cannot be updated while the program waits for 1 s.

In this project, an external interrupt-based event counter will be designed to count up by one and display on the seven-segment displays every time an external event is detected. In this project, external interrupt input INT0 (RB0) is used as the event input. An event is said to occur whenever the INT0 input goes from logic 0 to logic 1. The event count will be incremented inside an interrupt service routine (ISR). The displays will be refreshed inside a timer ISR so that the processor is free to do other tasks while the displays are refreshed.

In this project, Timer0 is used in the 8-bit mode (since the required delay is only several milliseconds) to refresh the displays. The time for a timer interrupt is given by the following:

$$\text{Time} = (4 \times \text{clock period}) \times \text{Prescaler} \times (256 - \text{TMR0L})$$

Where Prescaler is the selected prescaler value, and TMR0L is the value loaded into timer register TMR0L to generate timer interrupts every Time period. In our application, the clock frequency is 8 MHz, that is, the clock period $= 0.125$ μs, and Time $= 5$ ms. Selecting a prescaler value of 64, the number to be loaded into TMR0L can be calculated as follows:

$$\text{TMR0L} = 256 - \frac{\text{Time}}{4 * \text{clockperiod} * \text{prescaler}}$$

or

$$\text{TMR0L} = 256 - \frac{5000}{4 * 0.125 * 64} = 100$$

Thus, TMR0L should be loaded with 100. The value to be loaded into the TMR0 control register T0CON can then be found as follows:

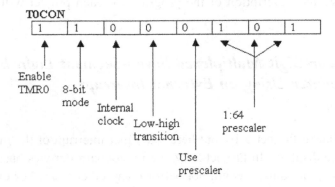

Thus, the T0CON register should be loaded with hexadecimal 0xC5. The next register to be configured is the interrupt control register INTCON where we will disable priority-based interrupts and enable the global interrupts and TMR0 and INT0 interrupts:

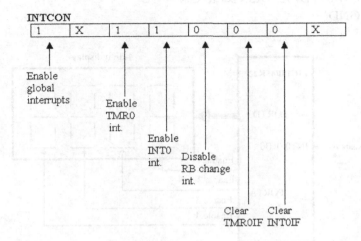

Taking the do not care entries (X) as 0, the hexadecimal value to be loaded into register INTCON is thus 0xB0.

In addition, we have to set bit 6 of register INTCON2 so that external interrupts are recognized on the low to high transition of the INT0 pin. Looking at the data sheet, we see that this bit is automatically set by default after power up (or reset).

In this project, the source of interrupt can either be from the timer or from an external event. Inside the ISR, we should check the interrupt flags of each source to determine the actual cause of the interrupt.

When a timer interrupt occurs, the TMR0L register should be reloaded inside the timer ISR. Interrupt flags of both interrupt sources must be cleared inside their ISR routines so that further interrupts can be accepted from these sources.

Figure 6.1 shows the block diagram of the project.

Project Hardware

The circuit diagram of the project is shown in Figure 6.2. The circuit is basically the same as in Figure 5.56, but here, additionally the external interrupt input is used for event triggering.

If you are using the EasyPIC V7 development board, make sure that the following jumpers are set correctly on the board:

SW4: DIS0, DIS1, DIS2, DIS3 set to ON
J17: set to GND

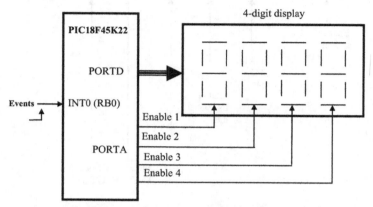

Figure 6.1: Block Diagram of the Project.

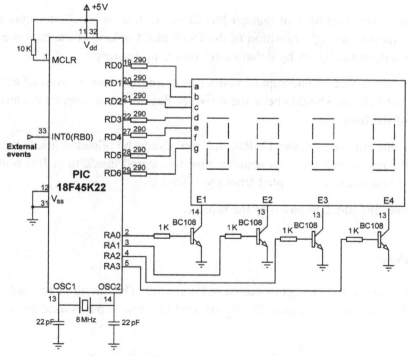

Figure 6.2: Circuit Diagram of the Project.

MAIN PROGRAM:

BEGIN
>   Configure PORTD as digital and output
>   Configure PORTA as digital and output
>   Configure PORTB as digital and RB0 input
>   Clear all digits
>   Configure TIMER0
>   Configure and enable external interrupts
>   **WAIT FOREVER**

END

INTERRUPT SERVICE ROUTINE:

BEGIN
>   **IF** the interrupt source = Timer0
>> Reload timer register
>> Clear timer interrupt flag
>> **CALL** Display to display digit data
>> Enable appropriate display digit
>   **ELSE IF** the interrupt source = INT0
>> Increment event count
>> Clear INT0 interrupt flag
>   **ENDIF**

END

BEGIN/DISPLAY
>   Extract the bit pattern to send to the port to display a number
>   Return the bit number to the calling programm

END/DISPLAY

Figure 6.3: PDL of the Project.

Project PDL

The PDL of the project is shown in Figure 6.3. The program is in two sections: the main program and the ISR. Inside the main program, TMR0L is configured to generate interrupts at every 5 ms. Also, input INT0 (RB0) is configured to accept external interrupts. Whenever an interrupt occurs, the source of the interrupt is detected. If the source is timer interrupt, then the display is refreshed. If the source of the interrupt is INT0, then it is assumed that an external event has occurred, and a counter is incremented by 1.

Project Program

mikroC Pro for PIC

The mikroC Pro for PIC program is called MIKROC-EVENT1.C and is shown in Figure 6.4. At the beginning of the main program, PORTD and PORTA are configured as

```
/*****************************************************************************
                4-DIGIT 7-SEGMENT DISPLAY EVENT COUNTER
                =========================================
```

In this project four common cathode 7-segment LED displays are connected to PORTD of a PIC18F45K22 microcontroller and the microcontroller is operated from an 8 MHz crystal. Four PORTA pins are used to enable/disable the LEDs. In addition, external interrupt input INT0 (RB0) is used to receive external events. An event is assumed to occur if pin INT0 goes from logic 0 to logic 1.

The program uses two ISR routines: the timer routine is used to refresh the Display every 5 ms. External interrupt ISR is used to increment the event count. The event count is displayed continuously on the 7-segment displays.

```
Author:     Dogan Ibrahim
Date:       August 2013
File:       MIKROC-EVENT1.C
*****************************************************************************/
#define DIGIT1 PORTA.RA0
#define DIGIT2 PORTA.RA1
#define DIGIT3 PORTA.RA2
#define DIGIT4 PORTA.RA3

unsigned int Cnt = 0;
unsigned char flag = 0;
unsigned int D1,D2,D3,D4,D5,D6;

//
// This function finds the bit pattern to be sent to the port to display a number
// on the 7-segment LED. The number is passed in the argument list of the function.
//
unsigned char Display(unsigned char no)
{
    unsigned char Pattern;
    unsigned char SEGMENT[] = {0x3F,0x06,0x5B,0x4F,0x66,0x6D,0x7D,0x07,0x7F,0x6F};

    Pattern = SEGMENT[no];                        // Pattern to return
    return (Pattern);
}

//
// Interrupt Service Routine
//
void interrupt (void)
{
    if(INTCON.TMR0IF == 1)                        // If Timer interrupt occured
    {
        TMR0L = 100;                              // Reload timer register
        INTCON.TMR0IF = 0;                        // Clear timer interrupt flag
        switch(flag)
```

Figure 6.4: mikroC Pro for PIC Program Listing.

```
          {
          case 0:
          {
          DIGIT1 = 0;                          // Disable digit 1
          D1 = Cnt/1000;                       // 1000s  digit
          PORTD = Display(D1);                 // Send to PORTD
          DIGIT4 = 1;                          // Enable digit 4
          flag = 1;
          break;
          }
          case 1:
          {
          DIGIT4 = 0;                          // Disable digit 4
          D2 = Cnt % 1000;
          D3 = D2/100;                         // 100s digit
          PORTD = Display(D3);                 // Send to PORTD
          DIGIT3 = 1;                          // Enable digit 3
          flag = 2;
          break;
          }
          case 2:
          {
          DIGIT3 = 0;                          // Disable digit 3
          D4 = D2 % 100;
          D5 = D4/10;                          // 10s digit
          PORTD = Display(D5);                 // Send to PORTD
          DIGIT2 = 1;                          // Enable digit 2
          flag = 3;
          break;
          }
          case 3:
          {
          DIGIT2 = 0;                          // Disable digit 2
          D6 = D4 % 10;
          PORTD = Display(D6);                 // Send to PORTD
          DIGIT1 = 1;                          // Enable digit 1
          flag = 0;
          break;
          }
          }
          }
          if(INTCON.INTOIF == 1)               // If external interrupt occurred
          {
            Cnt++;                             // Increment event count
            INTCON.INTOIF = 0;                 // Clear ext interrupt flag
          }
        }
      //
      // Start of MAIN Program
      //
      void main()
      {
```

Figure 6.4
cont'd

```
            ANSELA = 0;                          // Configure PORTA as digital
            ANSELD = 0;                          // Configure PORTD as digital
            ANSELB = 0;                          // Configure PORTB as digital
            TRISA = 0;                           // Configure PORTA as outputs
            TRISD = 0;                           // Configure PORTD as outputs
            TRISB = 1;                            // RB0 is event input

            DIGIT1 = 0;                          // Disable digit 1
            DIGIT2 = 0;                          // Disable digit 2
            DIGIT3 = 0;                          // Disable digit 3
            DIGIT4 = 0;                          // Disable digit 4
        //
        // Configure TIMER0 interrupts
        //
            T0CON = 0xC5;                        // TIMER0 in 8-bit mode
            TMR0L = 100;                         // Load Timer register
        //
        // Configure External interrupts and enable interrupts
        //
            INTCON = 0xB0;

            for(;;)                              // Endless loop
            {                                    // Wait and process interrupts
            }
        }
```

Figure 6.4
cont'd

digital outputs. Pin INT0 (RB0) is configured as the digital input since this is the event input.

Then, all the display digits are cleared, and register T0CON is loaded with 0xC5 to enable the TMR0 and set the prescaler to 64. The TMR0L register is loaded with 100 so that an interrupt can be generated every 5 ms. INTCON is then set to 0xB0 to enable global interrupts, timer interrupts, and external interrupts from input INT0. The remainder of the main program waits for interrupts to occur and does not do anything useful.

Inside the ISR, the program checks to determine the source of the interrupt. If the timer has caused the interrupt, then the timer register is reloaded, and the timer interrupt flag is cleared so that the processor can accept further interrupts from the timer. The display digits are then refreshed inside the timer ISR. Here, only one digit is enabled at any time.

```
/*****************************************************************************
              4-DIGIT 7-SEGMENT DISPLAY EVENT COUNTER
              =========================================
```

In this project four common cathode 7-segment LED displays are connected to PORTD of a
PIC18F45K22 microcontroller and the microcontroller is operated from an 8 MHz crystal.
Four PORTA pins are used to enable/disable the LEDs. In addition, external interrupt input
INT0 (RB0) is used to receive external events. An event is assumed to occur if pin INT0
goes from logic 0 to logic 1.

The program uses two ISR routines: the timer routine is used to refresh the Display every 5 ms.
External interrupt ISR is used to increment the event count. The event count is displayed
continuously on the 7-segment displays.

In this version of the program leading zeroes are blanked.

```
Author:      Dogan Ibrahim
Date:        August 2013
File:        MIKROC-EVENT2.C
*****************************************************************************/
#define DIGIT1 PORTA.RA0
#define DIGIT2 PORTA.RA1
#define DIGIT3 PORTA.RA2
#define DIGIT4 PORTA.RA3

unsigned int Cnt = 0;
unsigned char flag = 0;
unsigned int D1,D2,D3,D4,D5,D6;

//
// This function finds the bit pattern to be sent to the port to display a number
// on the 7-segment LED. The number is passed in the argument list of the function.
//
unsigned char Display(unsigned char no)
{
   unsigned char Pattern;
   unsigned char SEGMENT[] = {0x3F,0x06,0x5B,0x4F,0x66,0x6D,0x7D,0x07,0x7F,0x6F};

   Pattern = SEGMENT[no];                    // Pattern to return
   return (Pattern);
}

//
// Interrupt Service Routine
//
void interrupt (void)
{
   if(INTCON.TMR0IF == 1)                    // If Timer interrupt occured
   {
     TMR0L = 100;                            // Reload timer register
     INTCON.TMR0IF = 0;                      // Clear timer interrupt flag
```

Figure 6.5: mikroC Pro for the PIC-Modified Program.

```
switch(flag)
{
 case 0:
 {
  DIGIT1 = 0;                                    // Disable digit 1
  D1 = Cnt/1000;                                 // 1000s  digit
  if(D1 != 0)                                    // Check if blanking required
  {
    PORTD = Display(D1);                         // Send to PORTD
    DIGIT4 = 1;                                  // Enable digit 4
  }
  flag = 1;
  break;
 }
 case 1:
 {
  DIGIT4 = 0;                                    // Disable digit 4
  D2 = Cnt % 1000;
  D3 = D2/100;                                   // 100s digit
  if(D3 != 0 || D1 != 0)                         // Check if blanking required
  {
    PORTD = Display(D3);                         // Send to PORTD
    DIGIT3 = 1;                                  // Enable digit 3
  }
  flag = 2;
  break;
 }
 case 2:
 {
  DIGIT3 = 0;                                    // Disable digit 3
  D4 = D2 % 100;
  D5 = D4/10;                                    // 10s digit
  if(D5 != 0 || D3 != 0 || D1 != 0)             // Check if blanking is required
  {
    PORTD = Display(D5);                         // Send to PORTD
    DIGIT2 = 1;                                  // Enable digit 2
  }
  flag = 3;
  break;
 }
 case 3:
 {
  DIGIT2 = 0;                                    // Disable digit 2
  D6 = D4 % 10;
  PORTD = Display(D6);                           // Send to PORTD
  DIGIT1 = 1;                                    // Enable digit 1
  flag = 0;
  break;
 }
 }
}
```

Figure 6.5
cont'd

```
        if(INTCON.INT0IF == 1)                      // If external interrupt occurred
        {
           Cnt++;                                    // Increment event count
           INTCON.INT0IF = 0;                        // Clear ext interrupt flag
        }
}



//
// Start of MAIN Program
//
void main()
{

        ANSELA = 0;                                  // Configure PORTA as digital
        ANSELD = 0;                                  // Configure PORTD as digital
        ANSELB = 0;                                  // Configure PORTB as digital
        TRISA = 0;                                   // Configure PORTA as outputs
        TRISD = 0;                                   // Configure PORTD as outputs
        TRISB = 1;                                   // RB0 is event input

        DIGIT1 = 0;                                  // Disable digit 1
        DIGIT2 = 0;                                  // Disable digit 2
        DIGIT3 = 0;                                  // Disable digit 3
        DIGIT4 = 0;                                  // Disable digit 4
//
// Configure TIMER0 interrupts
//
        T0CON = 0xC5;                                // TIMER0 in 8-bit mode
        TMR0L = 100;                                 // Load Timer register
//
// Configure External interrupts and enable interrupts
//
        INTCON = 0xB0;

        for(;;)                                      // Endless loop
        {                                            // Wait and process interrupts
        }
}
```

Figure 6.5
cont'd

Figure 6.6: Example Display. (For color version of this figure, the reader is referred to the online version of this book.)

For each successive interrupt, data are sent to the corresponding digit, and its digit is enabled. A *switch* statement together with a *flag* variable is used to determine which digit should be refreshed.

If on the other hand the source of the interrupt is the external event, then the event count (Cnt) is incremented by 1, and the INT0 interrupt flag is cleared so that the processor can accept further interrupts from this source.

Function Display as before determines the bit pattern to be sent to the port to display a given number.

Modified Program

In Figure 6.4, the display shows leading zeroes, for example, number 14 is displayed as 0014. The program could easily be modified to blank leading zeroes. The modified program (called MIKROC-EVENT2.C) is shown in Figure 6.5.

Figure 6.6 shows an example display.

MPLAB XC8

The MPLAB XC8 version of the program is shown in Figure 6.7 (XC8-EVENT1.C). The operation of the program is the same as in Figure 6.5.

```
/*******************************************************************************
                4-DIGIT 7-SEGMENT DISPLAY EVENT COUNTER
                =============================================
```

In this project four common cathode 7-segment LED displays are connected to PORTD of a
PIC18F45K22 microcontroller and the microcontroller is operated from an 8 MHz crystal.
Four PORTA pins are used to enable/disable the LEDs. In addition, external interrupt input
INT0 (RB0) is used to receive external events. An event is assumed to occur if pin INT0 goes
from logic 0 to logic 1.

The program uses two ISR routines: the timer routine is used to refresh the Display every 5 ms.
External interrupt ISR is used to increment the event count. The event count is displayed
continuously on the 7-segment displays.

In this version of the program leading zeroes are blanked.

```
Author:     Dogan Ibrahim
Date:       August 2013
File:       XC8-EVENT2.C
*******************************************************************************/

#include <xc.h>
#pragma config MCLRE = EXTMCLR, WDTEN = OFF, FOSC = HSHP
#define _XTAL_FREQ 8000000

#define DIGIT1 PORTAbits.RA0
#define DIGIT2 PORTAbits.RA1
#define DIGIT3 PORTAbits.RA2
#define DIGIT4 PORTAbits.RA3

unsigned int Cnt = 0;
unsigned char flag = 0;
unsigned int D1,D2,D3,D4,D5,D6;

//
// This function finds the bit pattern to be sent to the port to display a number
// on the 7-segment LED. The number is passed in the argument list of the function.
//
unsigned char Display(unsigned char no)
{
   unsigned char Pattern;
   unsigned char SEGMENT[] = {0x3F,0x06,0x5B,0x4F,0x66,0x6D,0x7D,0x07,0x7F,0x6F};

   Pattern = SEGMENT[no];                          // Pattern to return
   return (Pattern);
}


//
// Interrupt Service Routine
//
```

Figure 6.7: MPLAB XC8 Program Listing.

```
void interrupt isr (void)
{
  if(INTCONbits.TMR0IF == 1)                          // If Timer interrupt occurred
  {
    TMR0L = 100;                                      // Reload timer register
    INTCONbits.TMR0IF = 0;                            // Clear timer interrupt flag
    switch(flag)
    {
      case 0:
      {
        DIGIT1 = 0;                                   // Disable digit 1
        D1 = Cnt/1000;                                // 1000s  digit
        if(D1 != 0)                                   // Check if blanking required
        {
          PORTD = Display(D1);                        // Send to PORTD
          DIGIT4 = 1;                                 // Enable digit 4
        }
        flag = 1;
        break;
      }
      case 1:
      {
        DIGIT4 = 0;                                   // Disable digit 4
        D2 = Cnt % 1000;
        D3 = D2/100;                                  // 100s digit
        if(D3 != 0 || D1 != 0)                        // Check if blanking required
        {
          PORTD = Display(D3);                        // Send to PORTD
          DIGIT3 = 1;                                 // Enable digit 3
        }
        flag = 2;
        break;
      }
      case 2:
      {
        DIGIT3 = 0;                                   // Disable digit 3
        D4 = D2 % 100;
        D5 = D4/10;                                   // 10s digit
        if(D5 != 0 || D3 != 0 || D1 != 0)             // Check if blanking is required
        {
          PORTD = Display(D5);                        // Send to PORTD
          DIGIT2 = 1;                                 // Enable digit 2
        }
        flag = 3;
        break;
      }
      case 3:
      {
        DIGIT2 = 0;                                   // Disable digit 2
        D6 = D4 % 10;
        PORTD = Display(D6);                          // Send to PORTD
```

Figure 6.7
cont'd

```
            DIGIT1 = 1;                                // Enable digit 1
            flag = 0;
            break;
          }
        }
      }
      if(INTCONbits.INTOIF == 1)                        // If external interrupt occurred
      {
          Cnt++;                                        // Increment event count
          INTCONbits.INTOIF = 0;                        // Clear ext interrupt flag
      }
    }

//
// Start of MAIN Program
//
    void main()
    {

        ANSELA = 0;                                     // Configure PORTA as digital
        ANSELD = 0;                                     // Configure PORTD as digital
        ANSELB = 0;                                     // Configure PORTB as digital
        TRISA = 0;                                      // Configure PORTA as outputs
        TRISD = 0;                                      // Configure PORTD as outputs
        TRISB = 1;                                      // RB0 is event input

        DIGIT1 = 0;                                     // Disable digit 1
        DIGIT2 = 0;                                     // Disable digit 2
        DIGIT3 = 0;                                     // Disable digit 3
        DIGIT4 = 0;                                     // Disable digit 4
//
// Configure TIMER0 interrupts
//
        T0CON = 0xC5;                                   // TIMER0 in 8-bit mode
        TMR0L = 100;                                    // Load Timer register
//
// Configure External interrupts and enable interrupts
//
        INTCON = 0xB0;

        for(;;)                                         // Endless loop
        {                                               // Wait and process interrupts
        }
    }
```

Figure 6.7
cont'd

Project 6.2—Calculator with a Keypad and Liquid Crystal Display
Project Description

Keypads are small keyboards that are used to enter numeric or alphanumeric data to microcontroller systems. Keypads are available in a variety of sizes and styles from 2×2 to 4×4 or even bigger.

In this project, a 4×4 keypad and a liquid crystal display (LCD) are used, and a simple calculator is designed. Figure 6.8 shows the picture of the keypad used.

Figure 6.9 shows the structure of the keypad used in this project, which consists of 16 switches, formed in a 4×4 array, and named **0–9**, **Enter**, $+,-,^*$ and /. Assuming that the keypad is connected to PORTC, the steps to detect which key is pressed is as follows:

- A logic 1 is applied to the first column via RC0.
- Port pins RC4–RC7 are read. If the data are nonzero, then a switch is pressed. If RC4 is 1, key 1 is pressed, if RC5 is 1, key 4 is pressed, if RC6 is 1, key 9 is pressed, and so on.
- A logic 1 is applied to the second column via RC1.
- Again Port pins RC4–RC7 are read. If the data are nonzero, then a switch is pressed. If RC4 is 1, key 2 is pressed, if RC5 is 1, key 6 is pressed, if RC6 is 1, key 0 is pressed, and so on.
- The above process is repeated for all the four columns continuously.

In this project, a simple integer calculator is designed. The calculator can add, subtract, multiply, and divide integer numbers and show the result on the LCD. The operation of

Figure 6.8: A 4 × 4 Keypad. (For color version of this figure, the reader is referred to the online version of this book.)

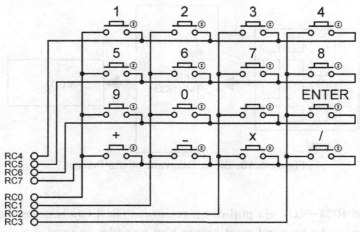

Figure 6.9: The 4 × 4 Keypad Structure.

the calculator is as follows: when power is applied to the system, the LCD displays text **CALCULATOR** for 2 s. Then, text **No1:** is displayed in the first row of the LCD, and the user is expected to type the first number and then press the **ENTER** key. Then, text **No2:** is displayed in the second row of the LCD where the user enters the second number and press the **ENTER** key. After this, the required operation key should be pressed. The result will be displayed on the LCD for 5 s, and then the LCD will be cleared, ready for the next calculation. The example below shows how numbers 12 and 20 can be added:

```
No1: 12 ENTER
No2: 20 ENTER
Op: +
Res = 32
```

In this project, the keypad is labeled as follows:

```
1   2   3   4
5   6   7   8
9   0       ENTER
+   -   X   /
```

One of the keys, between 0 and ENTER is not used in the project.

Project Hardware

The block diagram of the project is shown in Figure 6.10. The circuit diagram is given in Figure 6.11. A PIC18F45K22 microcontroller with an 8 MHz crystal is used in the project. Columns of the keypad are connected to port RC0—RC3, and rows are

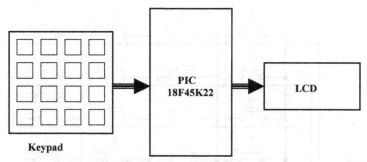

Figure 6.10: Block Diagram of the Project.

connected to port RC4—RC7 via pull-down resistors. The LCD is connected to PORTB in the default mode. An external reset button also provides to reset the microcontroller should it be necessary.

If you are using the EasyPIC V7 development board with the mikroElektronika 4×4 Keypad, then simply connect the keypad IDC10 ribbon cable connector to PORTC header at the edge of the board and enable the PORTC Pull-Down resistors by moving them downward (see Figure 6.11 for pull-down resistor requirements).

Project PDL

The project PDL is shown in Figure 6.12. The program consists of two parts: function **getkeypad** and the main program. Function **getkeypad** receives a key from the keypad. Inside the main program, PORTB is configured as the digital output, the LCD is initialized, and the heading "CALCULATOR" is displayed for 2 s. The program then executes in an endless loop. Inside this loop, two numbers and the required operation are received from the keypad. The microcontroller performs the required operation and displays the result on the LCD.

Project Program

The program listing (MIKROC-KEYPAD.C) is given in Figure 6.13. Each key is given a numeric value as follows:

```
0    1    2    3
4    5    6    7
8    9    10   11
12   13   14   15
```

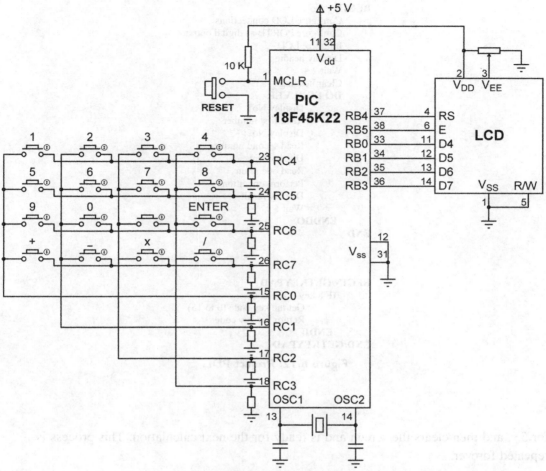

Figure 6.11: Circuit Diagram of the Project.

The program consists of a function called **getkeypad** that reads the pressed keys, and the main program. Variable **MyKey** stores the key value (0—15) pressed, variables **Op1** and **Op2** store the first and the second numbers entered by the user, respectively. All these variables are cleared to zero at the beginning of the program. A **while** loop is then formed to read the first number and store in variable **Op1**. This loop exits when the user presses the **ENTER** key. Similarly, the second number is read from the keyboard in a second **while** loop. Then, the operation to be performed is read and stored in variable **MyKey**, and a **switch** statement is used to perform the required operation and store the result in variable **Calc**. The result is converted into a string array using function **LongToStr** and leading blank characters are removed. The program displays the result on the LCD, waits

```
BEGIN
        Configure LCD connections
        Configure PORTB as digital output
        Initialize LCD
        Display heading
        Wait 2 s
        Clear heading
        DO FOREVER
                Display No1:
                Read first number
                Display No2:
                Read second number
                Display Op:
                Read operation
                Perform operation
                Display result
                Wait 5 s
        ENDDO
END


BEGIN/GETKEYPAD
        IF a key is pressed
                Get the key code (0 to 15)
                Return the key code
        ENDIF
END/GETKEYPAD
```

Figure 6.12: Project PDL.

for 5 s, and then clears the screen and is ready for the next calculation. This process is repeated forever.

Function **getkeypad** receives a key from the keypad. We start by sending a 1 to column 1, and then, we check all the rows. When a key is pressed, a logic 1 is detected in the corresponding row, and the program jumps out of the **while** loop. Then, a **for** loop is used to find the actual key pressed by the user as a number from 0 to 15.

It is important to realize that when a key is pressed or released, we get what is known as **contact noise** where the key output pulses up and down momentarily, and this produces a number of logic 0 and 1 pulses at the output. Switch contact noise is usually removed either in hardware or by programming, and this process is called **contact debouncing**. In software, the simplest way to remove the contact

```
/**********************************************************************
                    CALCULATOR WITH KEYPAD AND LCD
                    ================================
```

In this project a 4 x 4 keypad is connected to PORTC of a PIC18F45K22 microcontroller. Also an LCD is connected to PORTB. The project is a simple calculator which can perform integer arithmetic.

The keys are organised as follows:

```
 0   1   2   3
 4   5   6   7
 8   9  10  11
12  13  14  15
```

The keys are labeled as follows:

```
 1  2  3  4
 5  6  7  8
 9  0    Enter
 +  −  *  /
```

Author: Dogan Ibrahim
Date: August 2013
File: MIKROC-KEYPAD.C

```
***********************************************************/
// LCD module connections
sbit LCD_RS at RB4_bit;
sbit LCD_EN at RB5_bit;
sbit LCD_D4 at RB0_bit;
sbit LCD_D5 at RB1_bit;
sbit LCD_D6 at RB2_bit;
sbit LCD_D7 at RB3_bit;

sbit LCD_RS_Direction at TRISB4_bit;
sbit LCD_EN_Direction at TRISB5_bit;
sbit LCD_D4_Direction at TRISB0_bit;
sbit LCD_D5_Direction at TRISB1_bit;
sbit LCD_D6_Direction at TRISB2_bit;
sbit LCD_D7_Direction at TRISB3_bit;
// End LCD module connections

#define MASK 0xF0
#define Enter 11
#define Plus 12
#define Minus 13
#define Multiply 14
#define Divide 15

//
// This function gets a key from the keypad and returns it to calling program
//
```

Figure 6.13: mikroC Pro for PIC Program Listing.

```
unsigned char getkeypad()
{
    unsigned char i, Key = 0;

    PORTC = 0x01;                              // Start with column 1
    while((PORTC & MASK) == 0)                 // While no key pressed
    {
        PORTC = (PORTC << 1);                  // next column
        Key++;                                 // column number
        if(Key == 4)
        {
            PORTC = 0x01;                      // Back to column 1
            Key = 0;
        }
    }
    Delay_Ms(20);                              // Switch debounce

    for(i = 0x10; i !=0; i <<=1)               // Find the key pressed
    {
        if((PORTC & i) != 0)break;
        Key = Key + 4;
    }

    PORTC = 0x0F;
    while((PORTC & MASK) != 0);                // Wait until key released
    Delay_Ms(20);                              // Switch debounce

    return (Key);                              // Return key number
}



//
// Start of MAIN program
//
void main()
{
    unsigned char MyKey, i,j,op[12];
    unsigned long Calc, Op1, Op2;
    char *lcd;

    ANSELB = 0;                                // Configure PORTB as digital
    ANSELC = 0;                                // Configure PORTC as digital
    TRISB = 0;                                 // PORTB are outputs (LCD)
    TRISC = 0xF0;                              // RC4–RC7 are inputs

//
// Configure LCD
//
    Lcd_Init();                                // Initialize LCD
```

Figure 6.13
cont'd

```
                    Lcd_Cmd(_LCD_CLEAR);              // Clear LCD
                    Lcd_Out(1,1,"CALCULATOR");        // Display CALCULATOR
                    Delay_ms(2000);                   // Wait 2 s
                    Lcd_Cmd(_LCD_CLEAR);              // Clear display
          //
          // Program loop
          //
             for(;;)                                  // Endless loop
             {
               MyKey = 0;
               Op1 = 0;
               Op2 = 0;

               Lcd_Out(1,1,"No1: ");                  // Display No1:
               while(1)                               // Get first no
               {
                 MyKey = getkeypad();
                 if(MyKey == Enter)break;             // If ENTER pressed
                 MyKey++;                             // Key number pressed
                 if(MyKey == 10)MyKey = 0;            // If 0 key pressed
                 Lcd_Chr_Cp(MyKey + '0');
                 Op1 = 10 * Op1 + MyKey;              // First number in Op1
               }

               Lcd_Out(2,1,"No2: ");                  // Display No2:
               while(1)                               // Get second no
               {
                 MyKey = getkeypad();
                 if(MyKey == Enter)break;             // If ENTER pressed
                 MyKey++;
                 if(MyKey == 10)MyKey = 0;            // If 0 key pressed
                 Lcd_Chr_Cp(MyKey + '0');
                 Op2 = 10 * Op2 + MyKey;              // Second number in Op2
               }

               Lcd_Cmd(_LCD_CLEAR);                  // Clear LCD
               Lcd_Out(1,1,"Op: ");                   // Display Op:

               MyKey = getkeypad();                   // Get operation
               Lcd_Cmd(_LCD_CLEAR);
               Lcd_Out(1,1,"Res=");                   // Display Res=
               switch(MyKey)                          // Perform the operation
               {
                 case Plus:
                     Calc = Op1 + Op2;                // If ADD
                     break;
                 case Minus:
                     Calc = Op1 - Op2;                // If Subtract
                     break;
                 case Multiply:
                     Calc = Op1 * Op2;                // If Multiply
                     break;
```

Figure 6.13
cont'd

```
                case Divide:
                    Calc = Op1/Op2;                        // If Divide
                    break;
            }

            LongToStr(Calc, op);                           // Convert to string in op
            lcd = Ltrim(op);                               // Remove leading blanks

            Lcd_Out_Cp(lcd);                               // Display result
            Delay_ms(5000);                                // Wait 5 s
            Lcd_Cmd(_LCD_CLEAR);              /             / Clear LCD
        }
    }
```

Figure 6.13
cont'd

noise is to wait for about 20 ms after a switch key is pressed, and also after a switch key is released. In this project, **contact debouncing** is done in function **getkeypad**.

Program Using Built-in Keypad Function

In the program listing in Figure 6.13, a function called **getkeypad** has been developed to read a key from a keypad. mikroC Pro for PIC language has a built-in keypad library with functions **Keypad_Key_Press** and **Keypad_Key_Click** to read a key from a keypad when a key is pressed. The returned key has the code 1—16 (note that the returned key number is not from 0 to 15). Figure 6.14 shows a modified program (MIKROC-KEYPAD2.C) listing using the **Keypad_Key_Click** function to implement the calculator project. The circuit diagram is the same as in Figure 6.11.

Before using any keypad function, we have to call the **Keypad_Init** function to initialize the keypad library. Also, the connection port of the keypad must be declared at the beginning of the program. In this project, the keypad is connected to PORTC, and the following declaration must be made at the beginning of the program:

    char keypadPort at PORTC;

```
/********************************************************************************
                        CALCULATOR WITH KEYPAD AND LCD
                        ================================
```

In this project a 4 x 4 keypad is connected to PORTC of a PIC18F45K22 microcontroller. Also an
LCD is connected to PORTB. The project is a simple calculator which can perform integer
arithmetic.

The keys are labeled as follows:

```
1  2  3  4
5  6  7  8
9  0    Enter
+  −  *  /
```

In this version of the program built-in keypad libray is used.

```
Author:  Dogan Ibrahim
Date:    August 2013
File:    MIKROC-KEYPAD2.C
********************************************************************************/
// LCD module connections
sbit LCD_RS at RB4_bit;
sbit LCD_EN at RB5_bit;
sbit LCD_D4 at RB0_bit;
sbit LCD_D5 at RB1_bit;
sbit LCD_D6 at RB2_bit;
sbit LCD_D7 at RB3_bit;

sbit LCD_RS_Direction at TRISB4_bit;
sbit LCD_EN_Direction at TRISB5_bit;
sbit LCD_D4_Direction at TRISB0_bit;
sbit LCD_D5_Direction at TRISB1_bit;
sbit LCD_D6_Direction at TRISB2_bit;
sbit LCD_D7_Direction at TRISB3_bit;
// End LCD module connections

// Keypad module connections
char keypadPort at PORTC;
// End of keypad module connections

#define Enter 12
#define Plus 13
#define Minus 14
#define Multiply 15
#define Divide 16

//
// Start of MAIN program
//
void main()
```

Figure 6.14: Modified mikroC Pro for PIC Program Listing.

```
        {
            unsigned char MyKey, i,j,op[12];
            unsigned long Calc, Op1, Op2;
            char *lcd;

            ANSELB = 0;                                 // Configure PORTB as digital
            ANSELC = 0;                                 // Configure PORTC as digital
            TRISB = 0;                                  // PORTB are outputs (LCD)
            TRISC = 0xF0;                               // RC4–RC7 are inputs

            Keypad_Init();                              // Initialize keypad library
//
// Configure LCD
//
            Lcd_Init();                                 // Initialize LCD
            Lcd_Cmd(_LCD_CLEAR);
            Lcd_Out(1,1,"CALCULATOR");                  // Display CALCULATOR
            Delay_ms(2000);                             // Wait 2 s
            Lcd_Cmd(_LCD_CLEAR);                        // Clear display
//
// Program loop
//
            for(;;)                                     // Endless loop
            {
              MyKey = 0;
              Op1 = 0;
              Op2 = 0;

              Lcd_Out(1,1,"No1: ");                     // Display No1:
              while(1)                                  // Get first no
              {
                do
                   MyKey = Keypad_Key_Click();
                while(!MyKey);
                if(MyKey == Enter)break;                // If ENTER pressed
                if(MyKey == 10)MyKey = 0;               // If 0 key pressed
                Lcd_Chr_Cp(MyKey + '0');
                Op1 = 10 * Op1 + MyKey;                 // First number in Op1
              }

              Lcd_Out(2,1,"No2: ");                     // Display No2:
              while(1)                                  // Get second no
              {
                do
                   MyKey = Keypad_Key_Click();
                while(!MyKey);
                if(MyKey == Enter)break;                // If ENTER pressed
                if(MyKey == 10)MyKey = 0;               // If 0 key pressed
                Lcd_Chr_Cp(MyKey + '0');
                Op2 = 10 * Op2 + MyKey;                  // Second number in Op2
              }
```

Figure 6.14
cont'd

```
        Lcd_Cmd(_LCD_CLEAR);                    // Clear LCD
        Lcd_Out(1,1,"Op: ");                    // Display Op:

        do
            MyKey = Keypad_Key_Click();         // Get operation
        while(!MyKey);
        Lcd_Cmd(_LCD_CLEAR);
        Lcd_Out(1,1,"Res=");                    // Display Res=
        switch(MyKey)                           // Perform the operation
        {
          case Plus:
                Calc = Op1 + Op2;               // If ADD
                break;
          case Minus:
                Calc = Op1 – Op2;               // If Subtract
                break;
          case Multiply:
                Calc = Op1 * Op2;               // If Multiply
                break;
          case Divide:
                Calc = Op1/Op2;                 // If Divide
                break;
        }

        LongToStr(Calc, op);                    // Convert to string in op
        lcd = Ltrim(op);                        // Remove leading blanks

        Lcd_Out_Cp(lcd);                        // Display result
        Delay_ms(5000);                         // Wait 5 s
        Lcd_Cmd(_LCD_CLEAR);                    // Clear LCD
    }
}
```

Figure 6.14
cont'd

MPLAB XC8

The program listing for the MPLAB XC8 version of the program is shown in
Figure 6.15. Note here that the default LCD connections are slightly different when
using the MPLAB XC8 compiler, where RB4 is connected to E instead of RS, and RB5
is connected to RS instead of E. Also, the RW pin of the LCD is connected to RB6 of
the microcontroller.

```
/*******************************************************************************
                    CALCULATOR WITH KEYPAD AND LCD
                    ===============================
```

In this project a 4 x 4 keypad is connected to PORTC of a PIC18F45K22 microcontroller. Also an
LCD is connected to PORTB. The project is a simple calculator which can perform integer arithmetic.

The LCD is connected to the microcontroller as follows:

```
Microcontroller      LCD
=============        ===

  RB0               D4
  RB1               D5
  RB2               D6
  RB3               D7
  RB4               E
  RB5               R/S
  RB6               RW
```

The keys are organised as follows:

```
0   1    2    3
4   5    6    7
8   9   10   11
12  13  14   15
```

The keys are labeled as follows:

```
1 2 3 4
5 6 7 8
9 0  Enter
+ − * /
```

```
Author:   Dogan Ibrahim
Date:     August 2013
File:     XC8-KEYPAD.C
********************************************************************/

#include <xc.h>
#include <stdlib.h>
#include <plib/xlcd.h>
#include <plib/delays.h>
#pragma config MCLRE = EXTMCLR, WDTEN = OFF, FOSC = HSHP
#define _XTAL_FREQ 8000000

#define MASK 0xF0
#define Enter 11
#define Plus 12
#define Minus 13
```

Figure 6.15: MPLAB XC8 Program Listing.

```
#define Multiply 14
#define Divide 15


//
// This function creates seconds delay. The argument specifies the delay time in seconds.
//
void Delay_Seconds(unsigned char s)
{
  unsigned char i,j;

  for(j = 0; j < s; j++)
  {
    for(i = 0; i <  100; i++)__delay_ms(10);
  }
}


//
// This function creates 18 cycles delay for the xlcd library
//
void DelayFor18TCY( void )
{
Nop(); Nop(); Nop(); Nop();
Nop(); Nop(); Nop(); Nop();
Nop(); Nop(); Nop(); Nop();
Nop(); Nop();
return;
}



//
// This function creates 15 ms  delay for the xlcd library
//
void DelayPORXLCD( void )
{
   __delay_ms(15);
   return;
}



//
// This function creates 5 ms delay for the xlcd library
//
void DelayXLCD( void )
{
   __delay_ms(5);
   return;
}


//
// This function clears the screen
//
```

Figure 6.15
cont'd

```c
void LCD_Clear()
{
  while(BusyXLCD());
  WriteCmdXLCD(0x01);
}

//
// This function moves the cursor to position row,column
//
void LCD_Move(unsigned char row, unsigned char column)
{
 char ddaddr = 40 * (row–1) + column;
 while( BusyXLCD() );
 SetDDRamAddr( ddaddr );
}


//
// This function gets a key from the keypad and returns it to calling program
//
unsigned char getkeypad()
{
    unsigned char i, Key = 0;

    PORTC = 0x01;                                   // Start with column 1
    while((PORTC & MASK) == 0)                      // While no key pressed
    {
      PORTC = (PORTC << 1);                         // next column
      Key++;                                        // column number
      if(Key == 4)
      {
        PORTC = 0x01;                               // Back to column 1
        Key = 0;
      }
    }
    __delay_ms(20);                                 // Switch debounce

    for(i = 0x10; i !=0; i <<=1)                     // Find the key pressed
    {
      if((PORTC & i) != 0)break;
      Key = Key + 4;
    }

    PORTC = 0x0F;
    while((PORTC & MASK) != 0);                      // Wait until key released
    __delay_ms(20);                                 // Switch debounce

    return (Key);                                    // Return key number
}
```

Figure 6.15
cont'd

```
//
// Start of MAIN program
//
void main()
{
    unsigned char MyKey, i,j,op[10];
    unsigned long Calc, Op1, Op2;

    ANSELB = 0;                                     // Configure PORTB as digital
    ANSELC = 0;                                     // Configure PORTC as digital
    TRISB = 0;                                      // PORTB are outputs (LCD)
    TRISC = 0xF0;                                   // RC4–RC7 are inputs

    OpenXLCD(FOUR_BIT & LINES_5X7);                 // Initialize LCD
    while(BusyXLCD());                              // Wait if the LCD is busy
    WriteCmdXLCD(DON);                             // Turn Display ON
    while(BusyXLCD());                              // Wait if the LCD is busy
    WriteCmdXLCD(0x06);                            // Move cursor right
    putrsXLCD("CALCULATOR");                        // Display heading
    Delay_Seconds(2);                             // 2 s delay
    LCD_Clear();                                    // Clear display
//
// Program loop
//
    for(;;)                                         // Endless loop
    {
      MyKey = 0;
      Op1 = 0;
      Op2 = 0;

      LCD_Move(1,1);                                // Move to row = 1,column = 1
      putrsXLCD("No1: ");                           // Display No1:
      while(1)                                      // Get first no
      {
        MyKey = getkeypad();
        if(MyKey == Enter)break;                    // If ENTER pressed
        MyKey++;                                    // Key number pressed
        if(MyKey == 10)MyKey = 0;                   // If 0 key pressed
        putcXLCD(MyKey + '0');
        Op1 = 10 * Op1 + MyKey;                     // First number in Op1
      }

      LCD_Move(2,1);                                // Move to row = 2,column = 1
      putrsXLCD("No2: ");                           // Display No2:
      while(1)                                      // Get second no
      {
        MyKey = getkeypad();
        if(MyKey == Enter)break;                    // If ENTER pressed
        MyKey++;
```

Figure 6.15
cont'd

```
    if(MyKey == 10)MyKey = 0;              // If 0 key pressed
    putcXLCD(MyKey + '0');
    Op2 = 10 * Op2 + MyKey;                // Second number in Op2
}

LCD_Clear();                               // Clear LCD
LCD_Move(1,1);                             // Move to row = 1,column = 1
putrsXLCD("Op: ");                         // Display Op:

MyKey = getkeypad();                       // Get operation
LCD_Clear();
LCD_Move(1,1);
putrsXLCD("Res=");                         // Display Res=
switch(MyKey)                              // Perform the operation
{
  case Plus:
      Calc = Op1 + Op2;                    // If ADD
      break;
  case Minus:
      Calc = Op1 – Op2;                    // If Subtract
      break;
  case Multiply:
      Calc = Op1 * Op2;                    // If Multiply
      break;
  case Divide:
      Calc = Op1/Op2;                      // If Divide
      break;
}

Itoa(op, Calc, 10);                        // Convert to string in op
putsXLCD(op);                              // Display result
Delay_Seconds(5);                          // Wait 5 s
LCD_Clear();
  }
}
```

Figure 6.15
cont'd

Project 6.3—The High/Low Game
Project Description

This project uses a 4 × 4 keypad and an LCD to create the classical High/Low game. For those of you who do not know how to play the game, the rules for this version of the game are as follows:

- The computer will generate a secret random number between 1 and 32767.
- The top row of the LCD will display "Guess Now…".
- The player will try to guess what the number is by entering a number on the keypad and then pressing the ENTER key.

- If the guessed number is higher than the secret number, then the bottom row of the LCD will display "HIGH—Try Again".
- If the guessed number is lower than the secret number, then the bottom row of the LCD will display "LOW—Try Again".
- If the player guesses the number, then the bottom row will display "Well Done...".
- The program waits for 5 s, and the game restarts

Generating a Random Number

In our program, we will be generating a random integer number using the mikroC Pro for PIC library functions "srand" and "rand". Function "srand" must be called with an integer argument (or "seed") to prepare the random number generator library. Then, every time function "rand" is called a new random number will be generated between 1 and 32767. The *set of numbers* generated is the same if the program is restarted with the same "seed" applied to function "srand". Thus, if the game is restarted after resetting the microcontroller, the same set of numbers will be generated.

Block Diagram

The block diagram of the project is as in Figure 6.10.

The keys are organized on the keypad as shown below:

```
1 2 3 A
4 5 6 B
7 8 9 C
* 0 # D
```

The mikroC Pro for PIC returns the following numbers when a key is pressed on the keypad:

Key Pressed	Number Returned
1	1
2	2
3	3
A	4
4	5
5	6
6	7
B	8
7	9
8	10
9	11
C	12
*	13
0	14
#	15
D	16

We will be using key "C" as the ENTER key in our program. Also, we will be correcting the key numbering in our program so that, for example, when "7" is pressed on the keypad a 7 is returned and not a 9 as in the above table.

Circuit Diagram

The circuit diagram of the project is shown in Figure 6.16. The LCD is connected to PORTB as in the earlier projects. The rows and columns of the keypad are connected to PORTC.

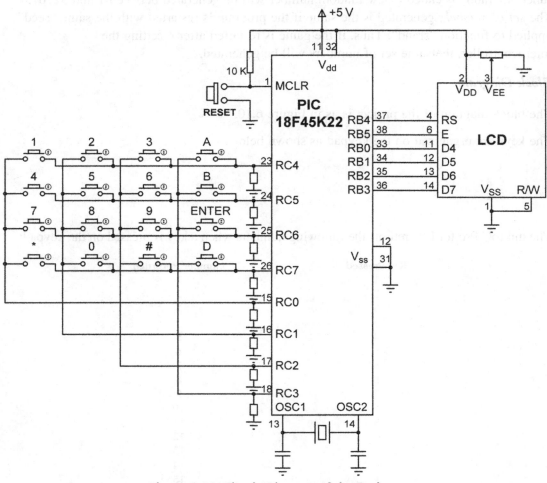

Figure 6.16: Circuit Diagram of the Project.

```
BEGIN
        Declare keypad port number
        Define LCD to microcontroller pin connections
        Configure PORTB as digital
        Initialize keypad library
        Initialize LCD
        Display heading "High/Low Game"
        Set new game flag
        Wait 2 s
        DO FOREVER
                IF new game flag is set THEN
                        Clear LCD
                        Turn OFF cursor
                        Generate a random number (secret number)
                        Display "Guess Now.." on row 1
                ENDIF
                Read and display (on row 2) numbers until ENTER is pressed
                IF entered number > secret number THEN
                        Display "HIGH—Try Again"
                        Wait 1 s
                        Clear second row of LCD
                ELSE IF entered number < secret number THEN
                        Display "LOW—Try Again"
                        Wait 1 s
                        Clear second row of LCD
                ELSE IF entered number = secret number THEN
                        Display "Well Done.."
                        Wait 5 s
                        Set new game flag
                ENDIF
        END
```

Figure 6.17: PDL of the Project.

Project PDL

The PDL of this project is given in Figure 6.17.

Project Program

mikroC Pro for PIC

The mikroC Pro for PIC program is named MIKROC-HILO.C, and the program listing of
the project is shown in Figure 6.18.

```
/*******************************************************************************
                    High/Low Game Using Keypad and LCD
                    ==================================
```

This project implements the High/Low game using the 4 x 4 keypad and an LCD. The program generates a random number between 1 and 32767 and expects the player to guess the number. The LCD displays "Guess Now.." on top row of the display.

The player then guesses the number by entering a number via the keypad and then pressing the ENTER key. If the guessed number is bigger than the generated number the message "HIGH—Try Again" will be displayed on the bottom row of the LCD.

If the guessed number is lower than the generated number then the message "LOW—Try Again" will be generated on the bottom row of the LCD. If on the other hand the player guesses the number correctly, the bottom row of the LCD will display the message "Well Done..".

The game will re-start after 5 s delay.

The microcontroller in this project is PIC18F45K22 and is operated from an 8 MHz crystal as before. An 4x4 keypad is connected to PORTC. The LCD is connected to PORTB.

```
Programmer:    Dogan Ibrahim
Date:          September 2013
File:          MIKROC-HILO.C


*******************************************************************************/
// Declare LCD connections
sbit LCD_RS at RB4_bit;
sbit LCD_EN at RB5_bit;
sbit LCD_D4 at RB0_bit;
sbit LCD_D5 at RB1_bit;
sbit LCD_D6 at RB2_bit;
sbit LCD_D7 at RB3_bit;

sbit LCD_RS_Direction at TRISB4_bit;
sbit LCD_EN_Direction at TRISB5_bit;
sbit LCD_D4_Direction at TRISB0_bit;
sbit LCD_D5_Direction at TRISB1_bit;
sbit LCD_D6_Direction at TRISB2_bit;
sbit LCD_D7_Direction at TRISB3_bit;
// End of LCD connections

// Declare keypad connection
char keypadPORT at PORTC;
// End of keypad connection

#define ENTER_KEY 12
```

Figure 6.18: Program Listing of the Project.

```
            unsigned char kp, new_game;
            unsigned int GuessNumber, PlayerNumber;
            int Diff;


            void main()
            {
                unsigned char Txt1[4];

                    ANSELB = 0;                                    // Configure PORT B as digital
                    ANSELC = 0;                                    // Configure PORTC as digital
                    Keypad_Init();                                 // Initialize keypad library
                    Lcd_Init();                                    // Initialize LCD
                    Lcd_Out(1, 1, "High/Low Game");                // Display heading
                    Delay_ms(2000);                                // Wait 2 s

                    new_game = 1;
                    srand(5);                                      // Random number seed

                for(;;)                                            // DO FOREVER
                    {
                        if(new_game == 1)
                        {
                            Lcd_Cmd(_LCD_CLEAR);                   // Clear LCD
                            Lcd_Cmd(_LCD_CURSOR_OFF);             // Turn OFF cursor
                            Lcd_Out(1, 1, "Guess Now..");          // Display "Guess Now.."
                            GuessNumber = rand();                  // Generate a random number
                        }
                        kp = 0;
                        PlayerNumber = 0;
                        Lcd_Out(2, 1, "");                         // Position cursor at 1,1
                        while(kp != ENTER_KEY)                     // Until ENTER pressed
                        {
                    do
                    {
                    kp = Keypad_Key_Click();                       // Look for key press
                    }while(!kp);

                    if(kp != ENTER_KEY)                            // If not ENTER key
                    {
                                    if(kp > 4 && kp <9)kp = kp = kp-1;   // 5 is 4, 6 is 5....
                    if(kp > 8 && kp < 12)kp = kp-2;                // 7 is 9, 8 is 10...
                    if(kp == 14)kp = 0;                            // 0 is14
                    PlayerNumber = 10 * PlayerNumber + kp;
                    ByteToStr(kp, Txt1);
                    Txt1[0] = Txt1[2];                             // Get the number
                    Txt1[1] = '\0';                                // Make a string
                    Lcd_Out_Cp(Txt1);                              // Display on LCD
                    }
                    }
```

Figure 6.18
cont'd

```
                    Delay_ms(1000);                        // Wait one s
                    Diff = PlayerNumber - GuessNumber;     // Find the diff

                    if(Diff > 0)                           // Greater ?
                    {
                            Lcd_Out(2, 1, "HIGH - Try Again");
                            new_game = 0;                  // Not a new game
                            Delay_ms(1000);
                            Lcd_Out(2,1,"            ");    // Clear second row
                    }

                    if(Diff < 0)                           // Less ?
                    {
                            Lcd_Out(2, 1, "LOW - Try Again");
                            new_game = 0;                  // Not a new game
                            Delay_ms(1000);
                            Lcd_Out(2,1,"           ");     // Clear second row
                    }

                    if(Diff == 0)                          // Equal ?
                    {
                            Lcd_Out(2,1, "Well Done..");
                            new_game = 1;                  // New game
                            Delay_ms(5000);                // Wait 5 s to
                    }                                      // re-start game
            }                                              // End of for
}                                                          // End of program
```

Figure 6.18
cont'd

At the beginning of the program, keypad PORT is declared as PORTC, and some other variables used in the program are also declared. Then, PORTB is configured as a digital output, keypad library is initialized, the LCD is initialized, and message "High/Low Game" is displayed on the LCD. After a 2 s delay, the program continues in an endless loop.

If this is a new game, the LCD is cleared, and message "Guess Now…" is displayed in the first row of the LCD. Then, a random number is generated between 1 and 32767 by calling library function "rand", and this number is stored in variable "GuessNumber". Note that the "srand" library function must be called with an integer number before calling "rand".

The keypad is then checked, and numbers are received until the ENTER key (key C) is pressed. The key numbers are then adjusted such that if, for example, 4 is pressed, number 4 is used by the program instead of 5. Similarly, if key 0 is pressed, number 0 is used by the program instead of 14 returned by the keypad library routine. The numbers entered by the player are displayed in the second row of the LCD as they are entered so that the player can see what he/she has entered. After the player presses the ENTER key, a 1 s delay is introduced. The number entered by the player is stored in variable "PlayerNumber" in decimal format.

The program then calculates the difference between the secret number in "GuessNumber" and the number entered by the player (in PlayerNumber). This difference is stored in variable "Diff".

If "Diff" is positive, that is, if the number entered by the player is greater than the secret number, then the program displays message "HIGH—Try Again", waits for a second, and clears the second row of the LCD, ready for the player to try another number.

Figure 6.19: Display from the Game—Start of the Game.

Figure 6.20: Display from the Game—User Guessed 258.

Figure 6.21: Display from the Game—The Guess was Low.

If "Diff" is negative, that is, if the number entered by the player is less than the secret number, then the program displays message "LOW—Try Again", waits for a second, and clears the second row of the LCD, ready for the player to try another number.

If "Diff" is 0, that is, if the number entered by the player is equal to the secret number, then the program displays message "Well Done..." waits for 5 s, and sets the "new_game" flag so that a new secret number can be generated by the program. The game continues as before.

Figures 6.19—6.21 show various displays from the game. Note that the keypad keys are not debounced in the keypad library, and sometimes, you may get double key strokes even though you press a key once. You should be firm and quick when pressing a key to avoid this from happening.

Project 6.4—Generating Waveforms
Project Description

This project demonstrates how various waveforms can be generated using a microcontroller. The following waveforms will be generated in this project: sawtooth wave, triangle wave, any arbitrary wave, sine wave, and square wave. The first three waveforms will be generated using a digital-to-analog converter (DAC).

Figure 6.22 shows the block diagram of a typical microcontroller-based waveform generation system. Here, the microcontroller generates the required waveform as a digital signal, and then, the DAC converts this signal into analog. In practical applications, a low-pass filter is used after the DAC to clean the signal and remove any high-frequency components.

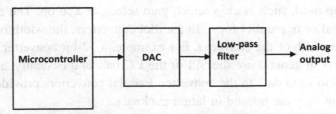

Figure 6.22: Block Diagram of Microcontroller-Based Waveform Generation.

Basically, two methods are used for waveform generation:

* The microcontroller calculates the waveform points in real-time and sends them to the DAC.
* The waveform points are stored in a look-up table. The microcontroller reads these points from the table and sends them to the DAC (this method is used to generate any arbitrary waveform, or to generate higher frequency waveforms).

As we shall see later, the rate at which the waveform points are sent to the DAC determines the frequency of the waveform.

Before going into the details of waveform generation, perhaps it is worthwhile to look at the operation of a DAC.

DAC Converter

A DAC converts a digital signal into an analog signal. The block diagram of a typical DAC is shown in Figure 6.23. This has a digital input, represented by D, analog output, represented by V_o, and a stable and accurate voltage reference, V_{ref}. In addition, some

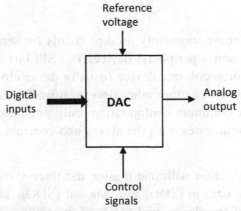

Figure 6.23: The Block Diagram of a Typical DAC.

control lines are also used, such as chip select, gain select, and so on. The digital input can either be in serial or in parallel form. In parallel converters, the width of the digital input is equal to the width of the converter. For example, a 12-bit converter has 12 input bits. Serial converters in general use the SPI or the I^2C bus, and basically, a clock and a data signal are used to send data to the converter. Parallel converters provide much faster conversion times, but they are housed in larger packages.

DACs are manufactured as either unipolar or bipolar as far as the output voltages are concerned. Unipolar converters can provide only positive output voltages, whereas bipolar converters provide both positive and negative voltages. In this book, we will only be using unipolar converters.

The relationship between the digital input—output and the voltage reference are given by

$$V_o = \frac{DV_{ref}}{2^n} \tag{6.1}$$

Where V_o is the output voltage, V_{ref} is the reference voltage, and n is the width of the converter. For example, in a 12-bit converter (resolution = 12 bits) with a +5 V reference voltage,

$$V_o = \frac{5D}{2^{12}} = 1.22D \text{ mV} \tag{6.2}$$

Thus, for example, if the input digital value is 1, the analog output voltage will be 1.22 mV, if the input value is 2, the analog output voltage will be 2.44 mV, and so on.

In this book, we shall be using a serial DAC for convenience and low cost. Most of the serial DACs use the SPI bus for communicating with a microcontroller. It is worth looking at the basic principles of the SPI bus before continuing with the project.

The SPI Bus

The SPI bus is one of the most commonly used protocols for serial communication between a microcontroller and a peripheral device. The SPI bus is a master—slave type bus protocol. In this protocol, one device (usually the microcontroller) is designated the *master*, and one or more other devices (usually sensors, converters, etc.) are designated *slaves*. In a minimum configuration, only one master and one slave are used. The master communicates with the slaves and controls all the activity on the bus.

Figure 6.24 shows a configuration with one master and three slaves. The SPI bus used three signals: clock (SCK), data in (SDI), and data out (SDO). The SDO of the master is connected to the SDIs of the slaves, and SDOs of the slaves are connected to the

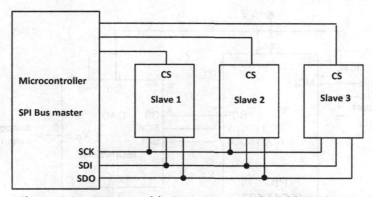

Figure 6.24: SPI Bus with One Master and Multiple Slaves.

SDI of the master. The master generates the SCK signals to enable data to be transferred on the bus. In every clock pulse, 1 bit of data is moved from the master to the slave, or from the slave to the master. The communication is only between a master and a slave, and the slaves cannot communicate with one an other. It is important to note that only one slave can be active at any time since there is no mechanism to identify the slaves. Thus, slave devices have enable lines (e.g. CS), which are normally controlled by the master. A typical communication between a master and several slaves can be as follows:

- Master enables slave 1.
- Master sends SCK signals to read or write data to slave 1.
- Master disables slave 1 and enables slave 2.
- Master sends SCK signals to read or write data to slave 2.
- The above process continues as required.

PIC18F microcontrollers provide one or more sets of special SPI bus compatible pins to enable the microcontroller to be connected to SPI slave peripheral devices. mikroC Pro for PIC and MPLAB XC8 compilers both provide SPI libraries to simplify programming and communication on the SPI bus.

Generating Sawtooth Waveform

In this part of the project, we will be generating a sawtooth waveform with the following specifications:

Output voltage	0 to +5 V
Frequency	100 Hz (period: 10 ms)
Step size	0.1 ms

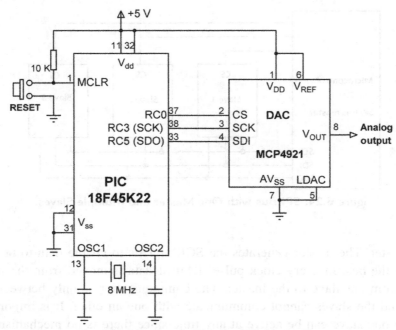

Figure 6.25: Circuit Diagram of the Project.

The circuit diagram of the project is shown in Figure 6.25. An MCP4921-type serial DAC is connected to the SPI port (PORTC) of a PIC18F45K22 microcontroller. The following connection is used between the microcontroller and the DAC:

Microcontroller Pin	DAC Pin
RC3 (SCK)	SCK
RC5 (SDO)	SDI
RC0	CS

If you are using the EasyPIC V7 development board and the 12-Bit DAC board, just plug in the DAC board to the PORTC IDC10 connector of the development board and configure the DAC board DIL jumper for the above connections. Also, set the reference voltage jumper on the DAC board to +5 V.

MCP4921 is a 12-bit serial DAC manufactured by Microchip Inc., having the following basic specifications:

- A 12-bit resolution,
- Up to a 20-MHz clock rate (SPI),
- Fast settling time of 4.5 μs,
- Unity or 2x gain output,
- External V_{ref} input,

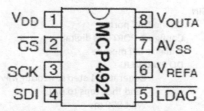

Figure 6.26: Pin Layout of MCP4921 DAC.

- A 2.7—5.5 V operation,
- Extended temperature range (−40 to +125 °C),
- An 8-pin DIL package.

Figure 6.26 shows the pin layout of the MCP4921. The pin definitions are as follows:

V_{DD}, AV_{SS}	power supply and ground
CS	chip select (LOW to enable the chip)
SCK, SDI	SPI bus clock and data in
V_{OUTA}	analog output
V_{REFA}	reference input voltage
LDAC	DAC input latch (transfers the input data to the DAC registers. Normally tied to ground so that CS controls the data transfer).

Data are written to the DAC in 2 bytes. The lower byte specified bits D0:D8 of the digital input data. The upper byte consists of the following bits:

D8:D11	bits D8:D11 of the digital input data
SHDN	1: output power down mode, 0: disable output buffer
GA	output gain control. 0: gain is 2x, 1: gain is 1x
BUF	0: input unbuffered, 1: input buffered
A/B	0: write to DAC_A, 1: write to DAC_B (MCP4921 supports only DAC_A)

Project PDL

The PDL of the project is shown in Figure 6.27.

Project Program

mikroC Pro for PIC

The mikroC Pro for the PIC program is called MIKROC-WAVE1.C and is given in Figure 6.28. At the beginning of the program, the CS enable input of the DAC is defined. PORTC is configured as digital, and the SPI bus module is initialized using built-in function SPI_Init. Inside the main program, an endless loop is formed, and the steps of the sawtooth waveform are sent out inside a *for* loop. Since there are 11 steps

```
BEGIN
        Define DAC port
        Configure PORTC as digital
        Initialize SPI module
        DO FOREVER
                Generate 11 step sawtooth wave
                Send the steps to DAC
                Wait 909 ms
        ENDDO
END


BEGIN/DAC
        Enable DAC
        Send high byte with 1x gain
        Send low byte
        Disable DAC
END/DAC
```

Figure 6.27: PDL of the Project.

(0–10) in the waveform and the required frequency is 100 Hz, that is, period 10 ms, then the duration of each step should be $10,000/11 = 909$ μs. The delay_us function is used to generate the required delay. Function DAC receives the digital data (0–4095) in its argument and sends the data to the DAC. The chip is enabled (CS = 0), and the high byte is sent out first by setting the output gain to 1x. Then, the low byte is sent through the SPI bus.

Figure 6.29 shows the output waveform obtained using the PSCGU250 digital oscilloscope. Here, the vertical axis is 1 V per division, and the horizontal axis is 5 ms per division. The graph is moved down the 0 V point for clarity. Note that the period of the waveform is around 13 ms (i.e. frequency of about 77 kHz, and not 100 Hz). This is because of the delay caused by the statements inside the *for* loop. We will see in the next section how to improve the frequency.

mikroC Pro for the PIC SPI library supports the following functions ("x" in these functions is either 1 or 2, and it designates the SPI module number to be used):

SPIx_Init: Initialize the SPI module in the master mode with $F_{OSC}/4$ clock, data transmitted on low to high edge, and data sampled at the middle of the interval.

SPIx_Init_Advanced: Similar to SPI_Init, but various initialization parameters can be selected.

SPIx_Read: Read a byte from the SPI bus.

SPIx_Write: Write a byte via the SPI bus.

```
/*****************************************************************************
                    Sawtooth Waveform Generation
                    ==============================

This project shows how a sawtooth wavefor with specified frequency can be generated using a
microcontroller. The PIC18F45K22 microcontroller is used with an 8 MHz crystal.

The generated sawtooth waveform has amplitude 0 to +5 V and frequency of 100 Hz (period: 10 ms).

The MCP4921 DAC chip is used to convert the generated signal into analog.  This is a 12-bit
converter controlled with the SPI bus signals.

Programmer:    Dogan Ibrahim
Date:          September 2013
File:          MIKROC-WAVE1.C

*****************************************************************************/
// DAC module connections
sbit Chip_Select at RC0_bit;
sbit Chip_Select_Direction at TRISC0_bit;
// End DAC module connections

//
// This function sends 12 bits digital data to the DAC. The data is passed
// through the function argument called "value"
//
void DAC(unsigned int value)
{
  char temp;

  Chip_Select = 0;                              // Enable DAC chip

  // Send High Byte
  temp = (value >> 8) & 0x0F;                   // Store bits 8:11 to temp
  temp |= 0x30;                                 // Define DAC setting (choose 1x gain)
  SPI1_Write(temp);                             // Send high byte via SPI

  // Send Low Byte
  temp = value;                                 // Store bits 0:7 to temp
  SPI1_Write(temp);                             // Send low byte via SPI

  Chip_Select = 1;                              // Disable DAC chip
}



void main()
{
    float i;
    unsigned int DAC_Value;
```

Figure 6.28: mikroC Pro for PIC Program Listing.

```
                        ANSELC = 0;                        // Configure PORTC as digital
                        Chip_select = 1;                   // Disable DAC
                        Chip_Select_Direction = 0;         // Set CS as output
                        SPI1_Init();                       // Initialize SPI module
            //
            // Generate the Sawtooth waveform
            //
                        for(;;)                            // Endless loop
                        {
                          for(i = 0; i <= 1; i = i + 0.1)  // Generate waveform steps
                          {
                            DAC_Value = i * 4095;
                            DAC(DAC_Value);                // Send to DAC converter
                            Delay_us(909);                 // Wait 909 ms
                          }
                        }
            }
```

Figure 6.28
cont'd

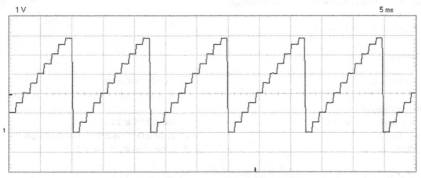

Figure 6.29: The Generated Sawtooth Waveform.

SPI_Set_Active: Sets the active SPI module to be used (for processors having more than one SPI module).

Modified Sawtooth Program

The program given in Figure 6.29 can be improved by using a timer to generate interrupts very close to 909 μs. Data can then be sent to the DAC inside the ISR.

Figure 6.30 shows the modified program (MIKROC-WAVE2.C). Timer0 is used in the 8-bit mode. To generate a delay of 909 μs, the value to be loaded into the timer register is calculated as (Project 6.1)

```
/*******************************************************************************
                        Sawtooth Waveform Generation
                        ============================

This project shows how a sawtooth waveform with specified frequency can be generated using a
microcontroller. The PIC18F45K22 microcontroller is used with an 8 MHz crystal.

The generated sawtooth waveform consists of 11 steps from 0 to +5 V and has a frequency of
100 Hz (period of 10 ms).

The MCP4921 DAC chip is used to convert the generated signal into analog. This is a 12-bit
converter controlled with the SPI bus signals.

This version of the program uses Timer0 interrupts to generate the waveform with the specified
frequency.

Programmer:     Dogan Ibrahim
Date:           September 2013
File:           MIKROC-WAVE2.C

*******************************************************************************/
// DAC module connections
sbit Chip_Select at RC0_bit;
sbit Chip_Select_Direction at TRISC0_bit;
// End DAC module connections

float Sample = 0.0;

//
// This function sends 12 bits digital data to the DAC. The data is passed
// through the function argument called "value"
//
void DAC(unsigned int value)
{
  char temp;

  Chip_Select = 0;                              // Enable DAC chip

  // Send High Byte
  temp = (value >> 8) & 0x0F;                   // Store bits 8:11 to temp
  temp |= 0x30;                                 // Define DAC setting (choose 1x gain)
  SPI1_Write(temp);                             // Send high byte via SPI

  // Send Low Byte
  temp = value;                                 // Store bits 0:7 to temp
  SPI1_Write(temp);                             // Send low byte via SPI

  Chip_Select = 1;                              // Disable DAC chip
}

//
```

Figure 6.30: Modified Program.

```
        // Timr interrupt service routine. Program jumps here at every 10 ms
        //
        void interrupt (void)
        {
            unsigned int DAC_Value;

            TMR0L = 28;                                 // Reload timer register
            INTCON.TMR0IF = 0;                          // Clear timer interrupt flag
        //
        // Generate the Sawtooth waveform
        //
            DAC_Value = Sample * 4095;
            DAC(DAC_Value);                             // Send to DAC converter
            Sample = Sample + 0.1;                      // Next sample
            if(Sample > 1.0)Sample = 0.0;
        }


        void main()
        {
          ANSELC = 0;                                   // Configure PORTC as digital
          Chip_select = 1;                              // Disable DAC
          Chip_Select_Direction = 0;                    // Set CS as output
          SPI1_Init();                                  // Initialize SPI module
        //
        // Configure TIMER0 interrupts
        //
          T0CON = 0xC2;                                 // TIMER0 in 8-bit mode
          TMR0L = 28;                                   // Load Timer register
          INTCON = 0xA0;                                // Enable global and Timer0 interrupts
          for(;;)                                       // Wait for interrupts
          {
          }
        }
```

Figure 6.30
cont'd

$$\text{Time} = (4 \times \text{clock period}) \times \text{Prescaler} \times (256 - \text{TMR0L})$$

Where Prescaler is the selected prescaler value, and TMR0L is the value loaded into timer register TMR0L to generate timer interrupts for every Time period. In our application, the clock frequency is 8 MHz, that is, clock period $= 0.125$ μs and Time $= 909$ μs. Selecting a prescaler value of 8, the number to be loaded into TMR0L can be calculated as follows:

$$\text{TMR0L} = 256 - \frac{\text{Time}}{4 * \text{clockperiod} * \text{prescaler}}$$

or

$$TMR0L = 256 - \frac{909}{4 * 0.125 * 8} = 28$$

Thus, TMR0L should be loaded with 28. The value to be loaded into the TMR0 control register T0CON can then be found as follows:

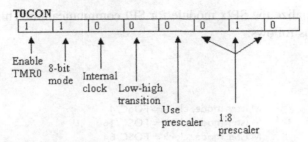

Thus, the T0CON register should be loaded with hexadecimal 0xC2. The next register to be configured is the interrupt control register INTCON:

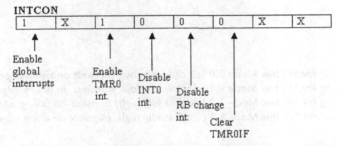

Taking the do not care entries (X) as 0, the hexadecimal value to be loaded into register INTCON is thus 0xA0.

Figure 6.31 shows the new waveform with the vertical axis of 1 V per division and the horizontal axis of 5 ms per division. Clearly, this waveform has the specified period of 10 ms.

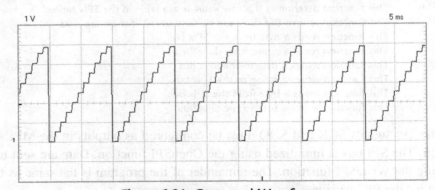

Figure 6.31: Generated Waveform.

MPLAB XC8

The MPLAB XC8 version of the program (XC8-WAVE2.C) is shown in Figure 6.32. MPLAB XC8 compiler supports the following SPI functions ("x" is 1 or 2, and it designates the SPI module number in multiple SPI processors. In single SPI processors, "x" can be omitted):

> OpenSPIx: Initialize the SPIx module for SPI communication. This function takes three arguments as follows:

sync_mode

SPI_FOSC_4	Master mode, clock = FOSC/4
SPI_FOSC_16	Master mode, clock = FOSC/16
SPI_FOSC_64	Master mode, clock = FOSC/64
SPI_FOSC_TMR2	Master mode, clock = TMR2 output/2
SLV_SSON	Slave mode, /SS pin control enabled
SLV_SSOFF	Slave mode, /SS pin control disabled

bus_mode

MODE_00	Setting for SPI bus Mode 0,0 (clock is idle low, transmit on rising edge)
MODE_01	Setting for SPI bus Mode 0,1 (clock is idle low, transmit on falling edge)
MODE_10	Setting for SPI bus Mode 1,0 (clock is idle high, transmit on falling edge)
MODE_11	Setting for SPI bus Mode 1,1 (clock is idle high, transmit on rising edge)

smp_phase

SMPEND	Input data sample at the end of data out
SMPMID	Input data sample at the middle of data out
CloseSPIx	This function disables the SPIx module
DataRdySPIx	This function determines if a new value is available in the SPIx buffer. The Function returns 0 if there are no data and 1 if there are data
getcSPIx	This function reads a byte from the SPIx bus
getsSPIx	This function reads a string from the SPIx bus
putcSPIx	This function writes a byte to the SPIx bus
putsSPIx	This function writes a string to the SPIx bus
ReadSPIx	This function reads a byte from the SPIx bus
WriteSPIx	This function writes a byte to the SPIx bus

Note that the SPI signals SCK and SDO must be configured as outputs in the MPLAB XC8 version. The SPI bus is initialized using the OpenSPI function. Data are sent to the SPI bus using the WriteSPI function. The remainder of the program is the same as the mikroC Pro for the PIC version.

```
/*****************************************************************************
                        Sawtooth Waveform Generation
                        ==============================
```

This project shows how a sawtooth waveform with specified frequency can be generated using a microcontroller. The PIC18F45K22 microcontroller is used with an 8 MHz crystal.

The generated sawtooth waveform consists of 11 steps from 0 to +5 V and has a frequency of 100 Hz (period of 10 ms).

The MCP4921 DAC chip is used to convert the generated signal into analog. This is a 12-bit converter controlled with the SPI bus signals.

This program uses Timer0 interrupts to generate the waveform with the specified frequency.

```
Programmer:     Dogan Ibrahim
Date:           September 2013
File:           XC8-WAVE2.C


*****************************************************************************/

#include <xc.h>
#include <plib/spi.h>
#pragma config MCLRE = EXTMCLR, WDTEN = OFF, FOSC = HSHP
#define _XTAL_FREQ 8000000

// DAC module connections
#define Chip_Select PORTCbits.RC0
#define Chip_Select_Direction TRISCbits.TRISC0
// End DAC module connections

// SPI bus signal directions
#define SCK_Direction TRISCbits.TRISC3
#define SDO_Direction TRISCbits.TRISC5
// End of SPI bus signal directions

float Sample = 0.0;

//
// This function sends 12 bits digital data to the DAC. The data is passed through the
// function argument called "value"
//
void DAC(unsigned int value)
{
  char temp;

  Chip_Select = 0;                                 // Enable DAC chip

  // Send High Byte
  temp = (value >> 8) & 0x0F;                       // Store bits 8:11 to temp
  temp |= 0x30;                                     // Define DAC setting (choose 1x gain)
```

Figure 6.32: MPLAB XC8 Program Listing.

```
        WriteSPI1(temp);                              // Send high byte via SPI

    // Send Low Byte
    temp = value;                                     // Store bits 0:7 to temp
    WriteSPI1(temp);                                  // Send low byte via SPI

    Chip_Select = 1;                                  // Disable DAC chip
}


//
// Timer interrupt service routine. Program jumps here at every 909 ms
//
void interrupt isr (void)
{
    unsigned int DAC_Value;

    TMR0L = 28;                                       // Reload timer register
    INTCONbits.TMR0IF = 0;                            // Clear timer interrupt flag
//
// Generate the Sawtooth waveform
//
    DAC_Value = (unsigned int)(Sample * 4095);
    DAC(DAC_Value);                                   // Send to DAC converter
    Sample = Sample + 0.1;                            // Next sample
    if(Sample > 1.0)Sample = 0.0;
}


void main()
{
    ANSELC = 0;
    Chip_Select_Direction = 0;
    SCK_Direction = 0;
    SDO_Direction = 0;

    Chip_Select = 1;                                  // Disable DAC
    OpenSPI1(SPI_FOSC_4, MODE_00, SMPMID);            // Initialize SPI module

//
// Configure TIMER0 interrupts
//
    T0CON = 0xC2;                                     // TIMER0 in 8-bit mode
    TMR0L = 28;                                       // Load Timer register
    INTCON = 0xA0;                                    // Enable global and Timer0 interrupts
    for(;;)                                           // Wait for interrupts
    {
    }
}
```

Figure 6.32
cont'd

Generating Triangle Waveform

In this part of the project, we will be generating a triangle waveform with the following specifications:

Output voltage	0 to +5 V
Frequency	100 Hz (period: 10 ms)
Step size	0.1 ms

The circuit diagram of the project is as in Figure 6.25. Since the required period is 10 ms, the rising and falling parts of the waveform will each be 454 μs. The value to be loaded into the timer register should therefore change to

$$TMR0L = 256 - \frac{Time}{4 * clockperiod * prescaler}$$

or

$$TMR0L = 256 - \frac{454}{4 * 0.125 * 8} = 142$$

Figure 6.33 shows the program listing (MIKROC-WAVE3.C). The ISR code is also changed to generate the required triangle waveform.

The generated waveform is shown in Figure 6.34 using a PGSCU250 PC-based oscilloscope. The vertical axis is 1 V per division and the horizontal axis 5 ms per division.

Generating an Arbitrary Waveform

In this part of the project, we will be generating an arbitrary waveform. One period of the shape of the waveform will be sketched, and values of the waveform at different points will be extracted and loaded into a look-up table. The program will output the data points at the appropriate times to generate the required waveform.

The shape of one period of the waveform to be generated is shown in Figure 6.35. Assume that the required period is 20 ms (50 Hz).

```
/**********************************************************************
                    Triangle Waveform Generation
                    ============================

This project shows how a triangle waveform with specified frequency can be generated using a
microcontroller. The PIC18F45K22 microcontroller is used with an 8 MHz crystal.

The generated triangle waveform consists of 11 steps rising and 11 steps falling from 0V to +5V and
has a frequency of 100 Hz (period of 10 ms, 5 ms rising and 5 ms falling).

The MCP4921 DAC chip is used to convert the generated signal into analog.  This is a 12-bit
converter controlled with the SPI bus signals.

This version of the program uses Timer0 interrupts to generate the waveform with the specified
frequency.

Programmer:    Dogan Ibrahim
Date:          September 2013
File:          MIKROC-WAVE3.C

**********************************************************************/
// DAC module connections
sbit Chip_Select at RC0_bit;
sbit Chip_Select_Direction at TRISC0_bit;
// End DAC module connections

float Sample = 0.0, Inc = 0.1;

//
// This function sends 12 bits digital data to the DAC. The data is passed
// through the function argument called "value"
//
void DAC(unsigned int value)
{
  char temp;

  Chip_Select = 0;                              // Enable DAC chip

  // Send High Byte
  temp = (value >> 8) & 0x0F;                   // Store bits 8:11 to temp
  temp |= 0x30;                                 // Define DAC setting (choose 1x gain)
  SPI1_Write(temp);                             // Send high byte via SPI

  // Send Low Byte
  temp = value;                                 // Store bits 0:7 to temp
  SPI1_Write(temp);                             // Send low byte via SPI

  Chip_Select = 1;                              // Disable DAC chip
}

//
```

Figure 6.33: mikroC Pro for the PIC Program Listing.

```
// Timer interrupt service routineProgram jumps here at every 10 ms
//
void interrupt (void)
{
    unsigned int DAC_Value;

    TMR0L = 142;                              // Reload timer register
    INTCON.TMR0IF = 0;                        // Clear timer interrupt flag
//
// Generate the Triangle waveform
//
    DAC_Value = Sample * 4095;
    DAC(DAC_Value);                           // Send to DAC converter
    Sample = Sample + Inc;                    // Next sample
    if(Sample > 1.0 || Sample < 0)
    {
      Inc = −Inc;
      Sample = Sample + Inc;
}    }


void main()
{
    ANSELC = 0;                               // Configure PORTC as digital
    Chip_select = 1;                          // Disable DAC
    Chip_Select_Direction = 0;                // Set CS as output
    SPI1_Init();                              // Initialize SPI module
//
 // Configure TIMER0 interrupts
 //
    T0CON = 0xC2;                             // TIMER0 in 8-bit mode
    TMR0L = 142;                              // Load Timer register
    INTCON = 0xA0;                            // Enable global and Timer0 interrupts
    for(;;)                                   // Wait for interrupts
    {
    }
}
```

Figure 6.33
cont'd

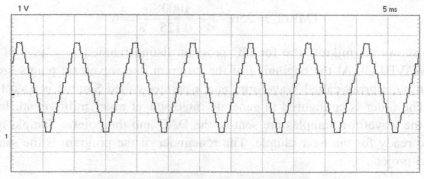

Figure 6.34: Generated Triangle Waveform.

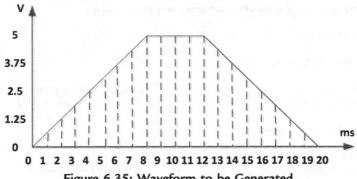

Figure 6.35: Waveform to be Generated.

The waveform takes the following values:

Time (ms)	Amplitude (V)	Time (ms)	Amplitude (V)
0	0	11	5.000
1	0.625	12	5.000
2	1.250	13	4.364
3	1.875	14	3.750
4	2.500	15	3.125
5	3.125	16	2.500
6	3.750	17	1.875
7	4.375	18	1.250
8	5.000	19	0.625
9	5.000	20	0
10	5.000		

The circuit diagram of the project is as given in Figure 6.25. Since the required period is 20 ms, each sample time is 1 ms. The value to be loaded into the timer register is found as follows:

$$ \text{TMR0L} = 256 - \frac{\text{Time}}{4 * \text{clockperiod} * \text{prescaler}} $$

or

$$ \text{TMR0L} = 256 - \frac{1000}{4 * 0.125 * 8} = 6 $$

Figure 6.36 shows the mikroC Pro for PIC program listing (named MIKROC-WAVE4.C). At the beginning of the program, the waveform points are stored in a floating point array called Waveform, and a pointer called Sample is used to index this array. The timer is configured to generate interrupts at every millisecond. Inside the timer ISR, the waveform samples are sent to the DAC and the pointer Sample is incremented ready for the next sample. The remainder of the program is the same as in the previous project.

```
/******************************************************************************
                        Arbitrary Waveform Generation
                        ============================
```

This project shows how an arbitrary waveform with specified frequency can be generated using a microcontroller. The PIC18F45K22 microcontroller is used with an 8 MHz crystal.

The generated waveform is first drawn, the waveform points extracted and stored in a floating point array. In this example the period of the Waveform is 20 ms, defined using 20 Waveform points.

The MCP4921 DAC chip is used to convert the generated signal into analog. This is a 12-bit converter controlled with the SPI bus signals.

This version of the program uses Timer0 interrupts to generate the waveform with the specified frequency.

```
Programmer:     Dogan Ibrahim
Date:           September 2013
File:           MIKROC-WAVE4.C

******************************************************************************/
// DAC module connections
sbit Chip_Select at RC0_bit;
sbit Chip_Select_Direction at TRISC0_bit;
// End DAC module connections

unsigned char Sample = 0;
//
// Store the waveform points in an array
//
float Waveform[] = {0, 0.625, 1.250, 1.875, 2.5, 3.125, 3.750, 4.375, 5, 5,
        5, 5, 5, 4.375, 3.750, 3.125, 2.5, 1.875, 1.250, 0.625};

//
// This function sends 12 bits digital data to the DAC. The data is passed
// through the function argument called "value"
//
void DAC(unsigned int value)
{
  char temp;

  Chip_Select = 0;                              // Enable DAC chip

  // Send High Byte
  temp = (value >> 8) & 0x0F;                   // Store bits 8:11 to temp
  temp |= 0x30;                                 // Define DAC setting (choose 1x gain)
  SPI1_Write(temp);                             // Send high byte via SPI

  // Send Low Byte
  temp = value;                                 // Store bits 0:7 to temp
```

Figure 6.36: mikroC Pro for PIC Program Listing.

```
        SPI1_Write(temp);                              // Send low byte via SPI

        Chip_Select = 1;                               // Disable DAC chip
    }

//
// Timer interrupt service routine. Program jumps here at every 10 ms
//
void interrupt (void)
{
    unsigned int DAC_Value;

    TMR0L = 6;                                          // Reload timer register
    INTCON.TMR0IF = 0;                                  // Clear timer interrupt flag
//
// Generate the arbitrary waveform
//
    DAC_Value = Waveform[Sample] * 4095/5;
    DAC(DAC_Value);                                     // Send to DAC converter
    Sample = Sample++;                                  // Next Sample
    if(Sample == 20)Sample = 0;
}


void main()
{
    ANSELC = 0;                                         // Configure PORTC as digital
    Chip_select = 1;                                    // Disable DAC
    Chip_Select_Direction = 0;                          // Set CS as output
    SPI1_Init();                                        // Initialize SPI module
//
// Configure TIMER0 interrupts
//
    T0CON = 0xC2;                                       // TIMER0 in 8-bit mode
    TMR0L = 6;                                          // Load Timer register
    INTCON = 0xA0;                                      // Enable global and Timer0 interrupts
    for(;;)                                             // Wait for interrupts
    {
    }
}
```

Figure 6.36
cont'd

Figure 6.37 shows the waveform generated by the program. Here, the vertical axis is 1 V per division and the horizontal axis 5 ms per division.

Generating Sine Waveform

In this part of the project, we will see how to generate a low-frequency sine wave using the built-in trigonometric **sin** function, and then send the output to a DAC.

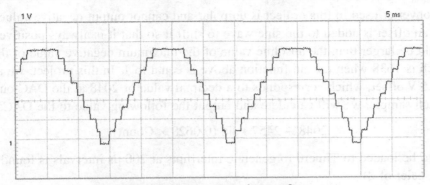

Figure 6.37: Generated Waveform.

The generated sine wave has an amplitude of 2 V, frequency of 50 Hz, and offset of 2.5 V.

The circuit diagram of the project is as in Figure 6.25 where the DAC is connected to PORTC of a PIC18F45K22 microcontroller.

The frequency of the sine wave to be generated is 50 Hz. This wave has a period of 20 ms, or 20,000 μs. If we assume that the sine wave will consist of 100 samples, then each sample should be output at 20,000/100 = 200 μs intervals. Thus, we will configure Timer0 to generate interrupts at every 200 μs, and inside the ISR, we will output a new sample of the sine wave. The sample values will be calculated using the trigonometric **sin** function of the compiler.

The sin function will have the following format:

$$\sin\left(\frac{2\pi \times \text{Count}}{T}\right)$$

where T is the period of the waveform and is equal to 100 samples. Count is a variable that ranges from 0 to 100 and is incremented by 1 inside the ISR every time a timer interrupt occurs. Thus, the sine wave is divided into 100 samples, and each sample is output at 200 μs. The above formula can be rewritten as follows:

$$\sin\left(0.0628 \times \text{Count}\right)$$

It is required that the amplitude of the waveform should be 2 V. With a reference voltage of +5 V and a 12-bit DAC converter (0−4095 quantization levels), 2 V is equal to decimal number 1638. Thus, we will multiply our sine function with the amplitude at each sample to give

$$1638 * \sin\left(0.0628 \times \text{Count}\right)$$

The D/A converter used in this project is unipolar and cannot output negative values. Therefore, an offset is added to the sine wave to shift it so that it is always positive. The offset should be larger than the absolute value of the maximum negative value of the sine wave, which is 1638 when the sin function above is equal to 1. In this project, we are adding a 2.5 V offset, which corresponds to a decimal value of 2048 at the DAC output. Thus, at each sample, we will calculate and output the following value to the DAC:

$$2048 + 2457 * \sin(0.0628 \times \text{Count})$$

The value to be loaded to Timer0 to generate interrupts at 200 μs intervals is found as (with a prescaler of 8):

$$\text{TMR0L} = 256 - \frac{\text{Time}}{4 * \text{clockperiod} * \text{prescaler}}$$

or

$$\text{TMR0L} = 256 - \frac{200}{4 * 0.125 * 8} = 206$$

mikroC Pro for PIC

Figure 6.38 shows the mikroC Pro for the PIC program listing (named MIKROC-WAVE5.C). At the beginning of the program, the chip select connection of the DAC chip is defined. Then, the sine wave amplitude is set to 1638, offset is set to 2048, and variable R is defined as $2\pi/100$. The chip select direction is configured as the output, DAC is disabled by setting its chip enable input, and SPI2 is initialized. The sine waveform values for a period are obtained offline outside the ISR using the following statement. The reason for calculating these values outside the ISR is to minimize the time inside the ISR so that higher frequency sine waves can be generated (it is also possible to generate higher frequency waveforms by increasing the clock frequency. For example, by enabling the clock PLL, the frequency can be multiplied by 4 to be 32 MHz):

```
for(i = 0; i < 100; i++)sins[i] = offset + Amplitude * sin(R * i);
```

The main program then configures Timer0 to generate interrupts at every 200 μs. Timer prescaler is taken as 8, and the timer register is loaded with 206. The main program then waits in an endless loop where the processing continues inside the ISR whenever the timer overflows.

Figure 6.39 shows the waveform generated by the program. It is clear from this figure that the generated sine waveform has period 20 ms as designed. Here, the vertical axis is 1 V per division, and the horizontal axis is 20 ms per division.

```
/*******************************************************************************
                            GENERATE SINE WAVE
                            ==================
```

In this project a DAC is connected to the microcontroller output and a 50 Hz sine wave with an amplitude of 2 V and an offset of 2.5 V is generated in real-time at the output of the DAC.

The MCP4921 12-bit DAC is used in the project. This converter can operate from +3.3 to +5 V, with a typical conversion time of 4.5 ms. Operation of the DAC is based on the standard SPI interface.

mikroC PRO for PIC trigonometric "sin" function is used to calculate the sine points. 50 Hz waveform has the period T = 20 ms, or, T = 20,000 us. If we take 100 points to sample the sine wave, then each Sample occurs at 200 us. Therefore, we need a timer interrupt service routine that will generate interrupts at every 200 us, and inside this routine we will calculate a new sine point and send it to the DAC. The result is that we will get a 50 Hz sine wave. Because the DAC is unipolar, we have to shift the output waveform to a level greater than its maximum negative value so that the waveform is always positive and can be output by the DAC.

```
Author:     Dogan Ibrahim
Date:       September 2013
File:       MIKROC-WAVE5.C
*******************************************************************************/
// DAC module connections
sbit Chip_Select at RC0_bit;
sbit Chip_Select_Direction at TRISC0_bit;
// End DAC module connections

#define T 100                               // 100 samples
#define R 0.0628                            // 2 * PI/T
#define Amplitude 1638                      // 2 V * 4096/5 V
#define offset 2048                         //2.5 * 4096/5 V

unsigned char temp, Count = 0;
float Sample;
unsigned int Value;
float sins[100];

//
// This function sends 12 bits digital data to the DAC. The data is passed
// through the function argument called "value"
//
void DAC(unsigned int value)
{
  char temp;

  Chip_Select = 0;                          // Enable DAC chip

  // Send High Byte
  temp = (value >> 8) & 0x0F;               // Store bits 8:11 to temp
```

Figure 6.38: mikroC Pro for PIC Program Listing.

```
        temp |= 0x30;                                // Define DAC setting (choose 1x gain)
        SPI1_Write(temp);                            // Send high byte via SPI

        // Send Low Byte
        temp = value;                                // Store bits 0:7 to temp
        SPI1_Write(temp);                            // Send low byte via SPI

        Chip_Select = 1;                             // Disable DAC chip
    }


//
// Timer ISR. Program jumps here at every 200 ms
//
void interrupt (void)
{
        TMR0L = 206;                                 // Reload timer register
//
// Get sine wave samples and send to DAC
//
        Value = sins[Count];
        DAC(Value);                                  // Send to DAC converter
        Count++;
        if(Count == 100)Count = 0;
        INTCON.TMR0IF = 0;                           // Clear timer interrupt flag
    }


void main()
{
    unsigned char i;
    ANSELC = 0;                                      // Configure PORTC as digital
    Chip_select = 1;                                 // Disable DAC
    Chip_Select_Direction = 0;                       // Set CS as output
    SPI1_Init();                                     // Initialize SPI module
//
// Generate the sine wave samples offline and load into an array called sins
//
    for(i = 0; i < 100; i++)sins[i] = offset + Amplitude * sin(R * i);
//
// Configure TIMER0 interrupts
//
    T0CON = 0xC2;                                    // TIMER0 in 8-bit mode
    TMR0L = 206;                                     // Load Timer register
    INTCON = 0xA0;                                   // Enable global and Timer0 interrupts
    for(;;)                                          // Wait for interrupts
    {
    }
}
```

Figure 6.38

cont'd

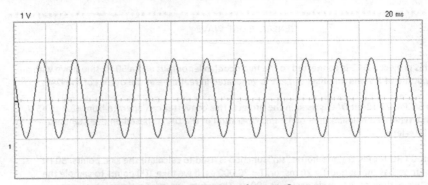

Figure 6.39: Generated Waveform.

Note that the code generated by the mikroC Pro for the PIC compiler is >2 K, and the licensed full version of the compiler is required to compile the program given in Figure 6.38.

MPLAB XC8

The MPLAB XC8 version of the program is shown in Figure 6.40 (XC8-WAVE5.C). In this program, the PLL clock multiplier is enabled by adding the PLLCFG = ON to the configuration register definition at the beginning of the program. The clock frequency (_XTAL_FREQ) is then changed to 32 MHz. The value to be loaded to Timer0 to generate interrupts at 200 μs intervals is calculated as (with a prescaler of 16):

$$TMR0L = 256 - \frac{Time}{4 * clockperiod * prescaler}$$

or

$$TMR0L = 256 - \frac{200}{4 * 0.03125 * 16} = 156$$

Note that the clock frequency is now 32 MHz, which has a period of 0.03125 μs.

Generating Square Waveform

The square wave is perhaps the easiest waveform to generate. If an accurate frequency is not required, then a loop can be formed to generate a square wave signal. For example, assuming we wish to generate 1 kHz (period = 1 ms) square wave signal on pin RB0 of

```
/********************************************************************
                      GENERATE SINE WAVE
                      ==================
```

In this project a DAC is connected to the microcontroller output and a 50 Hz sine wave with an amplitude of 2 V and an offset of 2.5 V is generated in real-time at the output of the DAC.

The MCP4921 12-bit DAC is used in the project. This converter can operate from +3.3 to +5 V, with a typical conversion time of 4.5 ms. Operation of the DAC is based on the standard SPI interface.

mikroC PRO for PIC trigonometric "sin" function is used to calculate the sine points. 50 Hz waveform has the period T = 20 ms, or, T = 20,000 us. If we take 100 points to sample the sine wave, then each Sample occurs at 200 us. Therefore, we need a timer interrupt service routine that will generate interrupts at every 200 us, and inside this routine we will calculate a new sine point and send it to the DAC. The result is that we will get a 50 Hz sine wave. Because the DAC is unipolar, we have to shift the output waveform to a level greater than its maximum negative value so that the waveform is always positive and can be output by the DAC.

```
Author:    Dogan Ibrahim
Date:      September 2013
File:      XC8-WAVE5.C
********************************************************************/

#include <xc.h>
#include <math.h>
#include <plib/spi.h>
#pragma config MCLRE = EXTMCLR, WDTEN = OFF, FOSC = HSHP, PLLCFG = ON
#define _XTAL_FREQ 32000000

// DAC module connections
#define Chip_Select PORTCbits.RC0
#define Chip_Select_Direction TRISCbits.TRISC0
// End DAC module connections

// SPI connections
#define SCK_Direction TRISCbits.RC3
#define SDO_Direction TRISCbits.RC5
//

#define T 100                          // 100 samples
#define R 0.0628                       // 2 * PI/T
#define Amplitude 1638                 // 2 V * 4096/5 V
#define offset 2048                    //2.5 * 4096/5 V

unsigned char temp, Count = 0;
float Sample;
unsigned int Value;
float sins[100];
```

Figure 6.40: MPLAB XC8 Program Listing.

```
//
// This function sends 12 bits digital data to the DAC. The data is passed
// through the function argument called "value"
//
void DAC(unsigned int value)
{
  char temp;

  Chip_Select = 0;                          // Enable DAC chip

  // Send High Byte
  temp = (value >> 8) & 0x0F;               // Store bits 8:11 to temp
  temp |= 0x30;                             // Define DAC setting (choose 1x gain)
  WriteSPI1(temp);                          // Send high byte via SPI

  // Send Low Byte
  temp = value;                             // Store bits 0:7 to temp
  WriteSPI1(temp);                          // Send low byte via SPI

  Chip_Select = 1;                          // Disable DAC chip
}


//
// Timer interrupt service routine. Program jumps here at every 200 us
//
void interrupt isr (void)
{
    TMR0L = 156;                            // Reload timer register
//
// Get sine wave samples and send to DAC
//
    Value = (unsigned int)sins[Count];
    DAC(Value);                             // Send to DAC converter
    Count++;
    if(Count == 100)Count = 0;
    INTCONbits.TMR0IF = 0;                  // Clear timer interrupt flag
}




void main()
{

    unsigned char i;
    ANSELC = 0;
    Chip_Select_Direction = 0;
    SCK_Direction = 0;
    SDO_Direction = 0;

    Chip_Select = 1;                        // Disable DAC
```

Figure 6.40
cont'd

```
        OpenSPI1(SPI_FOSC_4, MODE_00, SMPMID);              // Initialize SPI module


    //
    // Generate the sine wave samples offline and load into an array called sins
    //
        for(i = 0; i < 100; i++)sins[i] = offset + Amplitude * sin(R * i);
    //
    // Configure TIMER0 interrupts
    //
        T0CON = 0xC3;                            // TIMER0 in 8-bit mode
        TMR0L = 156;                             // Load Timer register
        INTCON = 0xA0;                           // Enable global and Timer0 interrupts
        for(;;)                                  // Wait for interrupts
        {
        }
    }
```

Figure 6.40
cont'd

the microcontroller with equal ON and OFF times, the following PDL shows how the
signal can easily be generated:

```
BEGIN
    Configure RB0 as digital output
    DO FOREVER
        Set RB0 = 1
        Wait 0.5 ms
        Set RB0 = 0
        Wait 0.5 ms
    ENDDO
END
```

The problem with this code is that the generated frequency is not accurate because of
two reasons: first, the built-in delay functions are not meant to be very accurate, and
second, the time taken to execute the other instructions inside the loop are not taken into
account.

An accurate square wave signal can be generated using a timer interrupt routine as
described in the previous waveform generation projects. Inside the ISR, all we have to do
is toggle the output pin where the waveform, is to be generated from. This is illustrated
with an example. In this example a square wave signal with frequency 1 kHz
(period = 1 ms) is generated from port pin RD0 of a PIC18F45K22 microcontroller.

The required mikroC Pro for the PIC program listing is shown in Figure 6.41
(MIKROC-WAVE6.C). Inside the main program, PORTD is configured as an analog

```
/*************************************************************************
                    GENERATE SQUARE WAVE
                    ====================

In this project a square wave signal with frequency 1 kHz is generated. The program uses timer
interrupts to send the signal to output.

Timer0 is configured to provide interrupts at every 0.5 ms. Inside the ISR the output pin is toggled,
thus generating a square wave signal.

Author:    Dogan Ibrahim
Date:      September 2013
File:      MIKROC-WAVE6.C
*************************************************************************/

#define OUT_PIN LATD.LATD0
//
// Timer ISR. Program jumps here at every 100 ms
//
void interrupt (void)
{
    TMR0L = 131;                              // Reload timer register
    OUT_PIN = ~OUT_PIN;                       // Toggle the output pin
    INTCON.TMR0IF = 0;                        // Clear timer interrupt flag
}


void main()
{
    ANSELD = 0;                               // Configure PORTC as digital
    TRISD = 0;
    OUT_PIN = 0;
//
// Configure TIMER0 interrupts
//
    T0CON = 0xC2;                             // TIMER0 in 8-bit mode
    TMR0L = 131;                              // Load Timer register
    INTCON = 0xA0;                            // Enable global and Timer0 interrupts
    for(;;)                                   // Wait for interrupts
    {
    }
}
```

Figure 6.41: mikroC Pro for PIC Program Listing

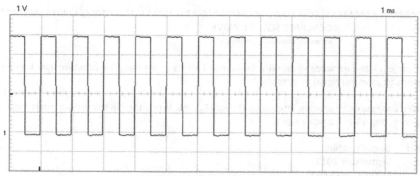

Figure 6.42: Generated Waveform.

output, and pin RD0 is configured as an output pin. Timer0 is configured to generate interrupts at every 500 μs. Inside the ISR, pin RD0 is toggled, and timer interrupt flag is cleared so that further interrupts can be accepted by the processor.

The value to be loaded to Timer0 to generate interrupts at 500 μs intervals is found as (with a prescaler of 8)

$$TMR0L = 256 - \frac{Time}{4 * clockperiod * prescaler}$$

or

$$TMR0L = 256 - \frac{500}{4 * 0.125 * 8} = 131$$

Figure 6.42 shows the waveform generated. Here, the vertical axis is 1 V per division, and the horizontal axis is 1 ms per division.

Another accurate method of generating a square wave is by using the PWM module of the microcontroller Chapter 2, Example 2.1. The PWM module makes use of ports CCP1, CCP2, CCP3, etc. on the microcontroller. But unfortunately, this module cannot be used to generate PWM signals with large periods such as 20 ms. The advantage of using the PWM module is that once configured this module works independent of the central processing unit (CPU), and thus, the CPU is free to do other tasks, while the PWM module is working.

An example is given in this section to show how to program the PWM module. In this program, it is required to generate a 20 kHz (period = 50 μs) square wave with equal ON and OFF times of 25 μs. The output signal will be available on pin RC2 (CCP1) of the microcontroller. The required program listing is given in Figure 6.43 (MIKROC-WAVE7.C). The PWM module is configured as follows:

The value to be loaded into Timer2 register can be calculated as follows:

$$PR2 = \frac{\text{PWM period}}{\text{TMR2PS} * 4 * T_{osc}} - 1$$

where

PR2 is the value to be loaded into the Timer2 register,
TMR2PS is the Timer2 prescaler value,
T_{osc} is the clock oscillator period (in seconds).

```
/***************************************************************************
                          GENERATE SQUARE WAVE
                          =====================

In this project a square wave signal with frequency 20 kHz (period 50 us) is generated. The
program uses the built-in PWM module to generate accurate square wave signal.

The output signal is available on pin RC2 (CCP1) of the microcontroller.

Author:    Dogan Ibrahim
Date:      September 2013
File:      MIKROC-WAVE7.C
***************************************************************************/

void main()
{

  ANSELC = 0;                          // Configure PORT C as digital
  TRISC = 0;                           // PORT C as output

  T2CON = 0b00000101;                  // Timer 2 with prescaler 4
  PR2 = 24;                            // Load PR2 register of Timer 2
  CCPTMRS0 = 0;                        // Enable PWM
  CCPR1L = 0X0C;                       // Load duty cycle
  CCP1CON = 0x2C;                      // Load duty cycle and enable PWM

  for(;;)                              // Wait here forever
  {
  }

}
```

Figure 6.43: mikroC Pro for PIC Program Listing.

Substituting the values into the equation, and assuming a prescaler of 4, an oscillator frequency of 8 MHz ($T_{osc} = 0.125$ μs), the PWM period of 50 μs, and duty cycle (ON time) of 25 μs, we get

$$PR2 = \frac{50 \times 10^{-6}}{4 \times 4 \times 0.125 \times 10^{-6}} - 1$$

Which gives PR2 = 24.

The 10-bit value to be loaded into PWM registers is given by (Chapter 2):

$$CCPR1L : CCP1CON\langle 5 : 4 \rangle = \frac{PWM\ duty\ cycle}{TMR2PS * T_{osc}}$$

Where the upper 8 bits will be loaded into register CCPR1L, and the two LSB bits will be loaded into bits 4 and 5 of CCP1CON.

or

$$CCPR1L : CCP1CON\langle 5 : 4 \rangle = \frac{25 \times 10^{-6}}{4 \times 0.125 \times 10^{-6}} = 50$$

This number in 12-bit binary is "00001100 10". Therefore, the value to be loaded into bits 4 and 5 of CCP1CON are the two LSB bits, that is, "10". Bits 2 and 3 of CCP1CON must be set to HIGH for PWM operation, and bits 6 and 7 are not used. Therefore, CCP1CON must be set to ("X" is do not care):

   XX101100 i.e. hexadecimal 0x2C

The number to be loaded into CCPR1L is the upper 8 bits, that is, "00001100", that is, hexadecimal 0x0C.

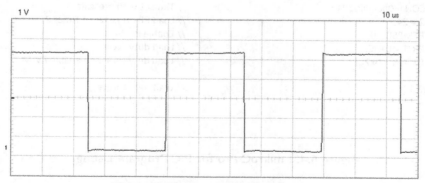

Figure 6.44: Generated Waveform.

At the beginning of the program, PORTC is configured as a digital output port. Then, Timer2 is configured and timer register PR2 is loaded to give the required period. PWM registers CCPR1L and CCP1CON are loaded with the duty cycle (ON time) and the PWM module is enabled. The main program then waits forever where the PWM works in the background to generate the required waveform.

Figure 6.44 shows the generated waveform, which has a period of exactly 50 μs. In this graph, the vertical axis is 1 V per division, and the horizontal axis is 10 μs per division.

The MPLAB XC8 version of the program is shown in Figure 6.45 (XC8-WAVE7.C).

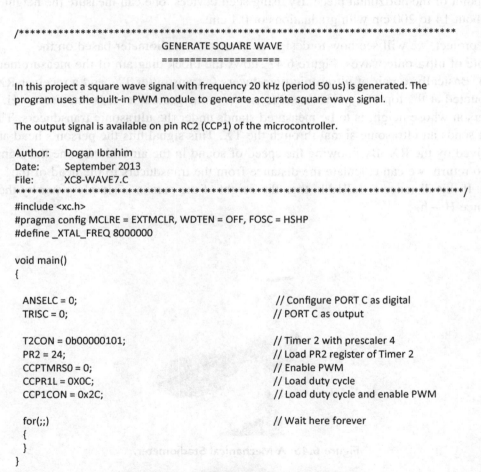

```
/****************************************************************************
                          GENERATE SQUARE WAVE
                          =====================

In this project a square wave signal with frequency 20 kHz (period 50 us) is generated. The
program uses the built-in PWM module to generate accurate square wave signal.

The output signal is available on pin RC2 (CCP1) of the microcontroller.

Author:    Dogan Ibrahim
Date:      September 2013
File:      XC8-WAVE7.C
*****************************************************************************/
#include <xc.h>
#pragma config MCLRE = EXTMCLR, WDTEN = OFF, FOSC = HSHP
#define _XTAL_FREQ 8000000

void main()
{

    ANSELC = 0;                              // Configure PORT C as digital
    TRISC = 0;                               // PORT C as output

    T2CON = 0b00000101;                      // Timer 2 with prescaler 4
    PR2 = 24;                                // Load PR2 register of Timer 2
    CCPTMRS0 = 0;                            // Enable PWM
    CCPR1L = 0X0C;                           // Load duty cycle
    CCP1CON = 0x2C;                          // Load duty cycle and enable PWM

    for(;;)                                  // Wait here forever
    {
    }
}
```

Figure 6.45: MPLAB XC8 Program Listing.

Project 6.5—Ultrasonic Human Height Measurement
Project Description

This project is about designing a microcontroller-based device to measure the human height using ultrasonic techniques. Having the correct height is very important especially during the child development ages. Human height is usually measured using a stadiometer. A stadiometer can either be mechanical or electronic. Most stadiometers are portable, although they can also be wall mounted. A mechanical stadiometer is commonly used in schools, clinics, hospitals, and doctors' offices.

As shown in Figure 6.46, a mechanical stadiometer consists of a long ruler, preferably mounted vertically on a wall, with a movable horizontal piece that rests on the head of the person whose height is being measured. The height is then read on the ruler corresponding to the point of the horizontal piece. By using such devices, one can measure the height from about 14 to 200 cm with graduations of 0.1 cm.

In this project, we will see how to design an electronic stadiometer based on the principle of ultrasonic waves. Figure 6.47 shows the block diagram of the measurement system. Basically, a pair of ultrasonic transducers (a transmitter TX, and a receiver RX) are mounted at the top of a pole whose height from the ground level is known, say H. The person whose height is to be measured stands under the ultrasonic transducers. The system sends an ultrasonic signal through the TX. This signal hits the person's head and is received by the RX. By knowing the speed of sound in the air and the time the signal takes to return, we can calculate the distance from the transducers to the head of the person. If this distance is called h, then the height of the person is simply given by the difference H − h.

Figure 6.46: A Mechanical Stadiometer.

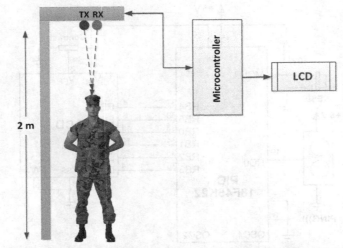

Figure 6.47: Block Diagram of the Height Measurement System.

In this project, the PING))) ultrasonic transducer pair (Figure 6.48), manufactured by Parallax is used. This transducer pair is mainly developed for distance measurement.

Project Hardware

The circuit diagram of the project is shown in Figure 6.49. The PING))) can be used to measure distances from 2 cm to 3 m. The specifications of this device are as follows:

- A 5 V supply voltage,
- A 30 mA supply current,
- A 40 kHz operation frequency,
- Small size ($22 \times 46 \times 16$ mm),
- Requirement of only one pin for connection to a microcontroller.

Figure 6.48: PING))) Ultrasonic Transducer Pair.

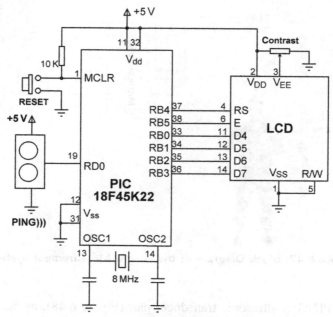

Figure 6.49: Circuit Diagram of the Project.

The device is connected to a microcontroller via the SIG pin, which acts as both an input and an output. In Figure 6.49, this I/O pin is connected to pin RD0 of a PIC18F45K22 microcontroller, operated from an 8 MHz crystal. The pin layout of the PING))) is shown in Figure 6.50. The operation of PING))) is as follows:

The device operates by emitting a short ultrasonic burst at 40 kHz from the TX output and then listening for the echo. This pulse travels in the air at the speed of sound, hits an

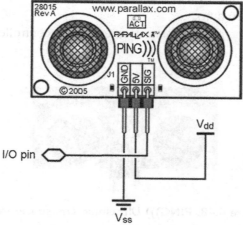

Figure 6.50: Pin Layout of PING))).

```
                          BEGIN
                              Define the LCD connections
                              Define PING))) connection
                              Configure PORTD as digital
                              Configure PORTB as digital
                              Initialize LCD
                              Send start-up message to the LCD
                              Configure Timer0 as 16-bit counter with count time of 1 μs
                          DO FOREVER
                                  Send a pulse to the ultrasonic module
                                  Start timer
                                  Wait until echo pulse is received
                                  Calculate the elapsed time
                                  Divide the time by 2 to find the time to the object
                                  Calculate the distance to the object (h)
                                  Calculate the height of the person (H–h)
                                  Display the height of the person on the LCD
                                  Wait 1 s
                          ENDDO
                      END
```

Figure 6.51: PDL of the Project.

object, and then bounces back and is received by the RX sensor. The PING))) provides an output pulse to the microcontroller that will terminate when the echo is detected, and thus, the width of this pulse is proportional to the distance to the target.

An LCD, connected to PORTB, is used to display the height of the person.

Project PDL

The PDL of the project is shown in Figure 6.51. Timer0 of the microcontroller is configured to operate as a counter in the 16-bit mode, and the count time is set to 1 μs using a prescaler value of 2, where with a clock period of 0.125 μs, the count time is given by

$$\text{Count time} = 4 \times 0.125 \times \text{Prescaler} = 4 \times 0.125 \times 2 = 1 \ \mu s.$$

For a prescaler of 2, the lower 2 bits of register T0CON must be loaded with "00".

Assuming that the total measured time is Tm, then the distance h to the object will be (assuming that the speed of sound in air is 340 m/s, or 34 cm/ms, or 0.034 cm/μs):

$T = Tm/2$ (time T to the object in microseconds)
$h = 0.034 * T$ (Distance $=$ Speed $\times$ Time, where h is in cm)

where T is the time it takes for the signal to echo back after hitting the object. Tm is the total time in microseconds measured by Timer0. The distance h in the above equation is in centimeters, and the time T is in microseconds.

The above operation requires floating point arithmetic. Instead, we can use long integer arithmetic to calculate the height with a good accuracy as follows:

$$h = 34 * T/1000$$

where T is a long integer.

Assuming that the PING))) device is mounted on a pole H centimeters high, the height of the person is simply given by

$$\text{Height of person} = H - h$$

The height is displayed on the LCD. After 1 s, the above process is repeated.

Project Program

mikroC Pro for PIC

The mikroC Pro for the PIC program is called MIKROC-HEIGHT.C and is shown in Figure 6.52. At the beginning of the main program, the LCD and ultrasonic module connections are defined, and PORTB and PORTD are configured as digital I/O ports. Then, the LCD is initialized, and the heading "HEIGHT" is displayed on the LCD for 2 s. Timer register T0CON is configured so that TIMER0 operates in a 16-bit counter mode with the count rate of 1 μs. Timer0 prescaler is set to 2.

The height calculation is carried out inside an endless loop formed using a *for* loop. Inside this loop, the counter is cleared, a pulse is sent to the ultrasonic module, counter is started, and the program waits until the echo signal is received. When the echo signal is received, the counter is stopped, and the total elapsed time is read and stored in variable Tm. Then, the height of the person is calculated and stored in variable Person_Height. This number is converted into a string in variable Txt and is displayed on the LCD. The program repeats after a 1 s delay. If, for example, the person's height is 150 cm, it is displayed in the following format:

> Height (cm)
>   150

The program given in Figure 6.52 can be improved by the following modifications:

- The height calculation can be done using floating point arithmetic to get more accurate results.
- The program assumes that the speed of sound in the air is fixed and is 340 m/s. In reality, the speed of sound in the air depends on several factors such as the ambient temperature and to a lesser extent the atmospheric pressure and relative humidity. A temperature sensor can be added to the project and more accurate results for the speed can be obtained.

```
/*******************************************************************************
                    ULTRASONIC HUMAN HEIGHT MEASUREMENT
                    =======================================
```

In this project the height of a person is found and displayed on a LCD. The project uses the PING))) Ultrasonic distance measuring transducer pair, connected to RD0 pin of a PIC18F45K22 microcontroller, operated from an 8 MHz crystal.

The LCD is connected to PORTB of the microcontroller.

The program assumes that the transducer pair is mounted on a pole 200 cm above the ground level where the person stands.

```
Author:     Dogan Ibrahim
Date:       September 2013
File:       MIKROC-HEIGHT.C
*******************************************************************************/

// LCD module connections
sbit LCD_RS at LATB4_bit;
sbit LCD_EN at LATB5_bit;
sbit LCD_D4 at LATB0_bit;
sbit LCD_D5 at LATB1_bit;
sbit LCD_D6 at LATB2_bit;
sbit LCD_D7 at LATB3_bit;
sbit LCD_RS_Direction at TRISB4_bit;
sbit LCD_EN_Direction at TRISB5_bit;
sbit LCD_D4_Direction at TRISB0_bit;
sbit LCD_D5_Direction at TRISB1_bit;
sbit LCD_D6_Direction at TRISB2_bit;
sbit LCD_D7_Direction at TRISB3_bit;
// End LCD module connections

// Ultrasonic module connection
sbit Ultrasonic at RD0_bit;
sbit Ultrasonic_Direction at TRISD0_bit;
// End of Ultrasonic module connections

#define Pole_Height 200

void main()
{
    unsigned long Tm;
    unsigned char Tl, Th;
    unsigned int h, Person_Height;
    char Txt[7];

    ANSELB = 0;                          // Configure PORTB as digital
    ANSELD = 0;                          // Configure PORTD as digital

    Lcd_Init();                          // Initialize LCD
```

Figure 6.52: mikroC Pro for PIC Program Listing.

```
        Lcd_Cmd(_LCD_CURSOR_OFF);              // Disable cursor
        Lcd_Out(1,1,"HEIGHT");                 // Display heading
        Delay_Ms(2000);                        // Wait 2 s
//
// Configure Timer 0 as a counter to operate in 16-bit mode with 1 ms
// count time. The prescaler is set to 2. The timer is stopped at this point.
//
        T0CON = 0x00;
//
// Start of program loop
// Send a pulse, start timer, get echo, stop timer, calculate distance and display
//
        for(;;)
        {
            Ultrasonic_Direction = 0;          // RD0 in output mode
            TMR0H = 0;                          // Clear high byte of timer
            TMR0L = 0;                          // Clear low byte of timer

            Ultrasonic = 0;
            Delay_us(3);
            Ultrasonic = 1;                     // Send a PULSE to Ultrasonic module
            Delay_us(5);
            Ultrasonic = 0;
            Ultrasonic_Direction = 1;          // RD0 in input mode
            while(Ultrasonic == 0);            // Wait until echo is received
            T0CON.TMR0ON = 1;                  // Start Timer0
            while(Ultrasonic == 1);
            T0CON.TMR0ON = 0;                  // Stop Timer0
            Tl = TMR0L;                         // Read timer low byte
            Th = TMR0H;                         // Read timer high byte
            Tm = Th * 256 + Tl;                // Timer as 16 bit value
            //
            // Now find the distance to person's head
            Tm = Tm/2;                          // Tm is half the time
            Tm = 34 * Tm;
            Tm = Tm/1000;                       // Divide by 1000
            h = (unsigned int)Tm;              // h is the distance to person's head
            Person_Height= Pole_Height – h;    // Person's height
            //
            // Now display the height
            //
            IntToStr(Person_Height, Txt);      // Convert into string to display
            Lcd_Cmd(_LCD_CLEAR);               // Clear LCD
            Lcd_Out(1,1, "Height (cm)");       // Display heading
            Lcd_Out(2,1, Txt);                 // Display the height
            Delay_Ms(1000);                    // Wait 1 s
        }
    }
```

Figure 6.52
cont'd

```
/*****************************************************************************
                    ULTRASONIC HUMAN HEIGHT MEASUREMENT
                    =======================================
```

In this project the height of a person is found and displayed on a LCD. The project uses the PING))) Ultrasonic distance measuring transducer pair, connected to RD0 pin of a PIC18F45K22 microcontroller, operated from an 8 MHz crystal.

The LCD is connected to the microcontroller as follows:

```
Microcontroller      LCD
============         ===

    RB0              D4
    RB1              D5
    RB2              D6
    RB3              D7
    RB4              E
    RB5              R/S
    RB6              RW
```

The program assumes that the transducer pair is mounted on a pole 200 cm above the ground level where the person stands.

```
Author:    Dogan Ibrahim
Date:      September 2013
File:      XC8-HEIGHT.C
*****************************************************************************/

#include <xc.h>
#include <plib/xlcd.h>
#include <stdlib.h>
#pragma config MCLRE = EXTMCLR, WDTEN = OFF, FOSC = HSHP
#define _XTAL_FREQ 8000000


// Ultrasonic module connection
#define Ultrasonic PORTDbits. RD0
#define Ultrasonic_Direction TRISDbits.TRISD0
// End of Ultrasonic module connections

#define Pole_Height 200

//
// This function creates seconds delay. The argument specifies the delay time in seconds.
//
void Delay_Seconds(unsigned char s)
{
    unsigned char i,j;

    for(j = 0; j < s; j++)
    {
```

Figure 6.53: MPLAB XC8 Program Listing.

```
        for(i = 0; i <  100; i++)__delay_ms(10);
    }
}


//
// This function creates 18 cycles delay for the xlcd library
//
void DelayFor18TCY( void )
{
Nop(); Nop(); Nop(); Nop();
Nop(); Nop(); Nop(); Nop();
Nop(); Nop(); Nop(); Nop();
Nop(); Nop();
return;
}


//
// This function creates 15 ms delay for the xlcd library
//
void DelayPORXLCD( void )
{
   __delay_ms(15);
   return;
}


//
// This function creates 5 ms delay for the xlcd library
//
void DelayXLCD( void )
{
   __delay_ms(5);
   return;
}


//
// This function clears the screen
//
void LCD_Clear()
{
   while(BusyXLCD());
   WriteCmdXLCD(0x01);
}


//
// This function moves the cursor to position row,column
//
void LCD_Move(unsigned char row, unsigned char column)
{
  char ddaddr = 40 * (row−1) + column;
```

Figure 6.53
cont'd

```
        while( BusyXLCD() );
        SetDDRamAddr( ddaddr );
    }


    void main()
    {
        unsigned long Tm;
        unsigned char Tl, Th;
        unsigned int h, Person_Height;
        char Txt[10];

        ANSELB = 0;                                 // Configure PORTB as digital
        ANSELD = 0;                                 // Configure PORTD as digital

        OpenXLCD(FOUR_BIT & LINES_5X7);              // Initialize LCD

        while(BusyXLCD());                          // Wait if the LCD is busy
        WriteCmdXLCD(DON);                          // Turn Display ON
        while(BusyXLCD());                          // Wait if the LCD is busy
        WriteCmdXLCD(0x06);                         // Move cursor right
        putrsXLCD("HEIGHT");                        // Display heading
        Delay_Seconds(2);                           // 2 s delay
        LCD_Clear();                                // Clear display
//
// Configure Timer 0 as a counter to operate in 16-bit mode with 1 ms
// count time. The prescaler is set to 2. The timer is stopped at this point.
//
        T0CON = 0x00;
//
// Start of program loop
// Send a pulse, start timer, get echo, stop timer, calculate distance and display
//
        for(;;)
        {
            Ultrasonic_Direction = 0;               // RD0 in output mode
            TMR0H = 0;                              // Clear high byte of timer
            TMR0L = 0;                              // Clear low byte of timer

            Ultrasonic = 0;
            __delay_us(3);
            Ultrasonic = 1;                         // Send a PULSE to Ultrasonic module
            __delay_us(5);
            Ultrasonic = 0;
            Ultrasonic_Direction = 1;               // RD0 in input mode
            while(Ultrasonic == 0);                 // Wait until echo is received
            T0CONbits.TMR0ON = 1;                   // Start Timer0
            while(Ultrasonic == 1);                 // Stop Timer0
            T0CONbits.TMR0ON = 0;                   // Stop Timer0
            Tl = TMR0L;                             // Read timer low byte
```

Figure 6.53

cont'd

```
Th = TMR0H;                              // Read timer high byte
Tm = Th * 256 + Tl;                      // Timer as 16 bit value
//
// Now find the distance to person's head
Tm = Tm/2;                               // Tm is half the time
Tm = 34 * Tm;
Tm = Tm/1000;                            // Divide by 1000
h = (unsigned int)Tm;                    // h is the distance to person's head
Person_Height= Pole_Height – h;          // Person's height
//
// Now display the height
//
itoa(Txt, Person_Height, 10);            // Convert into string to display
LCD_Clear();                             // Clear LCD
LCD_Move(1,1);
putrsXLCD("Height (cm)");                // Display heading
LCD_Move(2,1);
putrsXLCD(Txt);                          // Display the height
Delay_Seconds(1);                        // Wait 1 s
    }
}
```

Figure 6.53
cont'd

The temperature dependency of the speed of sound in dry air is given by

$$Speed = 331.4 + 0.6 T_C,$$

Where the speed is in meters per second and T_C is the ambient temperature in degrees centigrade. At 15 °C, the speed becomes nearly 340 m/s.

MPLAB XC8

The MPLAB XC8 program is called XC8-HEIGHT.C and is shown in Figure 6.53. Note that although the LCD is connected to PORTB of the microcontroller, the following pin connections differ from Figure 6.49:

- RB4 is connected to pin E of the LCD.
- RB5 is connected to pin RS of the LCD.
- RB6 is connected to pin RW of the LCD.

Project 6.6—Minielectronic Organ
Project Description

This project is about designing a microcontroller-based minielectronic organ using a 4 × 4 keypad with 16 keys to produce musical notes in one octave. Figure 6.54 shows the block diagram of the project.

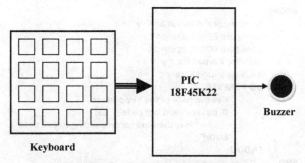

Figure 6.54: Block Diagram of the Project.

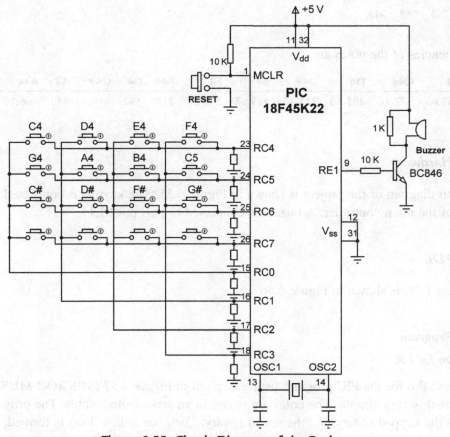

Figure 6.55: Circuit Diagram of the Project.

```
BEGIN
        Store musical notes in a table
        Configure PORTE as digital
        Configure PORTC as digital
        Initialize Keypad library
        Initialize Sound library
        DO FOREVER
                Get the code of the key pressed
                IF this is a valid key code (< 13)
                        Play the note corresponding to this note
                ENDIF
        ENDDO
END
```

Figure 6.56: PDL of the Project.

The musical notes are configured on the keypad as follows:

```
C4    D4    E4    F4
G4    A4    B4    C5
C4#   D4#   F4#   G4#
A4#
```

The frequencies of the notes are

Notes	C4	C4#	D4	D4#	E4	F4	F4#	G4	G4#	A4	A4#	B4
Hz	261.63	277.18	293.66	311.13	329.63	349.23	370	392	415.3	440	466.16	493.88

Project Hardware

The circuit diagram of the project is shown in Figure 6.55. The keypad is connected to PORTC of the microcontroller. A buzzer is connected to port pin RE1.

Project PDL

The project PDL is shown in Figure 6.56.

Project Program

mikroC Pro for PIC

The mikroC Pro for the PIC program listing is given in Figure 6.57 (MIKROC-MUSIC.C). The program is very simple. The notes are stored in an array called Notes. The program initializes the keypad library and the sound library. Then, an endless loop is formed, and

```
/*************************************************************************
                        MINI ELECTRONIC ORGAN
                        =====================

In this project a 4 x 4 keypad is connected to PORTC of a PIC18F45K22 microcontroller. Also a
buzzer is connected to port pin RE1. The keypad is organized such that pressing a key plays a
musical note. The notes on the keypad are organized as follows:

C4    D4    E4    F4
G4    A4    B4    C5
C4#   D4#   F4#   G4#
A4#

Author:  Dogan Ibrahim
Date:    September 2013
File:    MIKROC–MUSIC.C
*************************************************************************/

// Keypad module connections
char keypadPort at PORTC;
// End of keypad module connections

//
// Start of MAIN program
//
void main()
{
    unsigned char MyKey;
    unsigned Notes[] = {0,262,294,330,349,392,440,494,524,277,311,370,415,466};

    ANSELE = 0;                             // Configure PORTE as digital
    ANSELC = 0;                             // Configure PORTC as digital
    TRISC = 0xF0;                           // RC4–RC7 are inputs

    Keypad_Init();                          // Initialize keypad library
    Sound_Init(&PORTE, 1);                  // Initialize sound library
//
// Program loop
//
    for(;;)                                 // Endless loop
    {
        do
            MyKey = Keypad_Key_Press();     // Get code of pressed key
        while(!MyKey);

        if(MyKey <= 13)Sound_Play(Notes[MyKey], 100);   // Play the note
    }
}
```

Figure 6.57: mikroC Pro for PIC Program Listing.

inside this loop, the code of the pressed key is determined, and this is used as an index to
the Notes array to play the note corresponding to the pressed key. Notes are played for a
minimum of 100 ms when a key is pressed. Valid key codes are from 1 to 13, and keys
with codes >13 are not played.

Project 6.7—Frequency Counter with an LCD Display
Project Description

In this project, we look at the design of a simple frequency counter with an LCD display.

There are basically two methods used for the measurement of the frequency of an external signal.

Method I

This is perhaps the easiest method. The signal whose frequency is to be measured is connected to the clock input of a microcontroller counter (timer). The counter is enabled for a known time window, and the total count is read. Since each count corresponds to a clock pulse of the external signal, we can easily determine its frequency. For example, assuming that the time window is set to 1 s and the counter reaches 1000 at the end of 1 s, then the frequency of the signal is 1 kHz. Thus, we can simply display the counter value as the frequency of the signal in hertz. This way, using a 16-bit counter, we can measure frequencies up to 65,535 Hz. If using a 32-bit counter, the maximum frequency that we can measure will be 4,294,967,295 Hz. For the measurement of high frequencies, we can actually use a 16-bit counter and increment a variable each time the counter overflows (65535−0). The total count can then be found by multiplying this variable by 65536 and adding the current counter reading.

The nice thing about using a 1 s time window is that we can directly display the frequency in hertz by simply reading the counter value. This also means that the measurement resolution is 1 Hz, which is acceptable for most measurements. Increasing the time window to 10 s will increase the resolution to 0.1 Hz. On the other hand, decreasing the time window to 0.1 s will reduce the resolution to 10 Hz. Figure 6.58 illustrates this method.

Method II

In this method, the time between two or more successive edges of the signal is measured. The period and the frequency are then calculated easily. The counter (timer) is started as

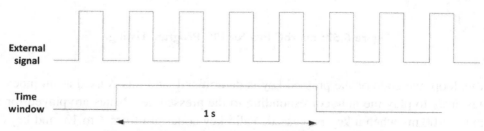

Figure 6.58: Frequency Measurement—Method I.

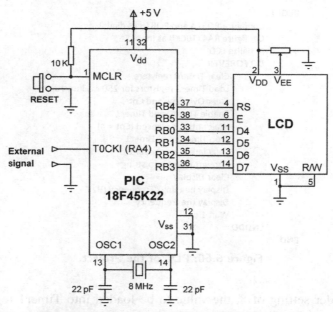

Figure 6.59: Circuit Diagram of the Project.

soon as the leading edge of the signal is detected. The counter is stopped when the next leading edge is detected. The elapsed time and hence the period of the signal and the frequency are then easily calculated. For example, assuming that the counter is configured to count using the internal clock at the rate of 1 μs, if the elapsed time between two edges is 1000 μs, then the period of the signal is 1 ms (1 kHz). The accuracy of the measurement depends on the accuracy of the internal clock and the frequency stability of the external signal. To improve the accuracy, sometimes more than two signal edges are taken.

In this project, the use of Method I will be illustrated by designing a frequency counter system.

Using Method I

Figure 6.59 shows the circuit diagram of the project. In this design, Timer0 is used in a 16-bit counter mode. The external signal whose frequency is to be measured is applied to Timer0 clock input T0CKI (RA4). An LCD is connected to PORTB of the microcontroller as in the previous LCD projects.

Timer1 is used to create the 1 s time window. It is not possible to generate a 1 s delay using Timer1 since the microcontroller clock frequency is high. Instead, the timer is configured to generate an interrupt every 250 ms, and when 4 interrupts are generated, it is assumed that 1 s has elapsed.

```
BEGIN
        Configure PORTA and PORTB as digital
        Configure RA4 (T0CKI) as input
        Initialize LCD
        DO FOREVER
                Clear Timer0 registers
                Load Timer1 registers for 250 ms interrupt
                Clear Overflow and Cnt
                Enable Timer0 and Timer1
                Wait until 1 s elapsed (Cnt = 4)
                Stop Timer0 and Timer1
                Calculate Timer0 count
                Convert count into string
                Clear Display
                Display heading "Frequency (Hz)"
                Display the frequency
                Wait 1 s
        ENDDO
END
```

Figure 6.60: PDL of the Project.

Assuming a prescaler setting of 8, the value to be loaded into Timer1 registers to generate interrupts at 250 ms (250,000 μs) intervals can be calculated from the following:

$$TMR0L = 65536 - \frac{\text{Time}}{4 * \text{clockperiod} * \text{prescaler}}$$

or

$$TMR0L = 65536 - \frac{250000}{4 * 0.125 * 8} = 3036$$

Decimal 3036 is equivalent to 0x0BDC in hexadecimals. Thus, TMR0H = 0x0B and TMR0L = 0xDC.

Project PDL

The project PDL is shown in Figure 6.60.

Project Program
mikroC Pro for PIC

The mikroC Pro for PIC program listing is given in Figure 6.61 (MIKROC-FREQ1.C). At the beginning of the program, PORTA and PORTB are configured as digital. RA4 is configured as an input, the LCD is initialized, and the cursor is disabled.

```
/*******************************************************************************
                              Frequency Counter
                              =================

This project is a frequency counter. The signal whose frequency is to be measured is applied
to pin RA4 (T0CKI) of a PIC18F45K22 microcontroller, operating from an 8 MHz crystal. The
project measures the frequency and displays on an  LCD in Hz.

The resolution of the measurement is 1 Hz. The project can measure frequencies from 1 Hz to
several MHz.

The project uses 2 timers, TIMER0 and TIMER1. TIMER1 is used to open a time window of 1 s
width. The pulses of the external signal increment the counter during this 1 s
window. At the end of 1 s both timers are stopped. The count in TIMER0 gives us
directly the frequency in Hz.

Both timers generate interrupts for higher accuracy. TIMER0 uses a variable called "Overflow"
to find out how many times it has overflowed (if any). This variable is used in calculating the
 total count. TIMER1 interrupts at every 250 ms and a variable called Cnt is incremented at
each interrupt. When Cnt = 4 then it is assumed that 1 s has elapsed.

Programmer:    Dogan Ibrahim
Date:          September 2013
File:          MIKROC-FREQ1.C

*******************************************************************************/
// LCD module connections
sbit LCD_RS at RB4_bit;
sbit LCD_EN at RB5_bit;
sbit LCD_D4 at RB0_bit;
sbit LCD_D5 at RB1_bit;
sbit LCD_D6 at RB2_bit;
sbit LCD_D7 at RB3_bit;

sbit LCD_RS_Direction at TRISB4_bit;
sbit LCD_EN_Direction at TRISB5_bit;
sbit LCD_D4_Direction at TRISB0_bit;
sbit LCD_D5_Direction at TRISB1_bit;
sbit LCD_D6_Direction at TRISB2_bit;
sbit LCD_D7_Direction at TRISB3_bit;
// End LCD module connections

#define PULSE PORTA.RA4

unsigned int Overflow;
unsigned char Cnt;
//
// Timer interrupt service routineProgram jumps here at every 10 ms
//
void interrupt (void)
{
```

Figure 6.61: mikroC Pro for PIC Program Listing.

```
        if(INTCON.TMR0IF == 1)                          // If TIMER0 interrupt
        {
          Overflow++;                                    // Increment Overflow count
          INTCON.TMR0IF = 0;                             // Clear Timer0 interrupt flag
        }
        if(PIR1.TMR1IF == 1)                             // If TIMER1 interrupt
        {
          TMR1H = 0x0B;                                  // Reload timer register
          TMR1L = 0xDC;
          Cnt++;                                         // Increment Cnt
          PIR1.TMR1IF = 0;                               // Clear Timer1 interrupt flag
        }
    }


    void main()
    {
        unsigned char Txt[11];
        unsigned long Elapsed;
        unsigned char L_Byte, H_Byte;

        ANSELA = 0;                                      // Configure PORTA as digital
        ANSELB = 0;
        TRISA.RA4 = 1;                                   // RA4 is input
        Lcd_Init();                                      // Initialize LCD
        Lcd_Cmd(_LCD_CURSOR_OFF);                        // Disable cursor
//
// Configure TIMER0 for 16-bit mode, no prescaler, clock provided by external
// signal into pin T0CKI. Timer is not started here
//
        T0CON = 0x28;                                    // TIMER0 16-bit,no prescaler,T0CKI clk
//
// Configure Timer1 for 250 ms Overflow. Timer1 is not started here
//
        T1CON = 0x36;

        PIE1 = 0x01;                                     // Enable TIMER1 interrupts
        PIR1.TMR1IF = 0;                                 // Clear TIMER1 interrupt flag
        INTCON = 0xE0;                                   // Enable TIMER0 and TIMER1 interrupts

        for(;;)                                          // Wait for interrupts
        {
          TMR0H = 0;                                     // Clear Timer0 registers
          TMR0L = 0;
          TMR1H = 0x0B;                                  // Load Timer1 registers
          TMR1L = 0xDC;
          //
          Overflow = 0;                                  // Clear Overflow count
          Cnt = 0;                                       // Clear Cnt
          //
```

Figure 6.61
cont'd

```
                // Start TIMER0. It will increment each time an external pulse is detected.
                // TIMER0 increments on the rising edge of the external clock
                //
                T0CON.TMR0ON = 1;
                //
                // Start Timer1 to count 4 × 250 ms = 1 s
                //
                T1CON.TMR1ON = 1;
                while(Cnt != 4);                              // Wait until 1 s has elapsed
                //
                // 1 s has elapsed. Stop both timers
                //
                T0CON.TMR0ON = 0;                             // Stop TIMER0
                T1CON.TMR1ON = 0;                             // Stop Timer1
                // Get TIMER0 count
                L_Byte = TMR0L;
                H_Byte = TMR0H;
                //
                // Store TIMER0 count is variable Elapsed
                //
                Elapsed = (unsigned long)256 * H_Byte + L_Byte;
                Elapsed = 65536 * Overflow + Elapsed;
                //
                // Convert into string and display
                //
                LongWordToStr(Elapsed, Txt);                 // Long to string conversion
                Lcd_Cmd(_LCD_CLEAR);                          // Clear LCD
                Lcd_Out(1,1,"Frequency (HZ)");               // Display heading
                Lcd_Out(2,1,Txt);                            // Display measured frequency

                Delay_ms(1000);                              // Wait 1 s and repeat
        }
}
```

Figure 5.61

cont'd

The program then configures Timer0 as a counter. Timer0 control register T0CON is loaded as follows:

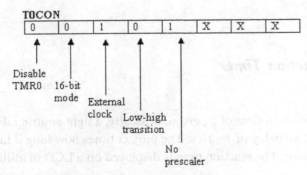

Thus, the T0CON register should be loaded with hexadecimal 0x28.

Similarly, Timer1 control register T1CON is loaded as follows:

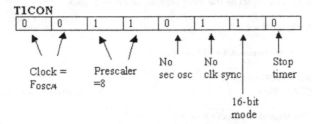

Thus, the T1CON register should be loaded with hexadecimal 0x36.

Then, Timer1 interrupts are enabled by setting PIE1 = 1 and also the timer1 interrupt flag is cleared by clearing PIR1.TMR1IF = 0. The next step is to configure interrupt register INTCON to enable Timer0 interrupts, unmasked interrupts (e.g. Timer1), and global interrupts. This is done by setting INTCON to hexadecimal 0xE0. The remainder of the program is executed in an endless loop. The following operations are carried out inside this loop:

- Timer0 registers cleared.
- Timer1 registers loaded for 250 ms interrupt.
- Variables Overflow and Cnt are cleared.
- Timer0 is enabled so that the counter counts every time external pulse is received.
- Timer1 is enabled so that interrupts are generated at every 250 ms.
- The program then waits until 1 s has elapsed (Cnt = 4).
- At this point, Timer0 and Timer1 are stopped.
- Timer0 high and low count is read.
- Total Timer0 count is calculated.
- Timer0 count is converted into string format and displayed on the LCD.
- Program waits for 1 s and above process is repeated.

If, for example, the frequency is 25 kHz it is displayed as follows:

        Frequency (Hz)
            25000

Project 6.8—Reaction Timer

Project Description

This project tests the reaction time of a person. Basically, a light emitting diode (LED) is turned ON after a random delay of 1—10 s. The project times how long it takes for the person to hit a switch in response. The reaction time is displayed on a LCD in milliseconds.

Figure 6.62 shows the block diagram of the project.

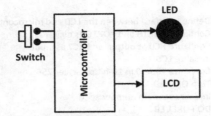

Figure 6.62: Block Diagram of the Project.

Project Hardware

Figure 6.63 shows the circuit diagram. The LED and the push-button switch are connected to port pins RC0 and RC7, respectively. The switch is configured in the active low mode so that when the switch is pressed logic 0 is sent to the microcontroller. The LCD is connected to PORTB as in the previous LCD projects.

Project PDL

The project PDL is shown in Figure 6.64. The program counts and displays the reaction time in milliseconds. Timer0 is used in the 16-bit mode to determine the reaction time. Using a prescaler of 256, Timer0 count rate is given by

$$4 \times 0.125 \times 256 = 128 \ \mu s$$

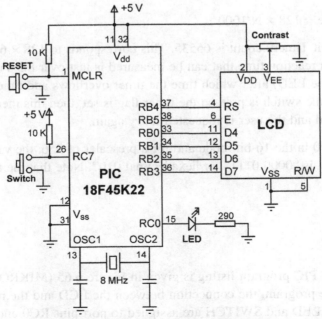

Figure 6.63: Circuit Diagram of the Project.

```
BEGIN
        Define interface between the LCD and microcontroller
        Configure PORTB and PORTC as digital
        Configure RC0 as output and RC7 as input
        Initialize LCD
        Configure Timer0 in 16-bit,prescaler 256
        Turn OFF LED
        Initialize random number seed
        DO FOREVER
                Generate random number 1 to 10
                Clear timer registers
                Turn ON LED
                Turn ON timer
                Wait until switch is pressed
                Stop timer
                Read timer count
                Convert count to milliseconds
                Convert count to string
                Display count on LCD
                Turn OFF LED
                Delay 2 s
                Clear LCD
        ENDDO
END
```

Figure 6.64: Project PDL.

We will start the timer as soon as the LED is turned ON and stop it when the switch is pressed. If the timer count is N, then the reaction time in milliseconds will be

Reaction time $= 128 \times N/1000$

The maximum 16-bit Timer0 count is 65535. This corresponds to $128 \times 65535 = 8388$ μs. Thus, the maximum reaction time that can be measured is just over 8.3 s (too long for anyone to react to the LED) after which time the timer overflows and timer flag TMR0IF is set to 1. If when the switch is pressed the timer flag is set, then this means that the timer has overflowed and the user is requested to try again.

For operating Timer0 in the 16-bit mode and for a prescaler of 256, the value to be loaded into register T0CON is "0000 0111", or hexadecimal 0x07. Note that the timer is stopped initially.

Project Program
mikroC Pro for PIC

The mikroC Pro for PIC program listing is given in Figure 6.65 (MIKRO-REACT.C). At the beginning of the program, the connection between the LCD and the microcontroller is defined; symbols LED and SWITCH are assigned to port pins RC0 and RC7,

```
/*****************************************************************************
                            Reaction Timer
                            ==============

This project is a reaction timer. Basically an LED is connected to RC0, a push-button switch
is connected to RC7 and an LCD is connected to PORTB.

The LED turns ON after a random delay (between 1 and 10 s) and the user is expected
to press the button as soon as he/she sees the LED. The elapsed time between the LED turning
ON and the button being pressed is displayed in milliseconds on the LCD

The project can measure reaction times from 1 ms to just over 8.3 s (In practice it is not
possible for any person to react in more than a few seconds). A message is sent to the LCD if
the maximum reaction time is exceeded.

Programmer:    Dogan Ibrahim
Date:          September 2013
File:          MIKROC-REACT.C

*****************************************************************************/
// LCD module connections
sbit LCD_RS at RB4_bit;
sbit LCD_EN at RB5_bit;
sbit LCD_D4 at RB0_bit;
sbit LCD_D5 at RB1_bit;
sbit LCD_D6 at RB2_bit;
sbit LCD_D7 at RB3_bit;

sbit LCD_RS_Direction at TRISB4_bit;
sbit LCD_EN_Direction at TRISB5_bit;
sbit LCD_D4_Direction at TRISB0_bit;
sbit LCD_D5_Direction at TRISB1_bit;
sbit LCD_D6_Direction at TRISB2_bit;
sbit LCD_D7_Direction at TRISB3_bit;
// End LCD module connections

#define LED PORTC.RC0
#define SWITCH PORTC.RC7

void main()
{
    unsigned char T_Low, T_High, Txt[11];
    unsigned int No;
    unsigned long Cnt;

    ANSELB = 0;                                    // Configure PORTB as digital
    ANSELC = 0;                                    // Configure PORTC as digital
    TRISC.RC0 = 0;                                 // Configure RC0 as output
    TRISC.RC7 = 1;                                 // Configure RC7 as input

    Lcd_Init();                                    // Initialize LCD
```

Figure 6.65: mikroC Pro for PIC Program Listing.

```
                Lcd_Cmd(_LCD_CURSOR_OFF);              // Disable cursor
                T0CON = 0x07;                          // Timer0, 16-bit, prescaler = 256
                LED = 0;                               // Turn OFF LED to start with
                srand(10);                             // Initialize random number seed
                INTCON.TMR0IF = 0;                     // Clear timer overflow flag

                for(;;)
                {
//
// Generate a random number between 1 and 32767 and change it to be between 1 and 10
//
                No = rand();                           // Random number between 1 and 32767
                No = No % 10 + 1;                      // Number between 1 and 10
                Vdelay_Ms(1000 * No);                  // Delay No seconds
                TMR0H = 0;                             // Clear Timer0 H register
                TMR0L = 0;                             // Clear Timer0 L register
                LED = 1;                               // Turn ON LED
//
// Turn ON Timer0 (counts in multiples of 128 microseconds)
//
                T0CON.TMR0ON = 1;                      // Turn ON Timer0
                while(SWITCH == 1);                    // Wait until the switch is pressed
//
// Switch is pressed. Stop timer and get the count
//
                T0CON.TMR0ON = 0;                      // Stop Timer0
                T_Low = TMR0L;                         // Read timer low byte
                T_High = TMR0H;                        // read timer high byte
                Cnt = (unsigned long)256 * T_High + T_Low;  // Get timer count
                Cnt = 128 * Cnt/1000;                  // Convert to milliseconds

                if(INTCON.TMR0IF == 1)                 // If timer overflow detected
                {
                  Lcd_Out(1,1,"Too long...");
                  Lcd_Out(2,1,"Try again...");
                  INTCON.TMR0IF = 0;                   // Clear timer overflow flag
                }
                else
                {
//
// Convert count to string and display on LCD
//
                  LongWordToStr(Cnt, Txt);             // Convert to string
                  Lcd_Out(1,1,"Reaction (ms)");        // Display heading
                  Lcd_Out(2,1,Txt);                    // Display reaction time in ms
                }
                LED = 0;                               // Turn OFF LED
                Delay_ms(2000);                        // Wait 2 s and repeat
                Lcd_Cmd(_LCD_CLEAR);                   // Clear LCD
            }
        }
```

Figure 6.65

cont'd

```
/*******************************************************************************
                              Reaction Timer
                              ============

This project is a reaction timer. Basically an LED is connected to RC0, a push-button switch is
connected to RC7 and an LCD is connected to PORTB.

The LED turns ON after a random delay (between 1 and 10 s) and the user is expected to
press the button as soon as he/she sees the LED. The elapsed time between the LED turning ON
and the button being pressed is displayed in milliseconds on the LCD

The project can measure reaction times from 1 ms to just over 8.3 s (In practice it is not
possible for any person to react in more than a few seconds). A message is sent to the LCD is
the maximum reaction time is exceeded.


Programmer:    Dogan Ibrahim
Date:          September 2013
File:          XC8-REACT.C


*******************************************************************************/


#include <xc.h>
#include <plib/xlcd.h>
#include <stdlib.h>
#pragma config MCLRE = EXTMCLR, WDTEN = OFF, FOSC = HSHP
#define _XTAL_FREQ 8000000

#define LED PORTCbits.RC0
#define SWITCH PORTCbits.RC7

//
// This function creates seconds delay. The argument specifies the delay time in seconds.
//
void Delay_Seconds(unsigned char s)
{
    unsigned char i,j;

    for(j = 0; j < s; j++)
    {
        for(i = 0; i <  100; i++)__delay_ms(10);
    }
}


//
// This function creates 18 cycles delay for the xlcd library
//
void DelayFor18TCY( void )
{
Nop(); Nop(); Nop(); Nop();
Nop(); Nop(); Nop(); Nop();
Nop(); Nop(); Nop(); Nop();
```

Figure 6.66: MPLAB XC8 Program Listing.

```
        Nop(); Nop();
        return;
    }


    //
    // This fucntion creates 15 ms delay for the xlcd library
    //
    void DelayPORXLCD( void )
    {
        __delay_ms(15);
        return;
    }


    //
    // This function creates 5 ms delay for the xlcd library
    //
    void DelayXLCD( void )
    {
        __delay_ms(5);
        return;
    }


    //
    // This function clears the screen
    //
    void LCD_Clear()
    {
        while(BusyXLCD());
        WriteCmdXLCD(0x01);
    }


    //
    // This function moves the cursor to position row,column
    //
    void LCD_Move(unsigned char row, unsigned char column)
    {
        char ddaddr = 40*(row − 1) + column;
        while( BusyXLCD() );
        SetDDRamAddr( ddaddr );
    }



    void main()
    {
        unsigned char T_Low, T_High, Txt[11];
        unsigned int No;
        unsigned long Cnt;
```

Figure 6.66
cont'd

```
    ANSELB = 0;                             // Configure PORTB as digital
    ANSELC = 0;                             // Configure PORTC as digital
    TRISCbits.TRISC0 = 0;                   // Configure RC0 as output
    TRISCbits.TRISC7 = 1;                   // Configure RC7 as input

    OpenXLCD(FOUR_BIT & LINES_5X7);         // Initialize LCD

    while(BusyXLCD());                      // Wait if the LCD is busy
    WriteCmdXLCD(DON);                      // Turn Display ON
    while(BusyXLCD());                      // Wait if the LCD is busy
    WriteCmdXLCD(0x06);                     // Move cursor right
    LCD_Clear();                            // Clear display

    T0CON = 0x07;                           // Timer0 for 16-bit, prescaler = 256
    LED = 0;                                // Turn OFF LED to start with
    srand(10);                              // Initialize random number seed
    INTCONbits.TMR0IF = 0;                  // Clear timer overflow flag

    for(;;)
    {
//
// Generate a random number between 1 and 32767 and change it to be between 1 and 10
//
        No = rand();                        // Random number between 1 and 32767
        No = No % 10 + 1;                   // Make the number between 1 and 10
        Delay_Seconds(No);                  // Delay No seconds
        TMR0H = 0;                          // Clear Timer0 H register
        TMR0L = 0;                          // Clear Timer0 L register
        LED = 1;                            // Turn ON LED
//
// Turn ON Timer0 (counts in multiples of 128 microseconds)
//
        T0CONbits.TMR0ON = 1;               // Turn ON Timer0
        while(SWITCH == 1);                 // Wait until the switch is pressed
//
// Switch is pressed. Stop timer and get the count
//
        T0CONbits.TMR0ON = 0;               // Stop Timer0
        T_Low = TMR0L;                      // Read timer low byte
        T_High = TMR0H;                     // read timer high byte
        Cnt = (unsigned long)256 * T_High + T_Low;    // Get timer count
        Cnt = 128 * Cnt/1000;               // Convert to milliseconds

        if(INTCONbits.TMR0IF == 1)          // If timer overflow detected
        {
          LCD_Move(1,1);
          putrsXLCD("Too long...");         // Display message
          LCD_Move(2,1);
          putrsXLCD("Try again...");
          INTCONbits.TMR0IF = 0;            // Clear timer overflow flag
        }
```

Figure 6.66

cont'd

```
              else
              {
//
// Convert count to string and display on LCD
//
                  utoa(Txt, Cnt, 10);                      // Convert to string
                  LCD_Move(1,1);
                  putrsXLCD("Reaction (ms)");              // Display heading
                  LCD_Move(2,1);
                  putrsXLCD(Txt);                          // Display reaction time in ms
              }
              LED = 0;                                     // Turn OFF LED
              Delay_Seconds(2);                            // Wait 2 s and repeat
              LCD_Clear();                                 // Clear LCD
          }
      }
```

Figure 6.66
cont'd

respectively. Then, PORTB and PORTC are configured as digital; RC7 is configured as an input pin. The LCD is initialized, Timer0 is configured to operate in the 16-bit mode with a prescaler of 256, and the count is disabled at this point. The timer overflow flag (INT0IF) is cleared, the LED is turned OFF, and the random number seed srand is loaded with an integer number. The remainder of the program is executed inside an endless loop.

Inside the endless loop, the LED is turned ON and Timer0 is turned ON to start counting the reaction time. When the SWITCH is pressed, the timer is stopped, and the count is read and converted into milliseconds. If the timer has overflowed (INT0IF is set), then the measurement is ignored since it is not correct anymore and message "Too long...Try again..." is sent to the LCD. Otherwise, the count is converted into a string and displayed on the LCD. The program repeats after a 2 s delay.

If, for example, the reaction time is 2568 ms, then it will be displayed as follows:

> Reaction (ms)
> 2568

The program can be modified if desired to measure the reaction time to sound. This will require the replacement of the LED with a buzzer and the generation of an audible sound with the required frequency.

MPLAB XC8

The MPLAB XC8 version of the program is shown in Figure 6.66 (XC8-REACT.C). The program operates as in the mikroC Pro for PIC version.

Project 6.9—Temperature and Relative Humidity Measurement
Project Description

This project demonstrates how the ambient temperature and relative humidity can be measured and then displayed on an LCD.

In this project, the SHT11 relative humidity and temperature sensor chip is used. This is a tiny eight-pin chip with dimensions 4.93 × 7.47 mm and thickness 2.6 mm, manufactured by Sensirion (http://www.sensirion.com). A capacitive sensor element is used to measure the relative humidity, while the temperature is measured by a band-gap sensor. A calibrated digital output is given for ease of connection to a microcontroller. The relative humidity is measured with an accuracy of ±4.5%RH, and the temperature accuracy is ±0.5 °C.

Because the sensor is very small, it is available as mounted on a small printed circuit board for ease of handling. Figure 6.67 shows a picture of the SHT11 sensor.

The sensor is operated with four pins. Pin 1 and pin 4 are the supply voltage and ground pins. Pin 2 and pin 3 are the data and clock pins, respectively. The clock pin synchronizes all the activities of the chip. It is recommended by the manufacturers that the data pin should be pulled high through a 10 K resistor, and a 100 nF decoupling capacitor should be connected between the power lines.

The SHT11 is based on serial communication where data are clocked in and out, in synchronization with the SCK clock. The communication between the SHT11 and a microcontroller consists of the following protocols (see the SHT11 data sheet for more detailed information).

RESET

At the beginning of data transmission, it is recommended to send a RESET to the SHT11 just in case the communication with the device is lost. This signal consists of sending nine

Figure 6.67: The SHT11 Temperature and Relative Humidity Sensor.

or more SCK signals while the DATA line is HIGH. Configuring the port pin as an input will force the pin to logic HIGH. A Transmission Start Sequence must follow the RESET.

The C code to implement the RESET sequence as a function is given below (SDA and SCK are the DATA and SCK lines, respectively). Note that the manufacturer's data sheet specifies that after SCK changes state it must remain in its new state for a minimum of 100 ns. Here, a delay of 1 μs is introduced between each SCK state change:

```
void Reset_Sequence()
{
  SCK = 0;                          // SCK low
  SDA_Direction = 1;                // Define SDA as input so that the SDA line
                                         becomes HIGH

  for (j = 0; j < 10; j++)          // Repeat 10 times
  {
    SCK = 1;                        // send 10 clocks on SCK line with 1 us delay
    Delay_us(1);                    // 1 us delay
    SCK = 0;                        // SCK is LOW
    Delay_us(1);                    // 1 us delay
  }
  Transmission_Start_Sequence();    // Send Transmission-start-sequence
}
```

Transmission Start Sequence

Before a temperature or relative humidity conversion command is sent to the SHT11, the transmission start sequence must be sent. This sequence consists of lowering the DATA line while SCK is HIGH, followed by a pulse on the SCK and rising DATA again while the SCK is still HIGH.

The C code to implement the transmission start sequence is given below:

```
void Transmission_Start_Sequence()
{
  SDA_Direction = 1;    // Set SDA HIGH
  SCK = 1;              // SCK HIGH
  Delay_us(1);          // 1 us delay
  SDA_Direction = 0;    // SDA as output
  SDA = 0;              // Set SDA LOW
  Delay_us(1);          // 1 us delay
  SCK = 0;              // SCK LOW
  Delay_us(1);          // 1 us delay
  SCK = 1;              // SCK HIGH
  Delay_us(1);          // 1 us delay
  SDA_Direction = 1;    // Set SDA HIGH
  Delay_us(1);          // 1 us delay
  SCK = 0;              // SCK LOW
}
```

Conversion Command

After sending the transmission start sequence, the device is ready to receive a conversion command. This consists of three address bits (only "000" is supported) followed by five command bits. The list of valid commands is given in Table 6.1. For example, the commands for relative humidity and temperature are "00000101" and "00000011", respectively. After issuing a measurement command, the sensor sends an ACK pulse on the falling edge of the eighth SCK pulse. The ACK pulse is identified by the DATA line going LOW. The DATA line remains LOW until the ninth SCK pulse goes LOW. The microcontroller then has to wait for the measurement to complete. This can take up to 320 ms. During this time, it is recommended to stop generating clocks on the SCK line and release the DATA line. When the measurement is complete, the sensor pulls the DATA line LOW to indicate that the data are ready. At this point, the microcontroller can restart the clock on the SCK line to read the measured data. Note that the data are kept in the SHT11 internal memory until they are read out by the microcontroller.

The data read out consists of 2 bytes of data and 1 byte of CRC checksum. The checksum is optional and if not used the microcontroller may terminate the communication by keeping the DATA line HIGH after receiving the last bit of the data (LSB). The data bytes are transferred with MSB first and are right justified. The measurement can be for 8, 12, or 14 bits wide. Thus, the fifth SCK corresponds to the MSB data for a 12-bit operation. For an 8-bit measurement, the first byte is not used. The microcontroller must acknowledge each byte by pulling the DATA line LOW, and sending an SCK pulse. The device returns to the sleep mode after all the data have been read out.

Acknowledgment

After receiving a command from the microcontroller, the sensor issues an acknowledgment pulse by pulling the DATA line LOW for one clock cycle. This takes place after the falling edge of the eighth clock on the SCK line, and the DATA line is pulled LOW until the end of the ninth clock on the SCK line.

The Status Register

The Status register is an internal 8-bit register that controls some functions of the device, such as selecting the measurement resolution, end of battery detection, and use of the internal

Table 6.1: List of Valid Commands

Command	Code
00011	Measure temperature
00101	Measure relative humidity
00111	Read status register
00110	Write status register
11110	Soft reset (reset interface, clear status register)

heater. To write a byte to the Status register, the microcontroller must send the write command ("00110"), followed by the data byte to be written. Note that the sensor generates acknowledge signals in response to receiving both the command and the data byte. Bit 0 of the Status register controls the resolution, such that when this bit is 1, both the temperature resolution and the relative humidity resolution are 12 bits. When this bit is 0 (the default state), the temperature resolution is 14 bits, and the relative humidity resolution is 12 bits.

The sensor includes an on-chip heating element that can be enabled by setting bit 2 of the Status register (the heater is off by default). By using the heater, it is possible to increase the sensor temperature by 5–10 °C. The heater can be useful for analyzing the effects of changing the temperature on humidity. Note that, during temperature measurements, the sensor measures the temperature of the heated sensor element and not the ambient temperature.

The steps for reading the humidity and temperature are summarized below:

Humidity (Assuming 12-Bit Operation with No CRC)

- Send Reset_Sequence.
- Send Transmission_start_sequence.
- Send "00000101" to convert relative humidity.
- Receive ACK from sensor on eighth SCK pulse going LOW. The ACK is identified by the sensor lowering the DATA line.
- Wait for the measurement to be complete (up to 320 ms), or until DATA line is LOW.
- Ignore first four SCK pulses.
- Get the four upper nibble starting with the MSB bit.
- Send ACK to sensor at the end of eighth clock by lowering the DATA line and sending a pulse on the SCK.
- Receive low 8 bits.
- Ignore the CRC by keeping the DATA line HIGH.
- The next measurement can start by repeating the above steps.

Temperature

The steps for reading the temperature are similar, except that the command "00000011" is sent instead of "00000101".

Conversion of Signal Output
Relative Humidity Reading (SO_{RH})

The humidity sensor is nonlinear, and it is necessary to perform a calculation to obtain the correct reading. The manufacturer's data sheet recommends the following formula for the correction:

$$RH_{linear} = C_1 + C_2 + SO_{RH} + C_3 \cdot SO_{RH}^2 (\%RH) \qquad (6.3)$$

Table 6.2: Coefficients for the RH Nonlinearity Correction

SO_{RH}	C_1	C_2	C_3
12 bits	-2.0468	0.0367	$-1.5955E\text{-}6$
8 bits	-2.0468	0.5872	$-4.0845E\text{-}4$

Where SO_{RH} is the value read from the sensor, and the coefficients are as given in Table 6.2.

For temperatures significantly different from 25 °C, the manufacturers recommend another correction to be applied to the relative humidity as follows:

$$RH_{TRUE} = (T - 25) \cdot (t_1 + t_2 \cdot SO_{RH}) + RH_{linear} \tag{6.4}$$

Where T is the temperature in degrees centigrade where the relative humidity reading is taken, and the coefficients are as given in Table 6.3.

Temperature Reading (SO_T)

The manufacturers recommend that the temperature reading of the SHT11 should be corrected according to the following formula:

$$T_{TRUE} = d_1 + d_2 \cdot SO_T \tag{6.5}$$

Where SO_T is the value read from the sensor, and the coefficients are as given in Table 6.4.

Block Diagram

The block diagram of the project is shown in Figure 6.68.

Circuit Diagram

The circuit diagram of the project is as shown in Figure 6.69. The SHT11 sensor is connected to PORTC of a PIC18F45K22 microcontroller, operated from an 8-MHz crystal. The DATA and SCK pins are connected to RC4 and RC3, respectively. The DATA pin is pulled up using a 10 K resistor as recommended by the manufacturers. Also, a 100 nF decoupling capacitor is connected between the V_{DD} pin and the ground.

Table 6.3: Coefficients for RH Temperature Correction

SO_{RH}	t_1	t_2
12 bits	0.01	0.00008
8 bits	0.01	0.00128

Table 6.4: Coefficients for Temperature Correction

V_{DD}	d_1 (°C)	d_1 (°F)
5	−40.1	−40.2
4	−39.8	−39.6
3.5	−39.7	−39.5
3	−39.6	−39.3
2.5	−39.4	−38.9

SO_T	D_2(°C)	D_2(°F)
14 bits	0.01	0.018
12 bits	0.04	0.072

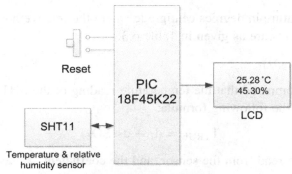

Figure 6.68: Block Diagram of the Project.

The project was tested using a plug-in SHT11 module (manufactured by mikroElektronika) together with the EasyPIC V7 development board, where the module was connected to the PORTC I/O connector located at the edge of the EasyPIC V7 development board (Figure 6.70).

Project PDL

The PDL of this project is given in Figure 6.71.

Project Program
mikroC Pro for PIC

The program listing of the project is shown in Figure 6.72 (MIKROC-SHT11.C). At the beginning of the program, the connections between the LCD and the microcontroller are defined. Then, the connections between the SHT11 sensor and the microcontroller are

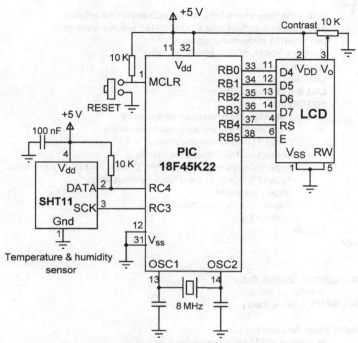

Figure 6.69: Circuit Diagram of the Project.

Figure 6.70: Use of the SHT11 Module with the EasyPIC V7 Development Board.

BEGIN
 Define the connections between the LCD and microcontroller
 Define the connections between the SHT11 and microcontroller
 Define SHT11 correction coefficients
 Configure PORTB, PORTC, PORTD as digital
 Initialise LCD
 Clear display
 CALL SHT11_Startup_Delay
 DO FOREVER
  CALL Measure to measure temperature
  CALL Measure to measure relative humidity
  Convert temperature to a string
  Convert relative humidity to a string
  Append degree symbol and letter "C" after the temperature value
  Append % sign after the relative humidity value
  Display temperature
  Display relative humidity
  Wait for 1 s
 ENDDO
END

BEGIN/SHT11_Startup_Delay
 Wait for 20 ms
END/SHT11_Startup_Delay

BEGIN/Reset_Sequence
 Implement SHT11 reset sequence
END/Reset_Seqeunce

BEGIN/Transmission_Start_Sequence
 Implement SHT11 transmission_start_sequence
END/Transmission_Start_Sequence

BEGIN/Send_ACK
 Send ACK signal to SHT11
END/Send_ACK

BEGIN/Measure
 Get type of measurement required
 Send Reset_Sequence
 Send Transmission_Start_Sequence
 Send address and temperature or humidity convert command to SHT11
 Send SCK pulse for the ACK signal
 Wait until measurement is ready (until DATA goes LOW)
 Read 8 bit measurement data
 Send ACK to SHT11
 Read remaining 8 bits
 Make corrections for temperature (or humidity)
END/Measure

Figure 6.71: PDL of the Project.

```
/*================================================================================
                TEMPERATURE AND RELATIVE HUMIDITY MEASUREMENT
                ===================================================
```

This project measures both the ambient temperature and the relative humidity and then displays the readings on an LCD.

The SHT11 single chip temperature and relative humidity sensor is used in this project. The sensor is connected as follows to a PIC18F45K22 type microcontroller, operating at 8 MHz:

Sensor   Microcontroller Port
DATA   RC4
SCK   RC3

A 10K pull-up resistor is used on the DATA pin. In addition, a 100 nF decoupling capacitor is used between the V_{DD} and the GND pins. The sensor is operated from a +5 V supply.

The connections between the LCD and the microcontroller is as in the earlier LCD based projects.

Author: Dogan Ibrahim
Date: September 2013
File: MIKROC-SHT11.C

```
================================================================================*/
// LCD module connections
sbit LCD_RS at RB4_bit;
sbit LCD_EN at RB5_bit;
sbit LCD_D4 at RB0_bit;
sbit LCD_D5 at RB1_bit;
sbit LCD_D6 at RB2_bit;
sbit LCD_D7 at RB3_bit;

sbit LCD_RS_Direction at TRISB4_bit;
sbit LCD_EN_Direction at TRISB5_bit;
sbit LCD_D4_Direction at TRISB0_bit;
sbit LCD_D5_Direction at TRISB1_bit;
sbit LCD_D6_Direction at TRISB2_bit;
sbit LCD_D7_Direction at TRISB3_bit;
// End LCD module connections

//SHT11 connections
sbit SHT11_SDA at RC4_bit;                              // SHT11 DATA pin
sbit SHT11_SCK at RC3_bit;                              // SHT11 SCK pin
sbit SHT11_SDA_Direction at TRISC4_bit;                 // DATA pin direction
sbit SHT11_SCK_Direction at TRISC3_bit;                 // SCK pin direction
//
// SHT11 Constants for calculating humidity (in 12 bit mode)
//
const float C1 = -2.0468;                               // -2.0468
const float C2 = 0.0367;                                // 0.0367
const float C3 = -1.5955E-6;                            // -1.5955* 10^-6
//
```

Figure 6.72: mikroC Pro for PIC Program Listing.

```
// SHT11 Constants for relative humidity temperature correction (in 12 bit mode)
//
const float t1 = 0.01;                                          // 0.01
const float t2 = 0.00008;                                       // 0.00008
//
// SHT11 temperature conversion coefficients (14 bit mode)
//
const float d1 = −40.1;                                         // −40.1
const float d2 = 0.01;                                          // 0.01

unsigned char i, mode;
unsigned int buffer;
float Res, Ttrue, RHtrue;
char T[] = "T=      ";
char H[] = "H=      ";

//
// Function to send the Transmission_Start_Sequence
//
void Transmission_Start_Sequence(void)
{
 SHT11_SDA_Direction = 1;                                       // Set SDA as input
 SHT11_SCK = 1;                                                 // SCK HIGH
 Delay_us(1);                                                   // 1 us delay
 SHT11_SDA_Direction = 0;                                       // Set SDA as output
 SHT11_SDA = 0;                                                 // SDA LOW
 Delay_us(1);                                                   // 1 us delay
 SHT11_SCK = 0;                                                 // SCK LOW
 Delay_us(1);                                                   // 1 us delay
 SHT11_SCK = 1;                                                 // SCK HIGH
 Delay_us(1);                                                   // 1 us delay
 SHT11_SDA_Direction = 1;                                       // Set SDA as input
 Delay_us(1);                                                   // 1 us delay
 SHT11_SCK = 0;                                                 // SCK low
}


//
// This function sends the Reset_Sequence
//
void Reset_Sequence()
{
    SHT11_SCK = 0;                                              // SCL low
    SHT11_SDA_Direction = 1;                                    // Define SDA as input
    for (i = 1; i <= 10; i++)                                   // Repeat 10 times
 {
    SHT11_SCK = 1;                                              // Send clock pulses
    Delay_us(1);
    SHT11_SCK = 0;
    Delay_us(1);
 }
 Transmission_Start_Sequence();
}
```

Figure 6.72
cont'd

```
//
// This function sends ACK
//
void Send_ACK()
{
 SHT11_SDA_Direction = 0;                          // Define SDA as output
 SHT11_SDA = 0;                                    // SDA low
 SHT11_SCK = 1;                                    // SCL high
 Delay_us(1);                                      // 1 us delay
 SHT11_SCK = 0;                                    // SCL low
 Delay_us(1);                                      // 1 us delay
 SHT11_SDA_Direction = 1;                          // Define SDA as input
}


//
// This function returns temperature or humidity depending on the argument
//
float Measure(unsigned char command)
{
  mode = command;                                  // Mode is 3 or 5
  Reset_Sequence();                                // Reset SHT11
  Transmission_Start_Sequence();                   // Start transmission sequence

  SHT11_SDA_Direction = 0;                         // Set SDA as output
  SHT11_SCK = 0;                                   // Set SCK as LOW
//
// Send address and command to SHT11 sensor. A total of 8 bits are sent
//
  for(i = 0; i < 8; i++)                           // Send address and command
  {
    if (mode.F7 == 1)SHT11_SDA_Direction = 1;      // If MSB (bit 7) is 1, Set SDA to 1
    else                                           // If MSB is 0
    {                                              // else MSB is 0
      SHT11_SDA_Direction = 0;                     // define SDA as output
      SHT11_SDA = 0;                               // Set SDA to 0
    }
    Delay_us(1);                                   // 1 us delay
    SHT11_SCK = 1;                                 // SCL high
    Delay_us(1);                                   // 1 us delay
    SHT11_SCK = 0;                                 // SCL low
    mode = mode << 1;                              / Move contents of j one place left
  }
//
// Give a SCK pulse for the ACK
//
  SHT11_SDA_Direction = 1;                         // Set SDA to input (to read ACK)
  SHT11_SCK = 1;                                   // SCL high
```

Figure 6.72
cont'd

```
        Delay_us(1);                                    // 1 us delay
        SHT11_SCK = 0;                                  // SCL low
        Delay_us(1);                                    // 1 us delay
    //
    // Now wait until the measurement is ready (SDA goes LOW when data becomes ready)
    //
        while (SHT11_SDA == 1)Delay_us(1);              // Wait until SDA goes LOW
    //
    // Now, the data is ready, read the data as 2 bytes. Read all 16 bits even though the
    // upper nibble may not be relevant
    //
      buffer = 0;
      for (i = 1; i <=16; i++)                          // DO 16 times
      {
        buffer = buffer << 1;                           // Move MSB one place left
        SHT11_SCK = 1;                                  // SCK HIGH
        if (SHT11_SDA == 1)buffer = buffer | 0x0001;    // Get the bit as 1 (OR with data)
        SHT11_SCK = 0;
        if (i == 8)Send_ACK();                          // If counter i = 8 then send ACK
      }

    //
    // Now make the corrections to the measured value. If mode = 3 then temperature, if on the
    // other hand, mode = 5 then relative humidity
    //
      if(command == 0x03)                               // Temperature correction
        Res = d1 + d2*buffer;
      else if(command == 0x05)                          // Relative humidity correction
      {
        Res = C1 + C2*buffer + C3*buffer*buffer;
        Res = (Ttrue – 25)*(t1 + t2*buffer) + Res;
      }
      return Res;                                       // Return temperature or humidity
    }


    //
    // This is the SHT11 startup delay (20 ms)
    //
    void SHT11_Startup_Delay()
    {
     Delay_ms(20);
    }



    //
    // Start of MAIN program
    //
    void main()
    {
```

Figure 6.72
cont'd

```
ANSELB = 0;                         // Configure PORT B as digital
ANSELC = 0;                         // Configure PORT C as digital
TRISB = 0;
TRISC = 0;
SHT11_SCK_Direction = 0;            // SCL is output

LCD_Init();                         // Initialise LCD
Lcd_Cmd(_LCD_CURSOR_OFF);           // Disable cursor
SHT11_Startup_Delay();              // SHT11 startup delay

for(;;)                             // DO FOREVER
{
  SHT11_SCK_Direction = 0;          // Define SCL1 as output
  Ttrue = Measure(0x03);            // Measure Temperature
  RHtrue = Measure(0x05);           // Measure Relative humidity
  SHT11_SCK_Direction = 1;          // Define SCK as input
  FloatToStr(Ttrue, T+3);           // Convert temperature to string
  FloatToStr(RHtrue, H+3);          // Convert rel humidity to string

  Lcd_Cmd(_LCD_CLEAR);              // Clear display
  T[8] = 178;                       // Insert Degree sign
  T[9] = 'C';                       // Insert C
  T[10] = 0x0;                      // Terminate with NULL
  H[8] = '%';                       // Insert %
  H[9]=0x0;                         // Terminate with NULL
  Lcd_Out(1,1,T);                   // Display temperature
  Lcd_Out(2,1,H);                   // Display humidity
  Delay_ms(1000);                   // Delay 1 s
 }
}
```

Figure 6.72
cont'd

defined. The temperature and relative humidity correction coefficients are then given as floating point numbers.

The main program then configures PORTB, PORTC, and PORTD as digital outputs, initializes the LCD, and clears the display. The program then enters an endless loop formed with a *for* statement. Inside this loop, the temperature is measured and corrected by calling function Measure with argument 3, and stored in a floating point variable T_{true}. Then, the relative humidity is read, corrected, and stored in floating point variable RH_{true}. The measured values are converted into strings by using built-in function FloatToStr. Finally, the degree symbol and letter "C" are appended to the temperature reading. Similarly, symbol "%" is appended to the relative humidity reading before it is displayed.

Function Measure is the most complicated function in the program. This function implements the measurement steps described earlier in the project. Argument *command* specifies the type of measurement required: 3 for temperature measurement and 5 for relative humidity measurement. After calling to functions Reset_Sequence and Transmission_Start_Sequence, the address and command are sent to the SHT11 device.

Figure 6.73: A Typical Display.

Bits of *mode* (3 or 5) are sent out through the MSB after shifting the data to the left in a loop. The program then waits until the conversion is ready, which is indicated by the DATA line going LOW. Once the data are ready, a loop is formed to read the 2 bytes from the sensor. At the end of the eighth clock pulse, an ACK signal is sent to the sensor. In the last part of this function, depending upon the type of conversion required, either the temperature or the relative humidity readings are corrected and returned to the calling program.

The program repeats after a delay of 1 s.

Figure 6.73 shows a typical display.

Project 6.10—Thermometer with an RS232 Serial Output
Project Description

Serial communication is a simple means of sending data to long distances quickly and reliably. The most commonly used serial communication method is based on the RS232 standard. In this standard, data are sent over a single line from a transmitting device to a receiving device in bit serial format at a prespecified speed, also known as the Baud rate, or the number of bits sent each second. Typical Baud rates are 4800, 9600, 19200, 38400 etc.

The RS232 serial communication is a form of asynchronous data transmission where data are sent character by character. Each character is preceded with a Start bit, seven or eight data bits, an optional parity bit, and one or more stop bits. The most commonly used format is eight data bits, no parity bit, and one stop bit. The least significant data bit is transmitted first, and the most significant bit is transmitted last.

A logic high is defined to be at −12 V, and a logic 0 is at +12 V. Figure 6.74 shows how character "A" (ASCII binary pattern 0010 0001) is transmitted over a serial line. The line is normally idle at −12 V. The start bit is first sent by the line going from high to low.

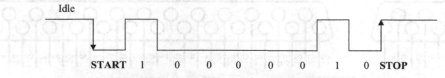

Idle

START 1   0   0   0   0   0   1   0 STOP

Figure 6.74: Sending Character "A" in Serial Format.

Then eight data bits are sent starting from the least significant bit. Finally, the stop bit is sent by raising the line from low to high.

In serial connection, a minimum of three lines are used for communication: transmit (TX), receive (RX), and ground (GND). Some high-speed serial communication systems use additional control signals for synchronization, such as CTS, DTR, and so on. Some systems use software synchronization techniques where a special character (XOFF) is used to tell the sender to stop sending, and another character (XON) is used to tell the sender to restart transmission. In this book, we will be using low-speed communication, and therefore, the basic pins shown in Table 6.5 will be used with no hardware or software synchronization.

Serial devices are connected to each other using two types of connectors: a nine-way connector and a 25-way connector. Table 6.5 shows the TX, RX, and GND pins of each types of connectors. The connectors used in RS232 serial communication are shown in Figure 6.75.

As described above, RS232 voltage levels are at ±12 V. On the other hand, microcontroller input—output ports operate at 0 to +5 V voltage levels. It is therefore necessary to translate the voltage levels before a microcontroller can be connected to an RS232 compatible device. Thus, the output signal from the microcontroller has to be converted into ±12 V, and the input from an RS232 device must be converted into 0 to +5 V before it can be connected to a microcontroller. This voltage translation is normally done using special RS232 voltage converter chips. One such popular chip is the MAX232.

Table 6.5: Minimum Required Pins for Serial Communication

Pin	Function
Nine-Pin Connector	
2	Transmit (TX)
3	Receive (RX)
5	Ground (GND)
Twenty Five-Pin Connector	
2	Transmit (TX)
3	Receive (RX)
7	Ground (GND)

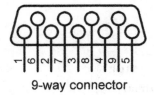
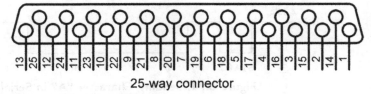

9-way connector   25-way connector

Figure 6.75: RS232 Connectors.

This is a dual converter chip having the pin configuration as shown in Figure 6.76. This particular device requires four external 1-μF capacitors for its operation.

In the PIC18F series of microcontrollers, serial communication can be handled either in the hardware or in the software. The use of the hardware option is easy. The PIC18F series of microcontrollers have built-in Universal Asynchronous Receiver Transmitter (USART) circuits providing special input–output pins for serial communication. For serial communication, all the data transmission is handled by the USART, but we have to configure the USART before receiving and transmitting data. With the software option, all the serial bit timing is handled in software and any input–output pin can be programmed and used for serial communication. In this book, we will only use the hardware UART functions.

In this project, a PC is connected to the microcontroller using an RS232 cable. The project receives the ambient temperature every second and then sends it to the PC where it is displayed on the PC screen.

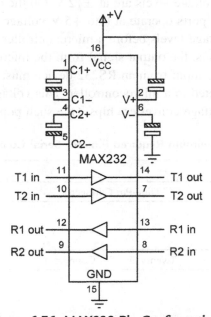

Figure 6.76: MAX232 Pin Configuration.

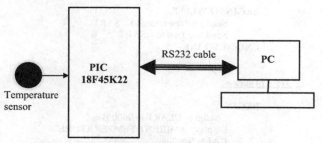

Figure 6.77: Block Diagram of the Project.

The block diagram of the project is shown in Figure 6.77.

Project Hardware

The circuit diagram of the project is shown in Figure 6.78. In this project, a PIC18F45K22 microcontroller is used with an 8 MHz crystal. The built-in USART of the microcontroller is used in this project. The serial communication lines of the microcontroller (RC6 and RC7) are connected to a MAX232 voltage translator chip and then to the serial input port (COM1) of a PC using a nine-pin connector. An LM35DZ-type analog temperature is connected to port pin AN0 (RA0) of the microcontroller.

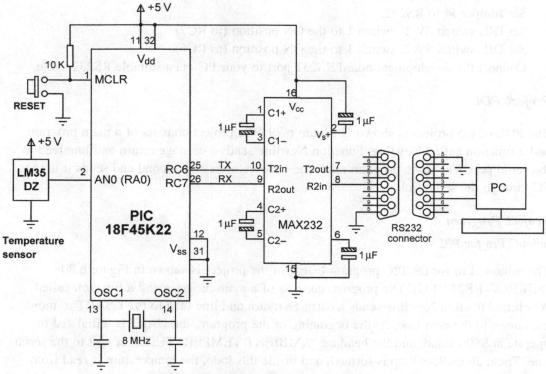

Figure 6.78: Circuit Diagram of the Project.

BEGIN/NEWLINE
Send carriage return to USART
Send line feed to USART
END/NEWLINE

<u>Main program:</u>

BEGIN
Configure USART to 9600 Baud
Display "AMBIENT TEMPERATURE"
CALL Newline
Display "===================="
CALL Newline
DO FOREVER
Read temperature from analog sensor
Convert temperature to Degrees C
Send temperature over the serial line
CALL Newline
Wait 1 s
ENDDO
END

Figure 6.79: Project PDL.

If you are using the EasyPIC V7 development board, the following jumpers must be configured on the board:
- Set Jumper J3 to RS232.
- Set Jumper J4 to RS232.
- Set DIL switch SW1, switch 1 to the ON position (to RC7).
- Set DIL switch SW2, switch 1 to the ON position (to RC6).
- Connect the development board RS232 port to your PC via a suitable RS232 cable.

Project PDL

The PDL of the project is shown in Figure 6.79. The project consists of a main program and a function called **Newline**. Function **Newline** sends a carriage return and line feed to the serial port. The main program reads the temperature every second and sends it to the PC through the serial link.

Project Program
mikroC Pro for PIC

The mikroC Pro for the PIC program listing of the project is shown in Figure 6.80 (MIKROC-RS232-1.C). The program consists of a main program and a function called **Newline**. Function **Newline** sends a carriage return and line feed to the USART to move the cursor to the next line. At the beginning of the program, the UART is initialized to operate at 9600 Baud, and the heading "AMBIENT TEMPERATURE" is sent to the serial line. Then, an endless loop is formed, and inside this loop, the temperature is read from

```
/*****************************************************************************
                   THERMOMETER WITH RS232 SERIAL OUTPUT
                   =======================================

In this project a PC is connected to a PIC18F452K2 microcontroller. The project reads the ambient
temperature using an LM35DZ type analog temperature sensor. The temperature is then sent to
a PC over the RS232 communications line. The PC displays the temperature on the screen.

This program uses the built in USART of the microcontroller. The USART is configured to operate
with 9600 Baud rate.

The serial TX pin is RC6 and the serial RX pin is RC7.

Author:    Dogan Ibrahim
Date:      September 2013
File:      MIKROC-RS232-1.C
*****************************************************************************/

//
// This functions send carriage-return and line-feed to USART
//
void Newline()
{
    Uart1_Write(0x0D);                          // Send carriage return
    Uart1_Write(0x0A);                          // Send line feed
}


//
// Start of MAIN program
//
void main()
{
    unsigned char Txt[14];
    unsigned char msg1[] = "AMBIENT TEMPERATURE";
    unsigned char msg2[] = "===================";
    unsigned int temp;
    float mV;
    char *res;

    ANSELA = 1;                                 // Configure RA0 as analog
    ANSELC = 0;                                 // Configure PORTC as digital
    TRISA = 1;                                  // Configure RA0 as input

    Uart1_Init(9600);                           // Initialize UART to Baud = 9600

    Uart1_Write_Text(msg1);                     // Display heading
    Newline();                                  // Newline
    Uart1_Write_Text(msg2);
    Newline();
```

Figure 6.80: mikroC Pro for PIC Program.

```
        for(;;)                              // Endless loop
        {

            temp = ADC_Read(0);              // Read temperature from channel 0
            mV = temp * 5000.0/1024;         // Convert to millivolts
            mV = mV/10.0;                    // Convert to Degrees Centigrade
            FloatToStr(mV, Txt);             // Convert into a string
            res = strchr(Txt, '.');          // Locate "."
            *(res+3) = '\0';                 // Insert NULL character 2 digits after "."
            Uart1_Write_Text(Txt);           // Send temperature over the serial line
            Newline();
            Delay_Ms(1000);                  // Wait 1 s
        }
    }
```

Figure 6.80
cont'd

channel 0 and is converted into millivolts after multiplying with 5000/1024. The result is then divided by 10 to give the temperature in Degrees Centigrade. The temperature is converted into a string using built-in function **FloatToStr**. The resulting string contains several digits after the decimal point as in the following example:

29.23687

The number of digits to display after the decimal point can be selected by finding the position of the decimal point and inserting a NULL character to terminate the string at the required point. In this program, the built-in string function **strchr** is used to find the address (pointer to) of the decimal point, and this is stored in character pointer **res**. Then, a NULL character is added to string **Txt** three digits after the decimal point. The result is that the string will be displayed with two digits after the decimal point. Thus, 29.23687 will be displayed as 29.23.

mikroC Pro for the PIC compiler supports the following hardware UART functions ("x" is 1 or 2 depending upon the UART module used in multiple UART processors):

UARTx_Init: This function initializes the UART module. The required Baud rate must be specified in the argument.

UARTx_Data_Ready: This function is used to check if data are available in the receive buffer. The function returns 1 if data are available for reading.

UARTx_Tx_Idle: This function is used to test if the transmit buffer is empty. The function returns 1 if the transmit buffer is empty.

UARTx_Read: This function reads a byte from the UART. Function UARTx_Data_-Ready should be used to make sure that data are available in the receive buffer.

UARTx_Read_Text: This function is used to read the text from UART. A delimiter and number of attempts to detect the delimiter should be specified to identify the end of data.

UARTx_Write: This function writes a byte to UART.

UARTx_Write_Text: This function is used to write zero-terminated text to the UART.

UART_Set_Active: Specify the UART to be used in processors having more than one UART.

Testing the Program

The program can be tested using a terminal emulator software such as **Hyperterminal**, which is distributed free of charge with the Windows operating systems. The steps to test the program are given below (these steps assume that serial port COM15 is used. If you are not sure which serial port is available on your PC, go to **Control Panel → System → Device Manager** and look for available Ports. Serial ports have identifiers COM, followed by a number):

- Connect the RS232 output from the microcontroller to the serial input port of a PC (e.g. COM15).
- Start Hyperterminal terminal emulation software and give a name to the session.
- Select File → New connection → Connect using and select COM15.
- Select the Baud rate as 9600, data bits as 8, no parity bits, 1 stop bit, and Flow Control None.
- Reset the microcontroller.

An example output from the Hyperterminal screen is shown in Figure 6.81.

Alternatively, the terminal emulator software included inside the mikroC Pro for PIC compiler can be used. The steps are given below (assuming that COM15 will be used):

- Click **Tools → USART Terminal**.
- Select COM15, 9600 Baud, 1 Stop bit, No parity, Flow control None, data Format ASCII (Figure 6.82).
- Click **Connect**. You should see the data being displayed on your terminal (Figure 6.83).

Using USB-RS232 Converter Cable

Most PCs nowadays do not have RS232 serial ports; instead, they are equipped with universal serial bus (USB) ports. If this is the case, you should be able to purchase and use a USB-RS232 converter cable. This is a special cable with built-in electronics to convert RS232 signals into USB and vice versa. When such a cable is connected to a PC, it creates a virtual serial communications port on your PC. You should go to **Control Panel → System → Device Manager** to see the name of the created COM port, and then use a terminal emulator software such as the ones given above, that is, Hyperterm or the mikroC Pro for PIC built-in terminal emulator.

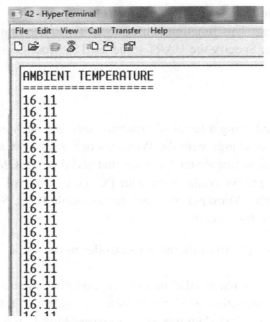

Figure 6.81: Hyperterminal Screen.

Figure 6.82: Use of the mikroC Pro for the PIC Built-in Terminal Emulator.

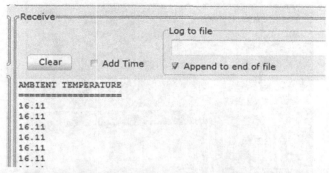

Figure 6.83: Displaying Data Using the mikroC Pro for the PIC Terminal Emulator.

Using the USB UART Port

The EasyPIC V7 development board has an on-board USB-RS232 converter module that enables you to connect the development board to the USB port on your PC. The steps to use this port are as follows:

- Set Jumper J3 to USB UART.
- Set Jumper J4 to USB UART.
- Set DIL switch SW1, switch 1 to ON position (to RC7).
- Set DIL switch SW2, switch 1 to ON position (to RC6).
- Connect the development board USB UART port to your PC via a USB cable.

Figure 6.84 shows the EasyPIC V7 development board configured for USB UART operation.

Note that the USB UART module on the development board uses the FT232RL chip to convert the signals. This chip requires the FTDI driver (known as VCP_DRIVERS) to be installed on your PC before you can use the USB UART communication. This driver is available on the EasyPIC V7 product DVD. Alternatively, it can be downloaded free of charge from the Internet.

MPLAB XC8

The mikroC Pro for PIC program listing of the project is shown in Figure 6.85 (MIKROC-RS232-1.C). Some of the important MPLAB XC8 hardware UART library functions are given below ("x" is either 1 or 2 and specifies the UART to be used in processors having more than one UART):

> BusyxUSART: This function returns 1 if the UART transmitter is busy.
> ClosexUSART: This function disables the UART.

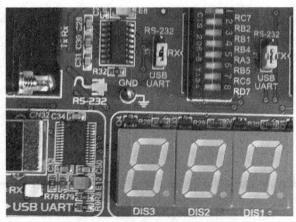

Figure 6.84: EasyPIC V7 Board Configured for the USB UART Operation. (For color version of this figure, the reader is referred to the online version of this book.)

DataRdyxUSART: This function returns 1 if data are available in the UART receive buffer.

getsxUSART: This function reads a fixed length of characters from UART. The number of characters to be read is specified in the function argument.

OpenxUSART: This function configures the specified UART. The function has two arguments. The first argument can be a bitwise AND of the following identifiers:

Interrupt on Transmission:

USART_TX_INT_ON   Transmit interrupt ON
USART_TX_INT_OFF   Transmit interrupt OFF

Interrupt on Receipt:

USART_RX_INT_ON   Receive interrupt ON
USART_RX_INT_OFF   Receive interrupt OFF

USART Mode:

USART_ASYNCH_MODE   Asynchronous Mode
USART_SYNCH_MODE   Synchronous Mode

```
/*******************************************************************************
                    THERMOMETER WITH RS232 SERIAL OUTPUT
                    =====================================
```

In this project a PC is connected to a PIC18F452K2 microcontroller. The project reads the ambient temperature using an LM35DZ type analog temperature sensor. The temperature is then sent to a PC over the RS232 communications line. The PC displays the temperature on the screen.

This program uses the built in USART of the microcontroller. The USART is configured to operate with 9600 Baud rate.

The serial TX pin is RC6 and the serial RX pin is RC7.

Author: Dogan Ibrahim
Date: September 2013
File: XC8-RS232-1.C
```
********************************************************************************/
#include <xc.h>
#include <string.h>
#include <plib/usart.h>
#include <plib/adc.h>
//#include <strings.h>
#include <stdlib.h>
#pragma config MCLRE = EXTMCLR, WDTEN = OFF, FOSC = HSHP
#define _XTAL_FREQ 8000000

//
// This function creates seconds delay. The argument specifies the delay time in seconds.
//
void Delay_Seconds(unsigned char s)
{
    unsigned char i,j;

    for(j = 0; j < s; j++)
    {
        for(i = 0; i <  100; i++)__delay_ms(10);
    }
}


void Newline()
{
    Write1USART(0x0D);                              // Send carriage return
    while(Busy1USART());                            // Wait while UART is busy
    Write1USART(0x0A);                              // Send line feed
}


//
// Start of MAIN program
```

Figure 6.85: MPLAB XC8 Program.

```
//
void main()
{
   char *Txt;
   unsigned char msg1[] = "AMBIENT TEMPERATURE";
   unsigned char msg2[] = "====================";
   unsigned int temp;
   float mV;
   char *res;
   int status;

   ANSELA = 1;                                    // Configure RA0 as analog
   ANSELC = 0;                                    // Configure PORTC as digital
   TRISA = 1;                                     // Configure RA0 as input

   Open1USART( USART_TX_INT_OFF    &
        USART_RX_INT_OFF    &
        USART_ASYNCH_MODE       &
        USART_EIGHT_BIT      &
        USART_CONT_RX           &
        USART_BRGH_LOW,
        12);

   putrs1USART(msg1);                             // Display heading
   Newline();                                     // Newline
   putrs1USART(msg2);
   Newline();
//
// Configure A/D converter
//
   OpenADC(ADC_FOSC_2 & ADC_RIGHT_JUST & ADC_12_TAD,
        ADC_CH0 & ADC_INT_OFF,
        ADC_TRIG_CTMU & ADC_REF_VDD_VDD & ADC_REF_VDD_VSS);


   for(;;)                                        // Endless loop
   {
     SelChanConvADC(ADC_CH0);                     // Select channel 0 and start conversion
     while(BusyADC());                            // Wait for completion
     temp = ReadADC();                            // Read converted data
     mV = temp * 5000.0/1024;                     // Convert to millivolt
     mV = mV/10.0;                                // Convert to Degrees Centigrade
     Txt = ftoa(mV, &status);                     // Convert to string

     res = strchr(Txt, '.');                      // Locate "."
     *(res+3) = '\0';                             // Insert NULL character 2 digits after "."
     putrs1USART(Txt);
     Newline();
     Delay_Seconds(1);                            // Wait 1 s
   }
}
```

Figure 6.85

cont'd

Transmission Width:

USART_EIGHT_BIT	8-Bit transmit/receive
USART_NINE_BIT	9-Bit transmit/receive

Slave/Master Select (synchronous mode only):

USART_SYNC_SLAVE	Synchronous Slave mode
USART_SYNC_MASTER	Synchronous Master mode

Reception mode:

USART_SINGLE_RX	Single reception
USART_CONT_RX	Continuous reception

Baud rate:

USART_BRGH_HIGH	High baud rate
USART_BRGH_LOW	Low baud rate

The second argument specifies the value to be written to the Baud rate generator register for the required Baud rate. The formula used to determine the Baud rate is (F_{OSC} is the oscillator frequency):

Asynchronous mode, high speed:

FOSC/(16 * (*spbrg* + 1))

Asynchronous mode, low speed:

FOSC/(64 * (*spbrg* + 1))

Synchronous mode:

FOSC/(4 * (*spbrg* + 1))

putrsxUSART: This function writes a string of characters to the UART, including the NULL character.

ReadxUSART: This function reads 1 byte from the UART.

WritexUSART: This function writes 1 byte to the UART.

Note that in Figure 6.85 the header file usart.h must be included at the beginning of the program. The UART module is initialized by using the following identifiers:

```
Open1USART(USART_TX_INT_OFF    &
           USART_RX_INT_OFF    &
           USART_ASYNCH_MODE   &
           USART_EIGHT_BIT     &
           USART_CONT_RX       &
           USART_BRGH_LOW,
           12);
```

The second argument specifies the Baud rate. Using a low speed with the required 9600 Baud and clock rate of 8 MHz $(F_{OSC}) = 8 \times 10^6$ Hz, we have

$$9600 = F_{OSC}/(64 * (spbrg + 1))$$

or

$$spbrg = [F_{OSC}/(64 * 9600)] - 1 = 12$$

Project 6.11—Microcontroller and a PC-Based Calculator
Project Description

In this project, a PC is connected to the microcontroller using an RS232 cable. The project operates as a simple integer calculator where numbers and operation to be performed are sent to the microcontroller via the PC keyboard, and the results are displayed on the PC screen.

A sample calculation is as follows:

```
CALCULATOR PROGRAM

 Enter First Number: 12
Enter Second Number: 2
    Enter Operation: +
           Result = 14
```

Project Hardware

The circuit diagram of the project is shown in Figure 6.86. The serial communication lines of the microcontroller (RC6 and RC7) are connected to an MAX232 voltage translator chip and then to the serial input port (COM1) of a PC using a nine-pin connector.

Project PDL

The PDL of the project is shown in Figure 6.87. The main program receives two numbers and the operation to be performed from the PC keyboard. The numbers are echoed on the PC monitor. The result of the operation is also displayed on the monitor.

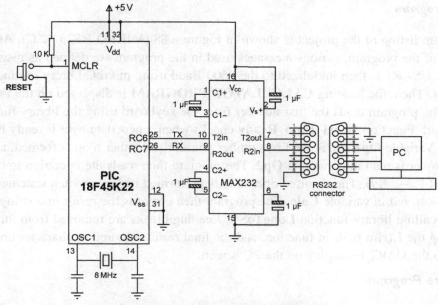

Figure 6.86: Circuit Diagram of the Project.

```
            BEGIN/NEWLINE
                    Send carriage return to USART
                    Send line feed to USART
            END/NEWLINE


      Main program:

            BEGIN
                    Configure USART to 9600 Baud
                    DO FOREVER
                            Display "CALCULATOR PROGRAM"
                            Display "Enter First Number:"
                            Read first number
                            Display "Enter Second Number:"
                            Read second number
                            Display "Operation:"
                            Read operation
                            Perform required operation
                            Convert result into string
                            Remove leading spaces from the result
                            Display "Result="
                            Display the result
                    ENDDO
            END
```

Figure 6.87: Project PDL.

Project Program

The program listing of the project is shown in Figure 6.88 (MIKRO-RS232-2.C). At the beginning of the program, various messages used in the program are defined as **msg1** to **msg5**. The USART is then initialized to the 9600-Baud using mikroC library routine **Uart1_Init**. Then, the heading **CALCULATOR PROGRAM** is displayed on the PC monitor. The program reads the first number from the keyboard using the library function **Uart1_Read**. Function **Uart1_Data_Ready** checks when a new data byte is ready before reading it. Variable **Op1** stores the first number. Similarly, another loop is formed, and the second number is read into variable **Op2**. The program then reads the operation to be performed (+, −, *, /). The required operation is performed inside a switch statement, and the result is stored in variable **Calc**. The program then converts the result into string format by calling library function **LongToStr**. Leading blanks are removed from this string using the **Ltrim** built-in function, and the final result is stored in character array **op** and sent to the UART to display on the PC screen.

Testing the Program

The program can be tested by using a terminal emulator program on the PC as described in the previous project. If you are using the mikroC Pro for a PIC terminal emulator, you should make the following settings before clicking the Connect button:

> New Line Settings: CR (0x0D)
> Append New Line

Keyboard data should be entered at the top of the terminal emulator window and the Send button should be clicked.

Figure 6.89 shows a sample run of the program using the Hyperterm terminal emulator software.

Project 6.12—GPS with an LCD Output
Project Description

This project is about designing a global positioning system-based system to display the latitude and longitude of current position on an LCD display.

The GPS is a satellite navigation system that provides time and location information anywhere on the Earth, 24 h a day, and in all weather conditions. Currently, the system is heavily used by motorists, ships, and in the air. The system is maintained by the US government and is freely available to anyone who has a GPS receiver.

```
/*******************************************************************************
                        CALCULATOR WITH PC INTERFACE
                        ============================
```

In this project a PC is connected to a PIC18F45K22 microcontroller. The project is a simple integer calculator. User enters the numbers using the PC keyboard. Results are displayed on the PC monitor.

The following operations can be performed:

  + – * /

This program uses the built in USART of the microcontroller. The USART is configured to operate with 9600 Baud rate.

The serial TX pin is RC6 and the serial RX pin is RC7.

```
Author:  Dogan Ibrahim
Date:    September 2013
File:    MIKRO-RS232-2.C
*******************************************************************************/

#define Enter 13
#define Plus '+'
#define Minus '−'
#define Multiply '*'
#define Divide '/'

//
// This functions send carriage-return and line-feed to USART
//
void Newline()
{
    Uart1_Write(0x0D);                          // Send carriage return
    Uart1_Write(0x0A);                          // Send line feed
}


void main()
{
    unsigned char MyKey, i,j,kbd[12];
    unsigned char *op;
    unsigned long Calc, Op1, Op2,Key;
    unsigned char msg1[] = "CALCULATOR PROGRAM";
    unsigned char msg2[] = "   Enter First Number: ";
    unsigned char msg3[]= "  Enter Second Nummber: ";
    unsigned char msg4[] = "    Enter Operation: ";
    unsigned char msg5[] = "          Result = ";
```

Figure 6.88: mikroC Pro for PIC Program.

```
                ANSELC = 0;
                //
                // Configure the USART
                //
                   Uart1_Init(9600);                        // Baud = 9600
                //
                // Program loop
                //
                   for(;;)                                  // Endless loop
                   {
                     MyKey = 0;
                     Op1 = 0;
                     Op2 = 0;

                     Newline();                             // Send newline
                     Newline();                             // Send newline
                     Uart1_Write_Text(msg1);                // Send TEXT
                     Newline();                             // Send newline
                     Newline();                             // Send newline

                //
                // Get the first number
                //
                     Uart1_Write_Text(msg2);                // Send TEXT to USART
                     do                                     // Get first number
                     {
                       if(Uart1_Data_Ready())               // If a character ready
                       {
                         MyKey = Uart1_Read();              // Get a character
                         if(MyKey == Enter)break;           // If ENTER key
                         Uart1_Write(MyKey);                // Echo the character
                         Key = MyKey – '0';
                         Op1 = 10*Op1 + Key;                // First number in Op1
                       }
                     }while(1);

                     Newline();

                //
                // Get the second character
                //
                     Uart1_Write_Text(msg3);                // Send TEXT to USART
                     do                                     // Get second number
                     {
                       if(Uart1_Data_Ready())
                       {
                         MyKey = Uart1_Read();              // Get a character
                         if(Mykey == Enter)break;           // If ENTER key
                         Uart1_Write(MyKey);                // Echo the character
                         Key = MyKey – '0';
                         Op2 = 10*Op2 + Key;                // Second number in Op2
                       }
                     }while(1);
```

Figure 6.88
cont'd

```
        Newline();
//
// Get the operation
//
        Uart1_Write_Text(msg4);
        do
        {
          if(Uart1_Data_Ready())
          {
            MyKey = Uart1_Read();          // Get a character
            if(MyKey == Enter)break;       // If ENTER key
            Uart1_Write(MyKey);            // Echo the character
            Key = MyKey;
          }
        }while(1);

//
// Perform the operation
//
        Newline();
        switch(Key)                        // Calculate
        {
          case Plus:
              Calc = Op1 + Op2;            // If ADD
              break;
          case Minus:
              Calc = Op1 – Op2;            // If Subtract
              break;
          case Multiply:
              Calc = Op1 * Op2;            // If Multiply
              break;
          case Divide:
              Calc = Op1/Op2;             // If Divide
              break;
        }

        LongToStr(Calc, kbd);             // Convert to string
        op = Ltrim(kbd);
        Uart1_Write_Text(msg5);
        Uart1_Write_Text(op);             // Display result

    }
}
```

Figure 6.88
cont'd

The GPS system consists of 24 satellites orbiting the Earth in six orbital planes at an altitude of 20,000 km. The orbital period is about 12 h, so each satellite orbits the earth twice a day. The orbits are organized such that normally six satellites are visible from any point on the Earth. To calculate the position, at least three satellites are required. The height can also be calculated if the fourth satellite is available. There are many books on the theory and operation of the GPS system. Also, there is a vast amount of information on the Internet on this topic.

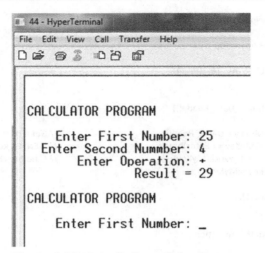

Figure 6.89: Sample Run of the Program.

GPS receivers are small handheld devices that receive information from the satellites and report the position (latitude and longitude) as well as the height of the user on a graphical display. GPS receivers usually provide ASCII output data to indicate the position and time information. These data, called the NMEA sentences, consist of a number of text, each one separated by a newline. We can read and decode these sentences to get information about our current coordinates.

In this project, we will be using a small GPS receiver board with an antenna and display our longitude and the latitude on an LCD display.

The block diagram of the project is shown in Figure 6.90.

Project Hardware

The project is based on the GPS Click board, manufactured by mikroElektronika (www.mikroe.com). Click boards are small peripheral modules designed for the

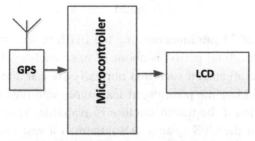

Figure 6.90: Block Diagram of the Project.

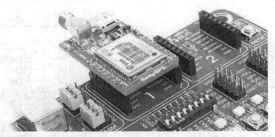

Figure 6.91: Connecting the GPS Click Board to EasyPIC V7 Development Board. (For color version of this figure, the reader is referred to the online version of this book.)

mikroBUS interface. There are Click boards available for many peripheral devices, such as LED, A/D converter, DAC, temperature sensor, GPS, relay, pressure sensor, light sensor, accelerator, WiFi, Ethernet, GSM, compass, and many more.

The mikroBUS is a 2 × 8-pin female header that provides interface signals, enabling many "Click Boards" to be connected to this bus. The mikroBUS provides interface signals for the following:

> Analog pin,
> Reset pin,
> SPI Bus pins (CS, clock, and data I/O),
> +3.3 V, +5 V, GND,
> PWM output line,
> Hardware interrupt pin,
> UART (TX and RX) pins,
> I^{2}C Bus pins (clock and data).

The EasyPIC V7 development board contains two identical mikroBUS connectors, enabling up to two Click boards to be connected directly to the board. Figure 6.91 shows the GPS Click board connected to one of the mikroBUS connectors on the EasyPIC V7 development board.

The GPS Click board uses the LEA-6S high performance GPS chip. The board can be interfaced with a microcontroller through the UART, I^{2}C bus, or through a USB connection. In this project, the UART interface is used for simplicity. An active or a passive antenna can be connected to the board to increase its sensitivity. Operation is with a 3.3 V power supply, and a power regulator should be used to provide 3.3 V if the board is used outside the EasyPIC V7 development board.

Figure 6.92 shows the circuit diagram of the project. In this diagram, only the UART interface to the GPS Click board is shown (see mikrolektronika web site for full circuit diagram of this Click board). The UART TX (RC6) and RX (RC7) pins of the microcontroller are connected directly to the GPS chip. PORTB is connected to an LCD as

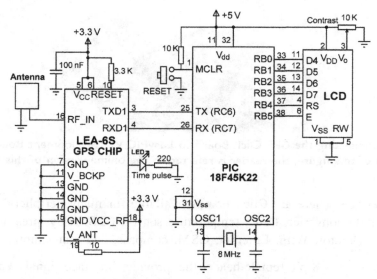

Figure 6.92: Circuit Diagram of the Project.

in the previous projects. An LED is connected to the Time_Pulse output of the GPS chip to see the device working. An antenna is connected to the RF_IN pin of the GPS chip.

Project PDL

The NMEA sentences generated by the GPS receivers start with the "$" character, followed by the name of the sentence and then its parameters. Each parameter is separated by a comma. A checksum byte is used at the end.

A sample of the NMEA sentences generated by the LEA-6 GPS every second are given below (different GPS chips may generate different sentences):

```
$GPRMC,101241.00,A,5127.36070,N,0003.12726,,0.118,,28813,,,A*7F
$GPVTG,,T,M,0.118,N0.218,K,A*0
$GPGGA,01241.00,527.36070,N00003.1272,E,1,06,1.0,40.8,M,4.4,M,,*63
$GPGSA,A,329,02,31,1,25,14,,,,,,2.57,1.6,2.01*0C
$GPGSV,,2,09,24,0,142,,25,7,072,21,2967,189,34,1,52,298,3*79
$GPGLL,517.35744,N,0003.13373E,103109.00,A,A*65
$GPRMC,101242.00,A,5127.36058,N,0003.12714,E0.326,,2806,1.6,40.8,M,454,M,,*6C
```

The above sentence list was obtained on a PC screen by the following program code after the microcontroller was connected to a PC and by running the terminal emulator software on the PC. See Project 6.11:

```
Uart1_Init(9600);
for(;;)
```

```
        BEGIN
                Define connection between LCD and microcontroller
                Configure PORTB and PORTC as digital output
                Initialize LCD
                Initialize UART to 9600 Baud
                DO FOREVER
                        Read until string "$GPGLL" is detected
                        Read until character "*" is detected
                        Extract latitude
                        Extract longitude
                        Display latitude and longitude
                        Wait 2 s
                ENDDO
        END
```

Figure 6.93: Project PDL.

```
{
    if(Uart1_Data_Ready() == 1)
    {
        c = Uart1_Read();
        Uart1_Write(c);
    }
}
```

The coordinates of the current position can be obtained from the $GPGLL sentence. This sentence decodes as follows:

```
            $GPGLL,517.35744,N,0003.13373,E,103109.00,A,A*65
```

5127.35744,N	Latitude 51 deg. 27.35744 min. North
00003.1276,E	Longitude 000 deg. 03.1276 min. East
10124100	Fix taken at 10:31:09 UTC
A	Data valid (V = data invalid)
A	Mode (A = autonomous, D = differential)
*65	Checksum

The PDL of the program is given in Figure 6.93. The program initially searches for characters $GPGLL, and then the remainder of the text is read until the newline character is obtained. The latitude and longitude are then extracted and displayed on the LCD.

Project Program

microC Pro for PIC

The mikroC Pro for the PIC program listing is shown in Figure 6.94 (MIKROC-GPS.C). At the beginning of the program, the connection between the LCD and the microcontroller is defined, and PORTB and PORTC are configured as digital. The LCD is then initialized, and message INVALID DATA is displayed to start with. Then, the UART is initialized, and the remainder of the program is executed in an endless loop.

```
/*****************************************************************************
                        GPS WITH LCD OUTPUT
                        ===================
```

In this project a GPS is connected to a PIC18F45K22 microcontroller. The coordinates (latitude and longitude) of the current location are read and displayed on an LCD.

The GPS Click Board (www.mikroe.com) is used in this project together with the EasyPIC V7 development board. The GPS board is connected to one of the mikroBUS connectors on the development board.

The LCD is connected to PORTB of the microcontroller as in the previous projects.

The communications between the GPS and the microcontroller is by using the serial port. TX pin (RC6) and the RX pin (RC7) of the microcontroller are connected to the corresponding serial pins of the GPS.

The latitude and longitude are determined by decoding the NMEA sentence $GPGLL. An example is given below:

    $GPGLL,5127.35744,N,00003.13373,E,103109.00,A,A*65

    5127.35744,N   Latitude 51 deg. 27.35744 min. North
    00003.1276,E   Longitude 000 deg. 03.1276 min. East
    103109.00   Fix taken at 10:31:09 UTC
    A   Data valid
    A   Data autonomous
    *65   Checksum

The program assumes fixed length NMEA sentence.

Author: Dogan Ibrahim
Date: September 2013
File: MIKRO-GPS.C
***/
// LCD module connections
sbit LCD_RS at RB4_bit;
sbit LCD_EN at RB5_bit;
sbit LCD_D4 at RB0_bit;
sbit LCD_D5 at RB1_bit;
sbit LCD_D6 at RB2_bit;
sbit LCD_D7 at RB3_bit;

sbit LCD_RS_Direction at TRISB4_bit;
sbit LCD_EN_Direction at TRISB5_bit;
sbit LCD_D4_Direction at TRISB0_bit;
sbit LCD_D5_Direction at TRISB1_bit;
sbit LCD_D6_Direction at TRISB2_bit;
sbit LCD_D7_Direction at TRISB3_bit;
// End LCD module connections
```

**Figure 6.94: mikroC Pro for PIC Program.**
```

```
void main()
{
    unsigned char buffer[50];
    unsigned char i,flag,c;
    unsigned char Lat[13], Lon[13];
    unsigned char gps[]="$GPGLL,";

    ANSELB = 0;
    ANSELC = 0;

    LCD_Init();                              // Initialize LCD
    Lcd_Cmd(_LCD_CURSOR_OFF);                // Disable cursor
    Lcd_Out(1,1,"INVALID DATA");             // Display INVALID DATA to start with

    Uart1_Init(9600);                        // Baud = 9600

    for(;;)                                  // Endless loop
    {
        for(i = 0; i < 50; i++)buffer[i] = 0;   // Clear the buffer

        i = 0;
        flag = 0;
//
// Read until "$GPGLL," is detected
//
        while(flag == 0)
        {
            if(Uart1_Data_Ready() == 1)
            {
                c = Uart1_Read();
                if(c == gps[i])
                {
                    i++;
                    if(i == 7)flag=1;
                }
                else i = 0;
            }
        }
//
// We come to this point when the string "$GPGLL," has been detected
//
        Uart1_Read_Text(buffer,"*",255);     // Read until "*" detected
        if(buffer[37] == 'A')                // If the sentence is valid
        {                                    // Get latitude Degrees
            Lat[0] = buffer[0];
            Lat[1] = buffer[1];
            Lat[2] = 178;                    // Degree character
            Lat[3] = ' ';
            Lat[4] = ' ';
            for(i = 0; i < 6; i++)Lat[5+i] = buffer[2+i];   // Get latitude minutes
            Lat[11] = buffer[11];            // Get latitude direction
            Lat[12] = 0x0;                   // Terminate the string
```

Figure 6.94

cont'd

```
                    Lon[0] = buffer[13];                    // Get longitude Degrees
                    Lon[1] = buffer[14];
                    Lon[2] = buffer[15];
                    Lon[3] = 178;                           // Degree character
                    Lon[4] = ' ';
                    for(i = 0; i < 6; i++)Lon[5+i] = buffer[16+i];   // Get longitude minutes
                    Lon[11] = buffer[25];                   // Get longitude direction
                    Lon[12] = 0x0;                          // Terminate the string
                    Lcd_Cmd(_LCD_CLEAR);                    // Clear LCD
                    Lcd_Out(1,1,"LAT=");                    // Display LAT=
                    Lcd_Out_Cp(Lat);                        // Display the latitude
                    Lcd_Out(2,1,"LON=");                    // Display LON=
                    Lcd_Out_Cp(Lon);                        // Display the longitude
                  }
                  else                                      // If invalid data
                  {
                    Lcd_Cmd(_LCD_CLEAR);                    // Clear LCD
                    Lcd_Out(1,1,"INVALID DATA");            // Display INVALID DATA
                  }
                  Delay_Ms(2000);                           // Wait 2 s
                }
          }
```

Figure 6.94
cont'd

At the beginning of the loop, the buffer that will hold the GPS data is cleared. The program then reads data from the GPS and looks for the string "$GPGLL,". When this string is detected, the Uart1_Read_Text function is used to read data from the GPS until the delimiting character "*" is detected (this character is at the end of the $GPGLL sentence just before the checksum). The buffer at this point contains all the parameters of the sentence $GPGLL, starting from the latitude parameter. The remainder of the program extracts the latitude and longitude parameters and loads into two string arrays called LAT and LON, respectively. These arrays are then displayed on the LCD after adding the degree sign and spaces at appropriate points.

Note that the code to detect the NMEA sentence "GPGLL," could have been done using the following two lines. Although the code seems to be smaller, it requires a very large buffer size (e.g. ≥1000 characters) since all the generated NMEA sentences will be read until the "$GPGLL," is detected as the delimiter:

```
Uart1_Read_Text(buffer, "$GPGLL,", 255);   // Read until "$GPGLL," detected
Uart1_Read_Text(buffer, "*",255);          // Read until "*" detected
```

The program checks the NMEA sentence validity every second, and if the sentence is not valid, the message INVALID DATA is displayed. The program assumes that the width of the parameters in the "$GPGLL" sentence have fixed sizes. Although this is the case with

Figure 6.95: Sample Display. (For color version of this figure, the reader is referred to the online version of this book.)

most GPS receivers, it may be necessary to locate the commas and then extract the parameters if this is not the case.

A sample display is shown in Figure 6.95.

Project 6.13—ON—OFF Temperature Control

Project Description

This project is about designing an ON—OFF type temperature control system for a small plant. Figure 6.96 shows the block diagram of the system to be designed.

The desired temperature setting is entered using a keypad. The temperature of the plant is measured using an analog temperature sensor. The microcontroller reads the temperature

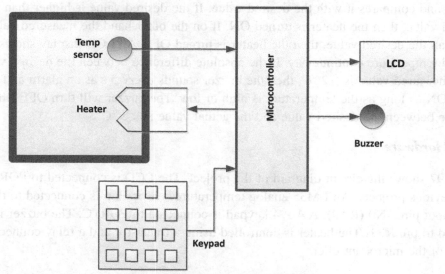

Figure 6.96: Block Diagram of the System.

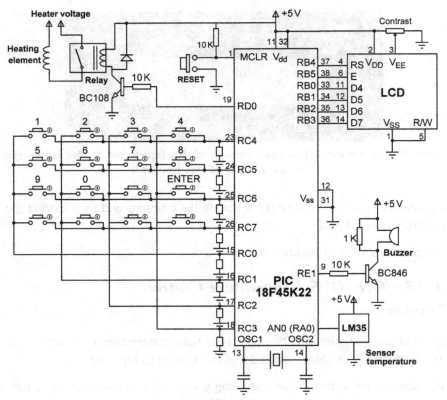

Figure 6.97: Circuit Diagram of the Project.

every 5 s and compares it with the desired value. If the desired value is higher than the measured value, then the heater is turned ON. If on the other hand the measured value is higher than the desired value, then the heater is turned OFF. An LCD display shows the measured temperature continuously. If the absolute difference between the desired value and the measured value is >2 °C, then the buzzer sounds every 5 s as an alarm and remains ON as long as the temperature is high or low. The buzzer will turn OFF when the difference between the desired value and the actual value is <2 °C.

Project Hardware

Figure 6.97 shows the circuit diagram of the project. The LCD is connected to PORTC as in the previous projects. An LM35 analog temperature sensor chip is connected to the analog input pin AN0 (RA0). A 4 × 4 keypad is connected to PORTC. The buzzer is connected to pin RE1. The heater is controlled using a transistor and a relay connected to pin RD0 of the microcontroller.

```
BEGIN
        Define LCD to microcontroller connections
        Define Keypad port
        Assign symbols BUZZER, HEATER, ON, OFF to port pins
        Configure PORTB, PORTC, PORTD, PORTE as digital
        Configure RA0 as input, RD0, RE1 and RC0:RC3 as output
        Initialize Keypad and Sound libraries
        Display heading "ON–OFF CONTROL"
        Wait 2 s
        Read desired temperature
        Wait until ENTER is pressed
        DO FOREVER
                Read plant temperature from channel 0
                Convert temperature to Degrees Centigrate
                Display the Set Point and actual pant temperatures
                IF Set Point > Plant temperature
                        Turn HEATER ON
                        IF Set Point – Plant temperature > 2
                                Sound alarm
                        ENDIF
                ELSE
                        Turn HEATER OFF
                        IF Set Point – Plant temperature > 2
                                Sound alarm
                        ENDIF
                ENDIF
                Wait 5 s
        ENDDO
END
```

Figure 6.98: Project PDL.

Project PDL

The project PDL is shown in Figure 6.98.

Project Program

mikroC Pro for PIC

The mikroC Pro for the PIC program listing is shown in Figure 6.99 (MIKROC-ON-OFF.C). At the beginning of the program, the connections between the LCD and the microcontroller are defined, the keypad port is defined, BUZZER, HEATER, and ON and OFF are defined and assigned as symbols to port pins. PORTB, PORTC, PORTD, and PORTE are configured as digital. RA0 is configured as an input pin. RD0, RE1, PORTB, and half of PORTC are configured as output pins. The program then initializes the Keypad and the Sound libraries. The LCD is cleared and message "ON–OFF CONTROL" is displayed for 2 s.

```
/***************************************************************************
                        ON-OFF TEMPERATURE CONTROL
                        ===========================
```

In this project the temperature of a plant is controlled using simple ON–OFF type controller.

The plant temperature is measured using an LM35DZ type analog temperature sensor. An LCD helps to enter the set point temperature, and also displays the set point as well as the actual plant temperature in real-time.

 A 4×4 keypad is used to set the desired temperature. If the set temperature is below the actual plant temperature then the heater relay is turned ON. If on the other hand the set temperature is above the actual plant temperature then the relay is turned OFF. A buzzer sounds if the absolute difference between the desired and the actual temperatures is more than 2 °C

```
        The LCD is connected to PORTB
        The buzzer is connected to pin RE1 through a transistor switch
        The keypad is connected to PORTC
        The heater relay is connected to pin RD0 through a transistor switch
        The LM35DZ temperature sensor is connected to pin AN0 (RA0)

        The program uses the built-in keypad library functions.
```

The control action is taken every 5 s

```
Author:  Dogan Ibrahim
Date:    September 2013
File:    MIKROC-ON-OFF.C
***************************************************************************/
// LCD module connections
sbit LCD_RS at RB4_bit;
sbit LCD_EN at RB5_bit;
sbit LCD_D4 at RB0_bit;
sbit LCD_D5 at RB1_bit;
sbit LCD_D6 at RB2_bit;
sbit LCD_D7 at RB3_bit;

sbit LCD_RS_Direction at TRISB4_bit;
sbit LCD_EN_Direction at TRISB5_bit;
sbit LCD_D4_Direction at TRISB0_bit;
sbit LCD_D5_Direction at TRISB1_bit;
sbit LCD_D6_Direction at TRISB2_bit;
sbit LCD_D7_Direction at TRISB3_bit;
// End LCD module connections

// Keypad module connections
char keypadPort at PORTC;
// End of keypad module connections

#define BUZZER PORTE.RE1
```

Figure 6.99: mikroC Pro for PIC Program.

```
#define HEATER PORTD.RD0
#define Enter 12
#define ON 1
#define OFF 0
//
// Start of MAIN program
//
void main()
{
    unsigned char MyKey,Txt[14];
    unsigned int SetPoint;
    unsigned char *op;
    unsigned int temp;
    float mV, PlantTemp;

    ANSELB = 0;                                 // Configure PORTB as digital
    ANSELC = 0;                                 // Configure PORTC as digital
    ANSELD = 0;                                 // Configure PORTD as digital
    ANSELE = 0;                                 // Configure PORTE as digital
    TRISA0_bit = 1;                             // Configure AN0 (RA0) as input
    TRISB = 0;                                  // PORTB are outputs (LCD)
    TRISC = 0xF0;                               // RC4–RC7 are inputs
    TRISD0_bit = 0;                             // RD0 is output
    TRISE1_bit = 0;                             // RE1 is output

    Keypad_Init();                              // Initialize keypad library
    Sound_Init(&PORTE,1);                       // Initialize Sound library
//
// Configure LCD
//
    Lcd_Init();                                 // Initialize LCD
    Lcd_Cmd(_LCD_CLEAR);
    Lcd_Out(1,1,"ON–OFF CONTROL");              // Display CALCULATOR
    Delay_ms(2000);                             // Wait 2 s
    Lcd_Cmd(_LCD_CLEAR);                        // Clear display

    BUZZER = OFF;                               // TURN OFF buzzer to start with
    HEATER = OFF;                               // Turn OFF heater to start with
//
// On startup read the set point temperature from the keypad
//
    Lcd_Out(1,1,"Enter Set Point");
    setPoint = 0;

    Lcd_Out(2,1,"SP: ");
    while(1)                                    // Get first no
    {
      do
         MyKey = Keypad_Key_Click();
      while(!MyKey);
      if(MyKey == Enter)break;                  // If ENTER pressed
```

Figure 6.99
cont'd

```
          if(MyKey == 10)MyKey = 0;              // If 0 key pressed
          Lcd_Chr_Cp(MyKey + '0');
          SetPoint = 10*SetPoint + MyKey;        // First number in Op1
        }

        Lcd_cmd(_LCD_CLEAR);                      // Clear LCD
        Lcd_Out(1,1,"SP = ");                     // Display SP=
        IntToStr(SetPoint,Txt);                   // Convert to string
        op = Ltrim(Txt);                          // Remove leading spaces
        Lcd_Out_CP(op);
        Lcd_Out(2,1,"ENTER to cont.");
//
// Wait until ENTER is pressed
//
        MyKey = 0;
        while(MyKey != Enter)
        {
          do
             MyKey = Keypad_Key_Click();
          while(!MyKey);
        }


// Program loop
//
    for(;;)                                       // Endless loop
      {
//
// Display the SetPoint and the Actual temperatures
//
        temp = ADC_Read(0);                       // Read from AN0 (RA0)
        mV = temp*5000.0/1024.0;                  // Convert to mV
        PlantTemp = mV/10.0;                      // Convert to degrees C
        Lcd_Cmd(_LCD_CLEAR);
        IntToStr(SetPoint,Txt);                   // Convert to string
        op = Ltrim(Txt);
        Lcd_Out(1,1,"SP=");                       // Display SP=
        Lcd_Out_Cp(op);                           // Display Set Point
        Lcd_Chr_Cp(178);
        Lcd_Chr_Cp('C');                          // Display C
        Lcd_Out(2,1,"AC=");                       // Display AC=
        FloatToStr(PlantTemp, Txt);               // Convert to string
        Lcd_Out_CP(Txt);
        Lcd_Chr_Cp(178);                          // Display Degree sign
        Lcd_Chr_Cp('C');                          // Display C character
//
// Implement the ON-OFF controller algorithm
//
        if(SetPoint > PlantTemp)                  // If SetPoint is bigger than actual
        {
          HEATER = ON;                            // Turn ON heater
```

Figure 6.99
cont'd

```
                    if((SetPoint − PlantTemp) > 2.0)        // If outside range
                       Sound_Play(1000,1000);               // Turn ON BUZZER
                 }
                 else                                        // If Set Point is not bigger than actual
                 {
                    HEATER = OFF;                            // Turn OFF heater
                    if((PlantTemp − SetPoint) > 2.0)         // Actual temp is bigger than setPoint
                       Sound_Play(1000,1000);                // Turn ON BUZZER
                 }
                 Delay_Ms(5000);                             // Wait 5 s and repeat
              }
        }
```

Figure 6.99
cont'd

The program then reads the desired temperature setting (Set Point temperature) from the keypad after displaying the following message and waiting for the user to enter the desired temperature:

> Enter Set Point
> SP:

After reading the desired temperature, the program displays a message to tell the user to press the ENTER key to continue. The remainder of the program is executed inside an endless loop that is repeated every 5 s.

Inside this loop, the program reads the plant temperature from analog channel 0 (AN0, or RA0) of the microcontroller and stores in the floating point variable PlantTemp in degrees centigrade. The Set Point and the actual plant temperature are then displayed in the following format (assuming the Set Point temperature is 25 °C, and the actual plant temperature is 20.45189 °C):

> SP = 25 °C
> AC = 20.45189 °C

The next part of the program implements the ON−OFF control algorithm. If the Set Point temperature is greater than the measured temperature, then the plant temperature is low, and HEATER is turned ON. If also the difference between the Set Point and the measured temperatures is >2 °C, then the BUZZER is sounded as an alarm. If the Set Point temperature is less than (or equal) to the measured temperature, then the HEATER is turned OFF. At the same time, if the difference between the Set Point and the measured values are >2 °C, then the BUZZER is sounded. The above process gets repeated every 5 s.

The program given in Figure 6.99 can be improved by the following modifications:

- The Set Point temperature can be stored in the electrically erasable programmable read-only memory of the microcontroller so that it does not need to be entered every time.

Figure 6.100: Sample Display 1. (For color version of this figure, the reader is referred to the online version of this book.)

Figure 6.101: Sample Display 2. (For color version of this figure, the reader is referred to the online version of this book.)

Figure 6.102: Sample Display 3. (For color version of this figure, the reader is referred to the online version of this book.)

Figure 6.103: Sample Display 4. (For color version of this figure, the reader is referred to the online version of this book.)

Figure 6.104: Sample Display 5. (For color version of this figure, the reader is referred to the online version of this book.)

- Instead of the simple ON−OFF, a more powerful control algorithm can be used (e.g. PID).
- Currently, the Set Point temperature must be an integer number. The keypad entry routine can be modified to accept floating point Set Point temperatures.

The program given in Figure 6.99 exceeds the free 2 K limit of the compiler, and users must have a licensed copy of the compiler to compile the program.

Figures 6.100−6.104 show sample displays from the project.

Advanced PIC18 Projects

Chapter Outline

In this chapter, we will be developing advanced projects using various peripheral devices and protocols such as Bluetooth, radiofrequency identification (RFID), WiFi, Ethernet, controller area network (CAN) bus, secure digital (SD) cards, universal serial bus (USB), and motor control. As in the previous chapter, the project description, hardware design, program description language (PDL), full program listing, and description of the program for each project will be given in detail.

Project 7.1—Bluetooth Serial Communication—Slave Mode

In this project, we shall be using a Bluetooth module to communicate with a personal computer (PC). The program will receive a text message from the PC and will display it on an liquid crystal display (LCD). In this project, our Bluetooth device is used as a slave device, and the PC is used as a master device.

Before going into the details of the project, it is worthwhile to review how Bluetooth devices operate.

The bluetooth is a form of digital communication standard for exchanging data over short distances using short-wavelength radiowaves in the industrial, scientific and medical (ISM) band from 2.402 to 2.489 GHz. The Bluetooth was originally conceived in 1994 as an alternative to the RS232 serial communications. Bluetooth communication occurs in the form of packets where the transmitted data are divided into packets, and each packet is transmitted using one of the designated Bluetooth channels. There are 79 channels, each with a 1-MHz bandwidth, starting from 2.402 GHz. The channels are hopped 1600 times per second using an adaptive frequency hopping algorithm. Because the communication is based on radiofrequency (RF), the devices do not have to be in the line of sight of each other in order to communicate.

Each Bluetooth device has a Media Access Control (MAC) address where communicating devices can recognize and establish a link if required.

Bluetooth communication operates in a master–slave structure, where one master can communicate with up to seven slaves. All the devices share the master's clock. Bluetooth devices in contact with each other form a *piconet*. At any time, data can be transferred between a master and a slave device. The master can choose which slave to communicate to. In the case of multiple slaves, the master switches from one slave to the next. Bluetooth is a secure way to connect and exchange data between various devices such as mobile phones, laptops, PCs, printers, faxes, global positioning system (GPS) receivers, and digital cameras.

Bluetooth's main characteristics can be summarized as follows:

- There are two classes of Bluetooth standards. The communication range is up to 100 m for Class 1, up to 10 m for Class 2, and up to 1 m for Class 3 devices.
- Class 1 devices consume 100 mW of power, Class 2 devices consume 2.5 mW, and Class 3 devices consume only 1 mW.
- The data rate is up to 3 Mbps.

The effective communication range depends on many factors, such as the antenna size and configuration, battery condition, and attenuation from walls.

Further information about Bluetooth communication standards can be obtained from many books, from the Internet, and from the Bluetooth Special Interest Group.

RN41 Bluetooth Module

In this project, the popular RN41 Bluetooth module, manufactured by Roving Networks, will be used. RN41 (Figure 7.1) is a Class 1 Bluetooth module delivering up to 3-Mbps data rate for distances up to 100 m. The module has been designed for easy interface to embedded systems. The basic features of the RN41 are as follows:

- Low-power operation (30 mA when connected, 250 μA in the sleep mode);
- Support for Universal Asynchronous Receiver Transmitter (UART) and USB data connection interfaces;
- On-board ceramic chip antenna;
- A 3.3-V operation;
- Baud rate from 1200 bps up to 921 kbps;
- A 128-bit encryption for secure communication;
- Error correction for guaranteed packet delivery.

RN41 is a 35-pin device. When operated using a UART interface, the following pins are of importance:

1:	GND
3:	GPIO6 (Set Bluetooth mode. 1 = auto master mode
4:	GPIO7 (Set Baud rate. 1 = 9600 bps, 0 = 115−kbps, or firmware setting)
5:	Reset (Active low)
11:	VDD (3.3-V supply)
12:	GND
13:	UART_RX (UART receive input)
14:	UART_TX (UART transmit output)

Figure 7.1: RN41 Bluetooth Module.

15: UART_RTS (UART RTS, goes high to disable host transmitter)
16: UART_CTS (UART CTS, if set high, it disables transmitter)
20: GPIO3 (autodiscovery = 1)
28: GND
29: GND

In low-speed interfaces, RTS and CTS pins are not used. Pin 3 is set to 1 for the auto master mode, pin 4 is set to 1 for 9600 Baud, Pin 5 is set to 1 for normal operation, and pin 20 is set to 1 for autodiscovery, GND and V_{DD} pins are connected to the ground and 3.3-V power supply lines. Thus, the module requires only two pins (pin 13 and pin 14) for interfacing to a microcontroller.

The PIC18F45K22 microcontroller operates with a +5-V power supply. The output logic high level from an I/O pin is minimum at +4.3 V. Similarly, the minimum input voltage to be recognized as logic high is +2.0 V.

The RN41 module operates with a +3.3-V power supply. The minimum output logic high level is $V_{DD} - 0.2 = +3.1$ V. Similarly, the maximum input logic high level is $V_{DD} + 0.4 = +3.7$ V.

The RN41 module cannot be connected directly to the PIC18F45K22 microcontroller. Although there is no problem with low logic levels, the output high voltage of +4.3 V of the microcontroller is much larger than the maximum allowable input high voltage of +3.7 V of RN41. Similarly, the minimum output high voltage of +3.1 V of the RN41 is just enough to provide the minimum high-level voltage required by the microcontroller inputs (minimum +2.0 V). In practice, +3.3- to +5.0-V and +5.0- to +3.3-V voltage converter chips are used in between the microcontroller and the RN41 module. A simple voltage converter circuit can be designed using a pair of transistors as switches to give both the required voltage level and also the correct logic polarity. Figures 7.2 and 7.3 show +5.0- to 3.3-V level converter and +3.3- to +5.0-V level converter circuits, respectively.

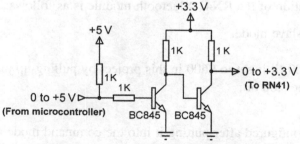

Figure 7.2: The +5- to +3.3-V Level Converter Circuit.

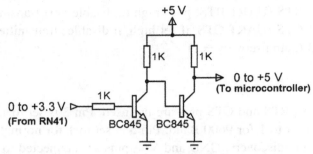

Figure 7.3: The +3.3- to +5.0-V Level Converter Circuit.

Project Hardware

The project uses the EasyBluetooth board (HW Rev. 1.02) manufactured by
mikroelektronica (www.mikroe.com). This is a small plug-in board having the following
specifications:

- RN41 Bluetooth module,
- Power select jumper (+5 or +3.3 V),
- A +5- to +3.3-V power regulator,
- Pull-up jumpers for pins GPIO3, GPIO4, GPIO6, and GPIO7,
- Voltage level converter transistor circuits,
- Dual in-line package switch for selecting signals for the microcontroller,
- IDC10 connector.

The RN41 Bluetooth module operates in two modes: data mode and command mode. The
data mode is the default mode, and in this mode, when the module receives data, it simply
strips the protocol headers and trailers and passes the raw data to the UART port.
Similarly, data to be sent out are famed by the addition of protocol headers and trailers
and are passed to the UART for transmission. Thus, the process of data communication is
transparent to the microcontroller.

The default configuration of the RN41 Bluetooth module is as follows:

- Bluetooth in the slave mode;
- Pin code: 1234;
- Serial port 115,200 (it is set to 9600 in this project by pulling-up pin GPIO7), eight data
  bits, no parity, one stop bit;
- No flow control.

The module can be configured after putting it into the command mode and sending
appropriate ASCII characters. There are two ways to put the module into the command

mode: Local communication with the module via the UART port and via the Bluetooth link. The new configuration takes effect after a reboot of the module.

In this book, we shall see how to configure the Bluetooth module via its UART port. This process requires the module ideally to be connected to a PC via its UART port and the use of a terminal emulator program (e.g. Hyperterm or mikroC pro for a PIC built-in terminal emulator). Once the module is rebooted, commands must be sent within 60 s (this can be changed if required).

If you are using the EasyPIC V7 development board, the Bluetooth module can be configured via the USB UART module of the development board. The steps are as follows:

- Configure the EasyBluetooth board jumpers as given below:
    DIL switch SW1: Set 1 to ON (this connects RN41 RX to UART TX)
    DIL switch SW1: Set 4 to ON (this connects RN41 TX to UART RX)
    DIL switch SW1: Set 7 to ON (this connects RN41 RESET to RC1)
    Set J1 to 5 V (EasyPIC V7 operates at +5 V)
    Set PI03 (GPI03) to Pull-Up
    Set PI07 (GPIO7) to Pull-Up (9600 baud)
- Configure the EasyPIC V7 development board as follows:
    SET Jumper J3 to USB UART
    Set Jumper J4 to USB UART
    DIL switch SW1: Set 1 to ON (RC7 to RX)
    DIL switch SW2: Set 1 to ON (RC6 to TX)
    J17 to GND (pin becomes low when button pressed)
    Insert a jumper at "Disable Protect" jumper position (near Button Press Level)
- When using the USB UART, the RX and TX pins should be reversed, and as a result, it is not possible to directly connect the EasyBluetooth board to PORTC of the EasyPIC V7 development board.

Connect an Insulation Displacement Connector (IDC) cable with two female connectors to PORTC and then make the following connections with wires between one end of the IDC cable and the EasyBluetooth board connector:

    Connect RC6 to pin P7 of EasyBluetooth board
    Connect RC7 to pin P6 of EasyBluetooth board
    Connect RC1 to pin P1 of EasyBluetooth board
    Connect VCC to pin VCC of EasyBluetooth
    Connect GND to pin GND of EasyBluetooth board

- Connect the USB UART port of the EasyPIC V7 development board to the PC USB port.
- Start the terminal emulator program on the PC and select the serial port, 9600 baud, 8 bits, no parity, 1 stop bit, and no flow control.

• Enter characters "$$$" on the terminal emulator window with no additional characters (e.g. no carriage return). The Bluetooth module should respond with characters "CMD". If there is no response from the Bluetooth module, reset the module by pressing button RC1 on the EasyPIC V7 development board and send the "$$$" characters again.

Figure 7.4 shows the response from the Bluetooth module, using the mikroC Pro for PIC terminal emulator. Note here that "Append New Line" box is disabled so that any other characters are not sent after the "$$$".

Now, enable the "Append New Line" by clicking the box and set the "New Line Setting" to CR (0x0D). Now, Enter character "h" to get a list of default configuration settings. Figure 7.5 shows a part of the display.

When a valid command is entered, the module returns string "AOK". If an invalid command is entered, the module returns "ERR", and "?" character for unrecognized commands. Some of the useful commands are given below. Command "—" exits from the command mode and returns to the data mode.

The various return codes have the following numeric values:

There are three types of commands: Set commands, Get commands, Change commands, Action commands, and General Purpose Input Output (GPIO) commands.

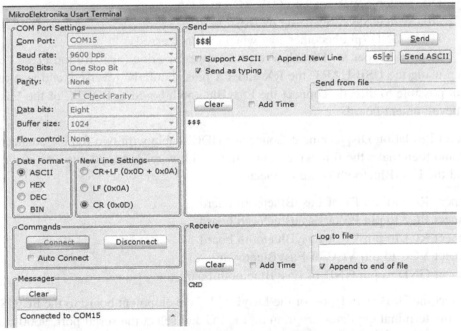

Figure 7.4: "CMD" Response from the Bluetooth Module.

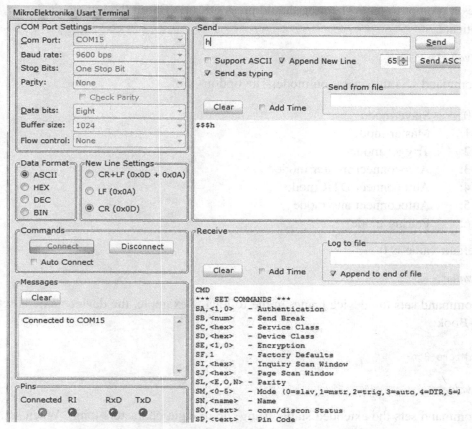

Figure 7.5: Part of the Default Configuration Settings.

Set commands: These commands store information to the flash memory. Changes take effect after a power cycle or reboot.

Get commands: These commands retrieve and display the stored information.

Change commands: These commands temporarily change the values of various settings.

Action commands: These commands perform action commands such as connections.

GPIO commands: These commands configure and manipulate the GPIO signals.

Some of the commonly used command examples are given below (details of the full command list can be obtained from the manufacturer's data sheet):

SA, <value>

This command forces authentication when a remote device attempts to connect. The <value> can be 0, 1, 2, or 4. The default value is 1 where the remote host receives a prompt to pair. The user should press OK or YES on the remote device to authenticate.

SM, <value>

This command sets the operation mode. The options are

  0:    Slave mode
  1:    Master mode
  2:    Trigger mode
  3:    Autoconnect master mode
  4:    Autoconnect DTR mode
  5:    Autoconnect any mode
  6:    Pairing mode

The default value is 0.

SN, <value>

This command sets the device name. In the following example, the device name is set to Micro-Book:

```
SN,Micro-Book
```

SO, <value>

This command sets the extended status string (up to eight characters long). When set, two status messages are sent to the local serial port: When a Bluetooth connection is established, the device sends the string <string>CONNECT. Also, when disconnecting, the device sends the string <string>DISCONNECT.

SP, <string>

This command sets the security pin code. The string can be up to 20 characters long. The default value is 1234.

SU, <value>

This command sets the Baud rate. Valid values are 1200, 2400, 4800, 9600, 19.2, 28.8, 38.4, 57.6, 115, 230, 460, or 921 K. The default value is either 115 K or 9600, set by the GPIO7 pin.

D

This command displays basic settings, such as the address, name, and pin code. Figure 7.6 shows an example.

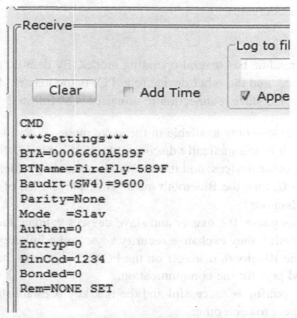

Figure 7.6: Displaying the Basic Settings with the D Command.

C, <address>

This command forces the device to connect to the specified remote address, where the address must be specified in hexadecimal format.

K

This command disconnects the current connection.

R,1

This command causes a reboot of the Bluetooth module.

I, <value1>, <value2>

This command performs a scan for a device. <value1> is the scan time, and 10 s is assumed if not specified. The maximum scan time is 48 s <value2> is the optional COD of the device.

SR, <hex value>

This command stores the remote address as a 12-digit hexadecimal code (6 bytes). Two additional characters can be specified with this command:

SR,I writes the last address obtained using the inquiry command. This option is useful when there is only one other Bluetooth device in the range.

SR,Z erases any stored addresses.

Making a Connection

The RN41 Bluetooth module has several operating modes. By default, the RN41 Bluetooth module is a slave device, and the other device (e.g. PC) is the master. Assuming that the device is configured as the slave, connection to a master is as follows:

- Discovery: This phase is only available in the slave mode. When we turn on the device in the slave mode, it is automatically discoverable. In this phase, the Bluetooth device is ready to pair with other devices, and it broadcasts its name, profile, and MAC address. If the master is a PC, then the Bluetooth manager displays a list of discoverable devices (if there is more than one).
- Pairing: During this phase, the master and slave devices validate the pin code, and if the validation is successful, they exchange security keys, and a link key is established. Double clicking the Bluetooth manager on the PC will pair with the device and create a virtual serial COM port for the communication.
- Connecting: If the pairing is successful and the link key is established, then the master and the slave connect to each other.

If the device is configured as the master, then the connection to a slave is as follows:

- The module makes connections when a connect command is received. The Bluetooth address of the remote node can be specified in the command string. The master mode is useful when we want to initiate connections rather than receive connections. Note that in the master mode the device is neither discoverable nor connectable.

The RN41 Bluetooth module also supports Autoconnect modes where the module makes connections automatically on power up and reconnects when the connection is lost.

Manual Connection Example

A manual connection example is given in this section to show the steps in connecting to a master device. Here, we assume that the master device is a PC, and the Bluetooth is enabled on the PC:

- Make sure that the Bluetooth on the PC is enabled to accept connections. Open Bluetooth Settings and enable the Discovery as shown in Figure 7.7.
- Configure and connect the EasyBluetooth board to PORTC as described earlier, configure the EasyPIC V7 development board, connect a cable from the USB UART on the development board to the PC USB port. Start the terminal emulation software on the PC. Get into the command mode (Press button RC1 and enter $$$ without any other characters). Name device as "Bluetooth-Slave" (optional, command SN), set extended

Figure 7.7: Enable the PC to Accept Connections.

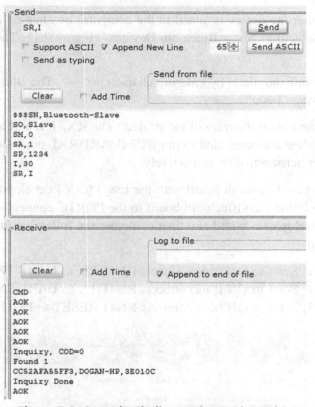

Figure 7.8: Steps in Finding a Bluetooth Device.

status string to "Slave" (optional, command SO), set the mode to slave (optional, command SM), enable authentication (optional, command SA), change the pass code to 1234 (command SP), look for Bluetooth devices (command I), and store the address of the found device (command SR). These steps are shown in Figure 7.8.

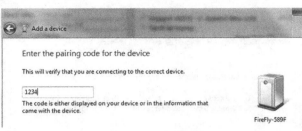

Figure 7.9: Entering the Pass Code.

- Connect (command C). Enter the pass code (default 1234) as shown in Figure 7.9. You should get a message to say that the device has been added to the computer (Figure 7.10).

You can check whether the device is added to the computer by clicking the "Bluetooth Devices" and then selecting "Show Bluetooth Devices" (Figure 7.11. The device name is "Bluetooth-Slave").

- Exit from the command menu by entering characters "—". The Bluetooth device should respond with string "END".

Figure 7.12 shows the circuit diagram of the project. The RX, TX, and RESET pins of the RN41 Bluetooth module are connected to pin RC7 (UART RX), pin RC6 (UART TX), and pin RC1 of the microcontroller, respectively.

If you are using the EasyBluetooth board with the EasyPIC V7 development board, you should directly plug in the EasyBluetooth board to the PORTC connector (no pin reversal necessary) of the development board and then configure the following jumpers on the EasyBluetooth board:

> DIL switch SW1: Set 1 to ON (this connects RN41 RX to UART TX)
> DIL switch SW1: Set 4 to ON (this connects RN41 TX to UART RX)
> DIL switch SW1: Set 7 to ON (this connects RN41 RESET to RC1)

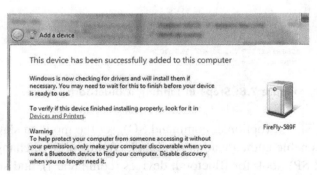

Figure 7.10: Device Added to the Computer.

Figure 7.11: Checking whether the New Device is Added to the Computer.

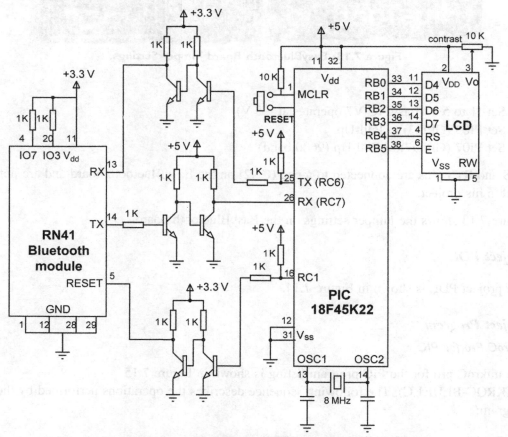

Figure 7.12: Circuit Diagram of the Project.

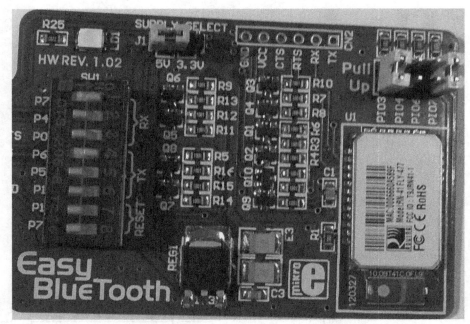

Figure 7.13: EasyBluetooth Board Jumper Settings.

Set J1 to 5 V (EasyPIC V7 operates at +5 V)
Set PI03 (GPI03) to Pull-Up
Set PI07 (GPIO7) to Pull-Up (9600 baud)

CTS and RTS pins are connected to a pad (CN2) on the EasyBluetooth board and are not used in his project.

Figure 7.13 shows the jumper settings on the EasyBluetooth board.

Project PDL

The project PDL is shown in Figure 7.14.

Project Program

mikroC Pro for PIC

The mikroC pro for the PIC program listing is shown in Figure 7.15 (MIKROC-BLUE1.C). The following sequence describes the operations performed by the program:

- Get into command mode (command $$$).

Main Program

BEGIN

Define the connection between the LCD and microcontroller
Configure PORTB and PORTC as digital
Configure RC1 as output
Enable UART interrupts
Initialize LCD
Initialize UART to 9600 baud
Reset the Bluetooth device
Get into command mode
Set device name
Set extended status
Set slave mode
Enable authentication
Set pass code
Wait until connected by the master
DO FOREVER
    IF message received flag is set
        Display message on the LCD
    ENDIF
ENDDO

END

BEGIN/INTERRUPT
    IF this is UART receive interrupt
        IF command mode
            Return command response
        ELSE
            Get message
            Set message received flag
        ENDIF
        Clear UART interrupt flag
    ENDIF
END/INTERRUPT

BEGIN/SEND_COMMAND
    DO WHILE correct response not received
        Send command to Bluetooth module
        Send carriage-return character
        Wait 500ms
    ENDDO
END

Figure 7.14: Project PDL.

- Configure the device by sending the following commands (these are optional and the defaults can be used if desired):

  Device name (command SN),

  Extended status (command SO),

```
/*******************************************************************************
                    BLUETOOTH COMMUNICATION
                    ==========================
```

This project is about using Bluetooth communication in a project. In this project a Bluetooth
module is used in slave mode. A PC is used in master mode. Messages sent to the slave module
are displayed on an LCD.

The Easy Bluetooth board (www.mikroe.com) is used in this project, connected to PORTC of an
EasyPIC V7 development board. An LCD is connected to PORTB of the microcontroller as in the
previous projects.

```
Author:  Dogan Ibrahim
Date:    September 2013
File:    MIKROC-BLUE1.C
*******************************************************************************/
// LCD module connections
sbit LCD_RS at RB4_bit;
sbit LCD_EN at RB5_bit;
sbit LCD_D4 at RB0_bit;
sbit LCD_D5 at RB1_bit;
sbit LCD_D6 at RB2_bit;
sbit LCD_D7 at RB3_bit;

sbit LCD_RS_Direction at TRISB4_bit;
sbit LCD_EN_Direction at TRISB5_bit;
sbit LCD_D4_Direction at TRISB0_bit;
sbit LCD_D5_Direction at TRISB1_bit;
sbit LCD_D6_Direction at TRISB2_bit;
sbit LCD_D7_Direction at TRISB3_bit;
// End LCD module connections

#define RESET PORTC.RC1

const CMD = 1;
const AOK = 2;
const CONN = 3;
const END = 4;

unsigned char Command_Mode, temp, Response, Data_Received;
unsigned char Buffer[6], Txt[16];
unsigned char Cnt = 0;
unsigned char i = 0;
//
// UART receive interrupt handler. In Command Mode we get the following responses:
// CMD, AOK, END, CONN. The interrupt handler returns responses in command mode
// and also stores the received data from the master
//
void interrupt(void)
{
  if(PIR1.RC1IF == 1)                           // Is this a UART receive interrupt ?
```

Figure 7.15: mikroC Pro for the PIC Program.

```
    {
      temp = UART1_Read();                                    // Read the received character
      if(Command_Mode == 1 && temp != 0x0)
      {
        Buffer[Cnt] = temp;
        Cnt++;
        if(Cnt == 4)
        {
          if(Buffer[0] == 'C' && Buffer[1] == 'O' && Buffer[2] == 'N' && Buffer[3] == 'N')
          {
            Response = CONN;
            Cnt = 0;
          }
        }

        if(Cnt == 5)
        {
          Cnt = 0;
          Response = 0;
          if(Buffer[0] == 'C' && Buffer[1] == 'M' && Buffer[2] == 'D' && Buffer[3] == 0x0D &&
             Buffer[4] == 0x0A)Response = CMD;
          if(Buffer[0] == 'A' && Buffer[1] == 'O' && Buffer[2] == 'K' && Buffer[3] == 0x0D &&
             Buffer[4] == 0x0A)Response = AOK;
          if(Buffer[0] == 'E' && Buffer[1] == 'N' && Buffer[2] == 'D' && Buffer[3] == 0x0D &&
             Buffer[4] == 0x0A)Response = END;
        }
      }
      else
      {
        if(temp == 0x0D)                                      // If END of data
        {
          Data_Received = 1;                                  // End of data received flag
          Txt[i] = 0x0;                                       // Terminate data with NULL
        }
        else
        {
          Txt[i] = temp;                                      // Store received data
          i++;                                                // Increment for next character
        }
      }
      PIR1.RC1IF = 0;                                         // Clear UART interrupt flag
    }
}


//
// Send a command to the Bluetooth Module. The first argument is the command string
// to be sent to the Bluetooth module. The second argument is the Response expected
// from the module (can be CMD, AOK, END, or CONN)
//
void Send_Command(char *msg, unsigned char Resp)
```

Figure 7.15
cont'd

```
  {
    do
    {
      UART1_Write_Text(msg);
      UART1_Write(0x0D);
      Delay_Ms(500);
    }while(Response != Resp);
  }



//
// Start of MAIN program
//
void main()
{
  ANSELB = 0;                              // Configure PORTB as digital
  ANSELC = 0;                              // Configure PORTC as digital
  TRISC1_bit = 0;                          // Configure RC1 as an output
//
// Enable UART receive interrupts
//
  PIE1.RC1IE = 1;                          // Clear UART1 interrupt flag
  INTCON.PEIE = 1;                         // Enable UART1 interrupts
  INTCON.GIE = 1;                          // Enable global interrupts

  LCD_Init();                              // Initialize LCD
  Lcd_Cmd(_LCD_CLEAR);                     // Clear LCD
  Lcd_Cmd(_LCD_CURSOR_OFF);                // Cursor off
  Lcd_Out(1,1,"Cmd Mode");                 // Display message

  Uart1_Init(9600);                        // Initialzie UART to 9600 Baud

  Command_Mode = 1;                        // We are getting into Command Mode
  RESET = 0;                               // Reset the Bluetooth module
  Delay_Ms(100);
  RESET = 1;
  Delay_Ms(1000);                          // End of Resetting the Bluetooth module

  do                                       // Send $$$ until in command mode
  {
    UART1_Write_Text("$$$");               // Get into command mode
    Delay_Ms(1000);
  }while(Response != CMD);

  Send_Command("SN,Bluetooth-Testing",AOK);  // Set device name
  Send_Command("SO,Slave", AOK);             // Set extended status
  Send_Command("SM,0", AOK);                 // Set into slave mode
  Send_Command("SA,1", AOK);                 // Enable authentication
  Send_Command("SP,1234", AOK);              // Set pass code
  Send_Command("---", END);                  // Exit command mode
```

Figure 7.15
cont'd

```
Lcd_Out(1,1,"Connecting");                // Display message
while(Response != CONN);                   // Wait until connected

Command_Mode = 0;                          // Now we are in Data mode
Data_Received = 0;

Lcd_Out(1,1,"Connected ");

for(;;)                                    // Display received messages on the LCD
{
  i = 0;
  while(Data_Received == 0);               // Wait until data up to CR is received
  Data_Received = 0;                       // Clear data receivd flag
  Lcd_Cmd(_LCD_CLEAR);                     // Clear LCD
  Lcd_Out(1,1,"Received Data:");           // Display "Received Data:" on first row
  Lcd_Out(2,1,Txt);                        // Displayed received data
  for(i = 0; i < 15; i++)Txt[i] = 0;       // Clear buffer for next time
  }
 }
}
```

Figure 7.15
cont'd

Slave mode (command SM),

Enable authentication (command SA),

Set pass code (command SP),

- Exit command mode (command "—").
- Wait for connection request from the master (PC).
- Read data from the master and display on the LCD.

At the beginning of the program, the connections between the LCD and the microcontroller are defined, symbol RESET is assigned to port pin RC1, and the various module responses are defined.

The main program configures PORTB and PORTC as digital, and RC1 pin is configured as an output pin. The LCD is initialized, the UART module is initialized to operate at 9600 baud, and the Bluetooth module is reset.

The program then puts the Bluetooth module into the command mode by sending characters "$$$" and waiting for the response string "CMD" to be received. This process is repeated until a response is received from the Bluetooth module. Once the correct response is received, commands are sent to the module to set the device name, extended status, mode, etc. Function Send_Command is used to send commands to the module. This function consists of the following code:

```
void Send_Command(char *msg, unsigned char Resp)
{
  do
  {
    UART1_Write_Text(msg);
```

```
        UART1_Write(0x0D);
        Delay_Ms(500);
    }while(Response != Resp);
}
```

The first argument is the command string, while the second argument is the expected response from the module. The last command to send is the "—", which exits from the command mode and waits for a connection from the master. Once a connection is established, the program waits until data are received from the master (until variable Data_Received becomes 1). The received text message in character array Txt is displayed on the second row of the LCD. Txt is then cleared, ready for then next message to be received.

Commands and data are received inside the UART data receive interrupt service routine (ISR). At the beginning of the ISR, the program checks to see whether or not the cause of the interrupt is actually the UART data reception. If this is the case, the received data are stored in variable temp. If we are in the command mode (Command_Mode = 1), then the received character must form part of the command response, and this character is stored in character array Buffer. After receiving four characters, the program checks to see if the received response is "CONN", and if so, the Response is set to CONN; otherwise, two more characters are received and the program checks to see if the response is one of "CMD", "AOK", or "END", followed by carriage-return (0x0D) and line-feed (0x0A) characters. The correct response is returned to the main program. If we are in the data mode (Command_Mode = 0), then the received character in temp is copied to character array Txt. If the end of data is detected (0x0D), then a NULL character is inserted at the end of Txt to turn it into a string.

The responses to various commands are as follows:

Command	Response
$$$	CMD<cr><lf>
SN<cr>	AOK<cr><lf>
SO<cr>	AOK<cr><lf>
SM<cr>	AOK<cr><lf>
SA<cr>	AOK<cr><lf>
SP<cr>	AOK<cr><lf>
—<cr>	END<cr><lf>

Testing the Program

In this section, we shall see how we can connect to a PC master device and receive a message from the PC and then display this message on the LCD. In this test, a Windows 7 PC is used.

- Make sure the Bluetooth adapter on the PC is turned on and the PC is set so that Bluetooth device can find the computer (open "Bluetooth Devices" in the hidden icons in the status bar and then "Open Settings" and configure as necessary if this is not the case).
- Compile and run the program. You should see the message "Connecting" on the LCD.

Figure 7.16: Device "Bluetooth Testing".

- Open "Bluetooth Devices" on the PC. Select "Add a Device". You should see the device name "Bluetooth Testing" (Figure 7.16). Double click on this device, and the program will ask you to enter the pairing code (Figure 7.17). Enter "1234" (Figure 7.18). The master should now connect to our slave device.

You are now ready to send messages to the slave device. The messages can be sent by finding the COM port that the Bluetooth is using on the PC for outgoing data. Open "Bluetooth Devices" on the PC. Then, "Open Settings", click "COM Ports" and see which port is assigned for outgoing data. In Figure 7.19, COM17 is the required port.

Now, we can send data via COM17 to our Bluetooth module. Start the Hyperterm terminal emulation software and type COM17 for "Connect using". Select 9600 Baud, 8 bits, no

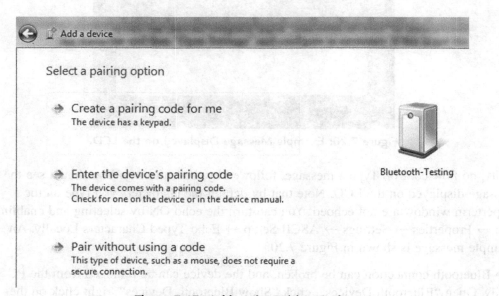

Figure 7.17: Asking the Pairing Code.

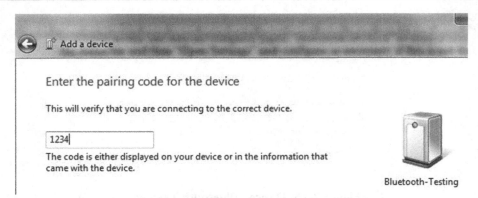

Figure 7.18: Entering the Pairing Code.

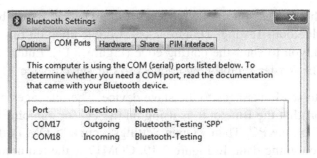

Figure 7.19: COM17 is Assigned for Outgoing Data.

Figure 7.20: Example Message Displayed on the LCD.

parity, no flow control. Type a message, followed by the Enter key. You should see the message displayed on the LCD. Note that by default the characters you type on the Hyperterm window are not echoed. You can turn the echo ON by selecting and enabling File → Properties → Settings → ASCII Setup → Echo Typed Characters Locally. An example message is shown in Figure 7.20.

The Bluetooth connection can be broken, and the device can be removed from the PC easily. Open "Bluetooth Devices", click "Show Bluetooth Devices", right click on the device, and select "Remove Device". Next time you connect you will need to enter the pass code again.

Modifications

The program in Figure 7.20 can be used for remote control applications. For example, the following modification shows how a message can be sent to the Bluetooth module to turn ON required light emitting diodes (LEDs) of PORTD. Here, for simplicity, we will assume that the message format is

```
PD = nnn
```

Where nnn is the three-digit decimal data (000−255) to be sent to PORTD.

Insert to the beginning of the program:

```
ANSELD = 0;    // Configure PORTD as digital
TRISD = 0;     // Configure PORTD as output
PORTD = 0;     // Clear PORTD to start with
```

Modify last part of the program as follows:

```
for(;)                              // Display received messages on the LCD
{
   i = 0;
   while(Data_Received == 0);       // Wait until data up to CR is received
   Data_Received = 0;               // Clear data received flag
   Lcd_Cmd(_LCD_CLEAR);             // Clear LCD
   Lcd_Out(1,1,"Received Data:");          // Display "Received Data:"
   Lcd_Out(2,1,Txt);                       // Displayed the received data
//
// Check for command PD = nnn and set PORTD accordingly
//
   if(Txt[0] == 'P' && Txt[1] == 'D' && Txt[2] == '=')
   {
      i = 100*(Txt[3]-'0') + 10*(Txt[4] - '0') + Txt[5] - '0';
      PORTD = i;
   }
   for(i = 0; i < 15; i++)Txt[i] = 0;      // Clear buffer for next time
}
```

As an example, entering command PD = 067 will turn ON LEDs 6, 1, and 0 of PORTD.

MPLAB XC8

The MPLAB XC8 program is similar, and the full program listing is shown in Figure 7.21 (XC8-BLUE1.C). Note that the LCD connections are different in the MPLAB XC8 version of the program as described in earlier MPLAB XC8 projects. In addition, since the MPLAB XC8 UART function putrsUSART sends the NULL character as well to the

```
/*********************************************************************************
                            BLUETOOTH COMMUNICATION
                            ===========================
```

This project is about using Bluetooth communication in a project. In this project a Bluetooth module is used in slave mode. A PC is used in master mode. Messages sent to the slave module are displayed on an LCD.

The Easy Bluetooth board (www.mikroe.com) is used in this project, connected to PORTC of an EasyPIC V7 development board. An LCD is connected to PORTB of the microcontroller as in the previous MPLAB XC8 projects.

```
Author:  Dogan Ibrahim
Date:    September 2013
File:    XC8-BLUE1.C
*********************************************************************************/
#include <xc.h>
#include <string.h>
#include <plib/usart.h>
#include <plib/xlcd.h>
#include <stdlib.h>
#pragma config MCLRE = EXTMCLR, WDTEN = OFF, FOSC = HSHP
#define _XTAL_FREQ 8000000

#define RST PORTCbits.RC1

unsigned char CMD = 1;
unsigned char AOK = 2;
unsigned char CONN = 3;
unsigned char END = 4;

unsigned char Command_Mode, temp, Response, Data_Received;
unsigned char Buffer[6], Txt[16];
unsigned char Cnt = 0;
unsigned char i = 0;
//
// UART receive interrupt handler. In Command Mode we get the following responses:
// CMD, AOK, END, CONN. The interrupt handler returns responses in command mode
// and also stores the received data from the master
//
void interrupt isr(void)
{
 if(PIR1bits.RC1IF == 1)                          // Is this a UART receive interrupt ?
 {
   temp = getc1USART();                           // Read the received character
   if(Command_Mode == 1 && temp != 0x0)
   {
     Buffer[Cnt] = temp;
     Cnt++;
     if(Cnt == 4)
     {
```

Figure 7.21: MPLAB XC8 Program.

```
            if(Buffer[0] == 'C' && Buffer[1] == 'O' && Buffer[2] == 'N' && Buffer[3] == 'N')
            {
              Response = CONN;
              Cnt = 0;
            }
          }

          if(Cnt == 5)
          {
            Cnt = 0;
            Response = 0;
            if(Buffer[0] == 'C' && Buffer[1] == 'M' && Buffer[2] == 'D' && Buffer[3] == 0x0D &&
               Buffer[4] == 0x0A)Response = CMD;
            if(Buffer[0] == 'A' && Buffer[1] == 'O' && Buffer[2] == 'K' && Buffer[3] == 0x0D &&
               Buffer[4] == 0x0A)Response = AOK;
            if(Buffer[0] == 'E' && Buffer[1] == 'N' && Buffer[2] == 'D' && Buffer[3] == 0x0D &&
               Buffer[4] == 0x0A)Response = END;
          }
        }
        else
        {
          if(temp == 0x0D)                            // If END of data
          {
            Data_Received = 1;                        // End of data received flag
            Txt[i] = 0x0;                             // Terminate data with NULL
          }
          else
          {
            Txt[i] = temp;                            // Store received data
            i++;                                      // Increment for next character
          }
        }
        PIR1bits.RC1IF = 0;                           // Clear UART interrupt flag
    }
}


//
// This function creates seconds delay. The argument specifies the delay time in seconds
//
void Delay_Seconds(unsigned char s)
{
  unsigned char i,j;

  for(j = 0; j < s; j++)
  {
    for(i = 0; i <  100; i++)__delay_ms(10);
  }
}

//
```

Figure 7.21
cont'd

```
// This function creates milliseconds delay. The argument specifies the delay time in ms
//
void Delay_Ms(unsigned int ms)
{
   unsigned int i;

   for(i = 0; i <  ms; i++)__delay_ms(1);
}


//
// This function creates 18 cycles delay for the xlcd library
//
void DelayFor18TCY( void )
{
Nop(); Nop(); Nop(); Nop();
Nop(); Nop(); Nop(); Nop();
Nop(); Nop(); Nop(); Nop();
Nop(); Nop();
return;
}


//
// This function creates 15 ms delay for the xlcd library
//
void DelayPORXLCD( void )
{
   __delay_ms(15);
   return;
}


//
// This function creates 5 ms delay for the xlcd library
//
void DelayXLCD( void )
{
   __delay_ms(5);
   return;
}


//
// This function clears the screen
//
void LCD_Clear()
{
   while(BusyXLCD());
   WriteCmdXLCD(0x01);
}
```

Figure 7.21
cont'd

```
//
// This function moves the cursor to position row,column
//
void LCD_Move(unsigned char row, unsigned char column)
{
 char ddaddr = 40*(row-1) + column;
 while( BusyXLCD() );
 SetDDRamAddr( ddaddr );
}


//
// This function sends commands to the Bluetooth module. The response is checked
// after sending a command
//
void Send_Command(const char msg[], unsigned char Resp)
{
   unsigned char i;

   do
   {
    i = 0;
    do
    {
       while(Busy1USART());                      // Check if UART is busy
       putc1USART(msg[i]);                        // Send to UART
       i++;
    }while(msg[i] != 0x00);                       // Until NULL terminator detected

       while(Busy1USART());
       putc1USART(0x0D);                          // Send carriage-return at the end
       Delay_Ms(500);
   }while(Response != Resp);
}



//
// Start of MAIN program
//
void main()
{
   ANSELB = 0;                                    // Configure PORTB as digital
   ANSELC = 0;                                    // Configure PORTC as digital
   TRISCbits.RC1 = 0;                             // Configure RC1 as an output
//
// Enable UART receive interrupts
//
   PIE1bits.RC1IE = 1;                            // Clear UART1 interrupt flag
   INTCONbits.PEIE = 1;                           // Enable UART1 interrupts
   INTCONbits.GIE = 1;                            // Enable global interrupts
```

Figure 7.21
cont'd

```
//
// Configure the LCD to use 4-bits, in multiple display mode
//
    Delay_Seconds(1);
    OpenXLCD(FOUR_BIT & LINES_5X7);

    Open1USART( USART_TX_INT_OFF    &                    // Initialize UART
            USART_RX_INT_ON    &
            USART_ASYNCH_MODE    &
            USART_EIGHT_BIT    &
            USART_CONT_RX    &
            USART_BRGH_LOW,
            12);
    while(BusyXLCD());                                    // Wait if the LCD is busy
    WriteCmdXLCD(DON);                                    // Turn Display ON
    while(BusyXLCD());                                    // Wait if the LCD is busy
    WriteCmdXLCD(0x06);                                   // Move cursor right
    LCD_Clear();                                          // Clear LCD
    LCD_Move(1,1);
    putrsXLCD("Cmd Mode");                                // Display message

    Command_Mode = 1;                                    // Getting into Command Mode
    RST = 0;                                             // Reset the Bluetooth module
    Delay_Ms(100);
    RST = 1;
    Delay_Seconds(1);                                    // End of Reset

    do                                                   // Get into command mode
    {
      putc1USART('$');    // Get into command mode
      while(Busy1USART());
      putc1USART('$');
      while(Busy1USART());
      putc1USART('$');
      Delay_Seconds(1);
    }while(Response != CMD);


    Send_Command("SN,BLUETOOTH2",AOK);                   // Set device name
    Send_Command("SO,SLAVE",AOK);                        // Set extended status
    Send_Command("SM,0",AOK);                            // Set into slave mode
    Send_Command("SA,1",AOK);                            // Enable authentication
    Send_Command("SP,1234",AOK);                         // Set pass code
    Send_Command("---", END);                            // Exit command mode
    LCD_Move(1,1);
    putrsXLCD("Connecting");                             // Display message
    while(Response != CONN);                             // Wait until connected

    Command_Mode = 0;                                    // Now we are in Data mode
    Data_Received = 0;
```

Figure 7.21
cont'd

```
            LCD_Move(1,1);
            putrsXLCD("Connected ");
//
// The received message is displayed inside this loop
//
        for(;;)                                              // Endless loop
        {
          i = 0;
          while(Data_Received == 0);                         // Wait until CR is received
          Data_Received = 0;                                 // Clear data received flag
          LCD_Clear();                                       // Clear LCD
          LCD_Move(1,1);
          putrsXLCD("Received Data:");                       // Display "Received Data:" on
          LCD_Move(2,1);
          putrsXLCD(Txt);                                    // Display the received data
          for(i = 0; i < 15; i++)Txt[i] = 0;                 // Clear buffer for next time
        }
}
```

Figure 7.21
cont'd

UART, we have used the putcUSART function to send characters to the Bluetooth module with no additional bytes.

Project 7.2—Bluetooth Serial Communication—Master Mode

In this project, we shall be using a Bluetooth module to communicate with a PC. The program will connect to a PC and then send a text message to the PC, which will be displayed on the screen. In this project, our Bluetooth device is used as a master device and the PC is used as a slave device.

This project is very similar to the previous project. Here, after a connection is established, the message "Bluetooth Test" will be sent to the master device every second. The LCD is used to display various messages and also the text sent to the slave device.

Project Hardware

The circuit diagram and the hardware configuration of this project is the same as the one given in the previous project.

Project PDL

The project PDL is given in Figure 7.22.

MAIN PROGRAM:

BEGIN

    Define connection between LCD and microcontroller
    Configure PORTB and PORTC as digital outputs
    Enable UART receive interrupts
    Initialize LCD
    Initialize UART to 9600 Baud
    Get into command mode
    Set Device name (command SN)
    Extended status (command SO)
    Slave mode (command SM)
    Enable authentication (command SA)
    Set pass code (command SP)
    Look for Bluetooth devices (command I)
    Store the address of the found Bluetooth device (command SR)
    Connect to the slave device (command C)
    DO FOREVER
        Send text "Bluetooth Test" to the slave device
        Display text "Bluetooth Test" on the LCD
        Wait 1 second
        Cleat LCD
    ENDDO
END

Interrupt Service Routine:

BEGIN

    IF this is a UART receive interrupt
        Read received character
        ELSE IF command mode
            Return CMD
        ELSE IF AOK
            Return AOK
        ELSE IF Done
            Return IDone
        ELSE IF CONNECT
            Return CONN
        ENDIF
    ENDIF
END

BEGIN/ Send_Command
    Send text to UART
    Send carriage-return character to UART
    Wait 500ms
END/Send_Command

Figure 7.22: Project PDL.

Project Program

mikroC Pro for PIC

The mikroC pro for PIC program listing is shown in Figure 7.23 (MIKROC-BLUE2.C). The following sequence describes the operations performed by the program:

- Get into command mode (command $$$).
- Configure the device by sending the following commands (these are optional and the defaults can be used if desired):
  - Device name (command SN);
  - Extended status (command SO);
  - Slave mode (command SM);
  - Enable authentication (command SA);
  - Set pass code (command SP);
  - Look for Bluetooth devices (command I);
  - Store the address of the found Bluetooth device (command SR);
  - Connect to the slave device (command C);
  - Send text "Bluetooth Test" to the slave device every second. Display the sent message on the LCD.

At the beginning of the program, the connections between the LCD and the microcontroller are defined, symbol RESET is assigned to port pin RC1, and the various module responses are defined.

The main program configures PORTB and PORTC as digital, and the RC1 pin is configured as an output pin. The LCD is initialized, the UART module is initialized to operate at 9600 baud, and the Bluetooth module is reset.

The program then puts the Bluetooth module into the command mode by sending characters "$$$" and waiting for the response string "CMD" to be received. This process is repeated until a response is received from the Bluetooth module. Once the correct response is received, commands are sent to the module to set the device name, extended status, mode, etc. Function Send_Command is used to send commands to the module as in the previous project. Command "I,30" looks for Bluetooth devices, and if a device is found, it gets its address. Command "SR,I" stores the address of the device just found. Command "C" connects to a slave device. The program then enters in a loop and sends message "Bluetooth Test" to the slave device every second.

Commands and data are received inside the UART data receive ISR. The ISR implemented here is slightly different from the one given in Figure 7.15. At the beginning of the ISR, the program checks to see whether or not the cause of the

```
/*******************************************************************
                    BLUETOOTH COMMUNICATION
                    ===========================

This project is about using Bluetooth communication in a project. In this project a Bluetooth
module is used in master mode. A PC is used in slave mode. Text "Bluetooth Test" is sent to
the slave device every second.

The Easy Bluetooth board (www.mikroe.com) is used in this project, connected to PORTC of
an EasyPIC V7 development board. An LCD is connected to PORTB of the microcontroller as in
the previous projects.

Author:  Dogan Ibrahim
Date:    September 2013
File:    MIKROC-BLUE2.C
*******************************************************************/
// LCD module connections
sbit LCD_RS at RB4_bit;
sbit LCD_EN at RB5_bit;
sbit LCD_D4 at RB0_bit;
sbit LCD_D5 at RB1_bit;
sbit LCD_D6 at RB2_bit;
sbit LCD_D7 at RB3_bit;

sbit LCD_RS_Direction at TRISB4_bit;
sbit LCD_EN_Direction at TRISB5_bit;
sbit LCD_D4_Direction at TRISB0_bit;
sbit LCD_D5_Direction at TRISB1_bit;
sbit LCD_D6_Direction at TRISB2_bit;
sbit LCD_D7_Direction at TRISB3_bit;
// End LCD module connections

#define RESET PORTC.RC1

const CMD = 1;
const AOK = 2;
const CONN = 3;
const IDone = 4;

unsigned char Response, temp;
unsigned char Buffer[60];
unsigned char i,Cnt = 0;
//
// UART receive interrupt handler. In Command Mode we get the following responses:
// CMD, AOK, Done, CONN. The interrupt handler returns responses in command mode
// and also stores the received data from the master
//
void interrupt(void)
{
  if(PIR1.RC1IF == 1)                          // Is this a UART receive interrupt ?
  {
```

Figure 7.23: mikroC Pro for the PIC Program.

```
            temp = UART1_Read();                      // Read the received character
            if(temp != 0x0)
            {   if(temp == 0x0A)
               {
                  if(Buffer[Cnt-4] == 'C' && Buffer[Cnt-3] == 'M' && Buffer[Cnt-2] == 'D' &&
                     Buffer[Cnt-1] == 0x0D)
                  {
                     Response = CMD;
                     Cnt = 0;
                  }
                  if(Buffer[Cnt-4] == 'A' && Buffer[Cnt-3] == 'O' && Buffer[Cnt-2] == 'K' &&
                     Buffer[Cnt-1] == 0x0D)
                  {
                     Response = AOK;
                     Cnt = 0;
                  }
                  if(Buffer[Cnt-5] == 'D' && Buffer[Cnt-4] == 'o' && Buffer[Cnt-3] == 'n' &&
                     Buffer[Cnt-2] == 'e' && Buffer[Cnt-1] == 0x0D)
                  {
                     Response = IDone;
                     Cnt = 0;
                  }
                  if(Buffer[Cnt-8] == 'C' && Buffer[Cnt-7] == 'O' && Buffer[Cnt-6] == 'N' &&
                     Buffer[Cnt-5] == 'N' && Buffer[Cnt-4] == 'E' && Buffer[Cnt-3] == 'C' &&
                     Buffer[Cnt-2] == 'T')
                  {
                     Response = CONN;
                     Cnt = 0;
                  }
                  Cnt=0;
               }
               else
               {
                  Buffer[Cnt] = temp;
                  Cnt++;
               }
            }
         }
         PIR1.RC1IF = 0;                               // Clear UART interrupt flag
      }


//
// Send a command to the Bluetooth Module. The first argument is the command string
// to be sent to the Bluetooth module. The second argument is the Response expected
// from the module (can be CMD, AOK, END, or CONN)
//
void Send_Command(char *msg, unsigned char Resp)
{
  do
  {
```

Figure 7.23
cont'd

```
            UART1_Write_Text(msg);
            UART1_Write(0x0D);
            Delay_Ms(500);
    }while(Response != Resp);
}


//
// Start of MAIN program
//
void main()
{
    ANSELB = 0;                                  // Configure PORTB as digital
    ANSELC = 0;                                  // Configure PORTC as digital
    TRISC1_bit = 0;                              // Configure RC1 as an output
//
// Enable UART receive interrupts
//
    PIE1.RC1IE = 1;                              // Clear UART1 interrupt flag
    INTCON.PEIE = 1;                             // Enable UART1 interrupts
    INTCON.GIE = 1;                              // Disable global interrupts

    LCD_Init();                                  // Initialize LCD
    Lcd_Cmd(_LCD_CLEAR);                         // Clear LCD
    Lcd_Cmd(_LCD_CURSOR_OFF);                    // Cursor off
    Lcd_Out(1,1,"Cmd Mode");                     // Display message
    Uart1_Init(9600);                            // Initialize UART to 9600 Baud

    RESET = 0;                                   // Reset the Bluetooth module
    Delay_Ms(100);
    RESET = 1;
    Delay_Ms(1000);                              // End of Resetting the Bluetooth module

    do                                           // Send $$$ for command mode
    {
        UART1_Write_Text("$$$");                 // Get into command mode
        Delay_Ms(1000);
    }while(Response != CMD);

    Lcd_Out(1,1,"Send Commands");
    Send_Command("SN,Bluetooth-Master",AOK);     // Set device name
    Send_Command("SO,Master", AOK);              // Set extended status
    Send_Command("SM,1", AOK);                   // Set into master mode
    Send_Command("SA,1", AOK);                   // Enable authentication
    Send_Command("SP,1234", AOK);                // Set pass code
    Lcd_Out(1,1,"Look For Devices");
    Lcd_Out(2,1,"Wait 30 secs");

    UART1_Write_Text("I,30");                    // Look for Bluetooth devices (wait 30 s)
    UART1_Write(0x0D);                           // Send carriage-return
```

Figure 7.23
cont'd

```
            while(Response != IDone);                        // Wait up to 30 s

            Lcd_Cmd(_LCD_CLEAR);
            Lcd_Out(1,1,"Device Found ");                    // Display message
            Send_Command("SR,I", AOK);                       // Store address of the device just found

            UART1_Write_Text("C");
            UART1_Write(0x0D);
            while(Response != CONN);                          // Connect to the slave just found

            for(;;)
            {
              Uart1_Write_Text("Bluetooth Test");             // Send text to the slave
              Uart1_Write(0x0D);                              // Send carriage-return
              LCD_Out(1,1,"Sent Message:");
              Lcd_Out(2,1,"Bluetooth Test");
              Delay_Ms(1000);
              Lcd_Cmd(_LCD_CLEAR);
            }
        }
```

Figure 7.23
cont'd

interrupt is actually the UART data reception. The command responses are terminated with the line-feed character 0x0A. The program stores the received characters in array Buffer until the line-feed character is received. Then, the program checks to see what type of response this is and sets variable Response accordingly. Response is set to CMD, AOK, Done, or CONNECT.

The Bluetooth device returns the following responses when a command is sent:

Command	Response
$$$	CMD<cr><lf>
SN<cr>	AOK<cr><lf>
SO<cr>	AOK<cr><lf>
SM<cr>	AOK<cr><lf>
SA<cr>	AOK<cr><lf>
SP<cr>	AOK<cr><lf>
—<cr>	END<cr><lf>
I<cr>	Inquiry, COD=0<cr><lf>
	Found n<cr><lf>
	<address>,<name>,<COD>,<serial port><cr><lf>
	Inquiry Done<cr><lf>
SR<cr>	AOK<cr><lf>
C<cr>	TRYING<cr><lf>
	MasterCONNECT<cr><lf>

Testing the Program

In this section, we shall see how we can connect to a PC slave device and send a message from our Bluetooth device. In this test, a Windows 7 PC is used.

- Make sure the Bluetooth adapter on the PC is turned on and the PC is set so that the Bluetooth device can find the computer (open "Bluetooth Devices" in hidden icons in the status bar and then "Open Settings" and configure as necessary if this is not the case). You should also enable the option "Alert me when a new Bluetooth device wants to connect".
- Compile and run the program.
- You should see these messages on the LCD: "CMD Mode", "Send Commands", "Look For Devices", "Wait 30 s". The program will look for Bluetooth devices. This may take up to 30 s, and you should wait until the message "Device Found" is displayed on the LCD.
- At this stage, the PC will display an alert message to say that a new Bluetooth device wants to connect (Figure 7.24). Click the message and then enter the pass code 1234 (Figure 7.25). Open a terminal emulation session and enter the COM port number (in Bluetooth settings), and connect with 9600 baud, 8 bits, no parity, no flow control. The two devices will connect, and you should see the message "Bluetooth Test" displayed on the terminal emulation window.

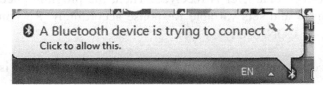

Figure 7.24: Alert from the PC.

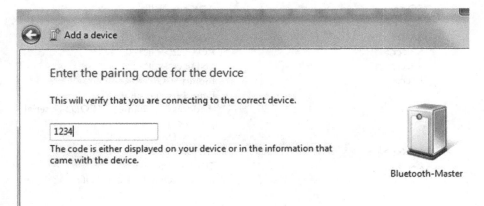

Figure 7.25: Entering the Pass Code.

You should now try as an exercise to use two EasyBluetooth boards, one configured as the master and one as the slave, and connect them to each other to exchange data.

Project 7.3—Using the RFID

In this project, we shall be using an RFID receiver to read the unique identifier (UID) number stored on an RFID tag.

Before going into the details of the project, it is worthwhile to review the basic principles of the RFID.

Radiofrequency Identification

RFID involves the use of RF electromagnetic waves to transfer data without making any contact to the source of the data. Generally, an RFID system consists of two parts: A Reader and one or more transponders (also known as Tags) that carry the data. RFID systems evolved from barcode labels as a means of identifying an object. A comparison of RFID systems with barcodes reveals the following:

- Barcodes are read only. Most RFID systems are read write.
- Barcodes are based on optical technology and may be affected from environmental lighting, making the reliable reading distance not more than several feet. RFID systems are based on RF waves and are not affected by environmental lighting. The reading distance of an RFID system can be $\geq 20-40$ ft.
- Barcode images are normally printed on paper, and their readability is affected by aging and the state of the paper, for example, dirt on the paper and torn paper. RFID systems do not suffer from environmental lighting, but their operation may be affected if attached to metals.
- Barcodes can be generated and distributed electronically, for example, via e-mail and mobile phone. For example, boarding passes with barcodes can be printed.
- Barcode reading is more labor intensive as it requires the light beam to be directed onto them. RFID readers on the other hand can read the data wirelessly and without touching the tags.
- Barcodes are much cheaper to produce and use than RFID systems.

Some common uses of RFID systems are as follows:

- Tracking of goods,
- Tracking of persons and animals,
- Inventory systems,
- Public transport and airline tickets,
- Access management,

- Passports,
- Hospitals and healthcare,
- Libraries,
- Museums,
- Sports,
- Defense,
- Shoplifting detection.

There are three types of RFID tags: Active, Passive, or Semipassive. Passive tags have no internal power sources, and they draw their power from the electromagnetic field generated by the RFID reader (Figure 7.26). Passive tags have no transmitters; they simply alter the electromagnetic field emitted by the reader such that the reader can detect. Because the reader has no transmitter, the range of passive tags is limited to several feet. Passive tags have the advantage that their cost is low compared to other types of tags.

Active tags have their own power sources (small batteries). These tags also have their own transmitters. As a result, active RFID systems have much greater detection ranges, usually a few hundred feet. To extend the battery life, these tags are normally in a low-power state until they detect the electromagnetic waves transmitted by the RFDI receiver. After they leave the electromagnetic field, they return back to the low-power mode.

Semipassive tags have their own power sources (small batteries). But the power source is used just to power the microchip embedded inside the tag. Like the passive tags, these tags do not have any transmitters, and they rely on the same principle as the passive tags for transferring their data. Semipassive tags have greater detection ranges than the passive tags.

RFID tags can be read only where a fixed serial number is written on the tag, and this number is used by the reader to identify the tag, or they may be read write where data can be written onto the tag by the user. Some tags are write once, where the user can only write once onto the tag but read as many times as required.

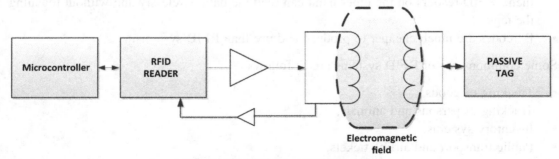

Figure 7.26: Passive RFID System.

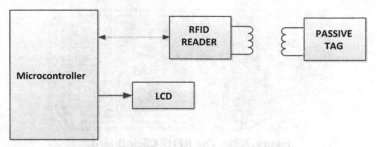

Figure 7.27: Block Diagram of the Project.

The data on the tag are stored in a nonvolatile memory. This may be a small memory that, for example, stores just a serial number, or larger memory capable of storing, for example, product-related information, or a person's details.

An RFID tag contains at least two parts: an antenna for receiving (and transmitting on some tags) the signal and a microchip with memory for controlling all operations of the tag and for storing the data.

There are tags that operate in the LF, HF, or the UHF bands. Usually, every country sets its own rules and the allocation of frequencies. LF tags are generally passive, operating in the 120- to 150-kHz band, and their detection range is not >10 cm. The 13.56-MHz ISM HF band is very popular for passive tags, offering good detection ranges up to 1 m. The 433-MHz ultra high frequency (UHF) band tags are active, offering detection ranges of over several hundreds of meters but having higher costs. The 865—868 UHF ISM band tags are passive with ranges up to 10 m, and having very low costs. The microwave band tags are active with high data rates and usually high ranges, but their cost is high.

There are several standards that control and regulate the design and development of RFID based products. It is important that the tag we use is compatible with the RFID receiver we are using. Also, the receivers generally support various standards and the selection of a particular standard is generally programmable. We should also make sure that the receiver and the tag use the same standard.

The block diagram of the project is shown in Figure 7.27. An RFID receiver chip is connected to the microcontroller. The microcontroller reads the serial number (UID) on the tag and displays on the LCD.

Project Hardware

The project uses the RFID Click board manufactured by mikrolektronika (Figure 7.28). This is a small mikroBUS compatible board, featuring the CR95HF 13.56-MHz RFID

Figure 7.28: The RFID Click Board.

transceiver as well as the trace antenna. The board is designed to operate with a 3.3-V supply voltage, and it can communicate with the microcontroller using one of several busses, such as UART and serial peripheral interface (SPI). The basic specifications of the CR95HF chip are as follows:

- Support for reading and writing;
- A 13.56-MHz frequency, supporting the following standards:
    ISO/IEC 14443 Types A and B
    ISO/IEC 15693
    ISO/IEC 18092
- Host interface for UART, SPI, and INT;
- A 32-pin VFQFPN package;
- A 3.3-V operation.

In this project, we shall be using an RFID tag compatible with the standard ISO/IEC 14443 Type A. We shall be using the UART interface for simplicity. The CR95HF pins used while operating in the UART mode are as follows:

Pin	Pin name	Pin Description
1	TX1	Driver output to the coil
2	TX2	Driver output to the coil
5	RX1	Receiver input from the coil
6	RX2	Receiver input from the coil
8	GND	Ground
12	UART_RX/IRQ_IN	UART receive pin + Interrupt input
13	VPS	Power supply
14	UART_TX/IRQ_OUT	UART transmit pin + interrupt output
19	SSI_0	Select comms interface
20	SSI_1	Select comms interface
22	GND	Ground
29	XIN	Crystal input
30	XOUT	Crystal output
31	GND	Ground
32	VPS_TX	Power supply

The CR95HF chip requires a 27.12-MHz crystal to be connected between its XIN-XOUT pins together with a pair of 10-pF capacitors. These are included on the RFID Click board.

Figure 7.29 shows the circuit diagram of the project. The UART pins of the CR95HF chip, UART_RX and UART_TX are connected to UART pins RC6 (TX) and RC7 (RX) of the microcontroller, respectively. The antenna is connected to pins TX1-TX2 and RX1-RX" of the CR95HF. An LCD is connected to PORTB as in the earlier projects.

If you are using the RFID Click board together with the EasyPIC V7 development board, then configure the RFID board as follows (see the RFID Click board schematic):

Solder the jumpers in position B (to use the UART interface).
Connect the RFID board to mikroBUS 1 connector on the development board.

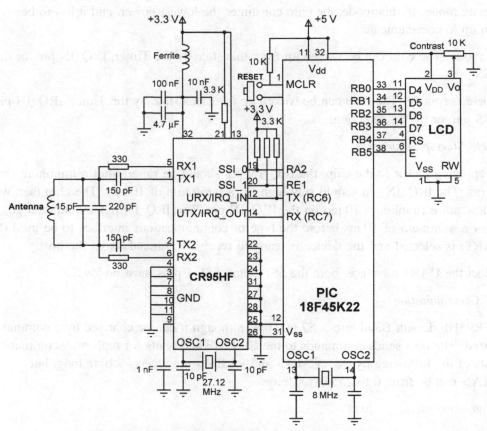

Figure 7.29: Circuit Diagram of the Project.

The connections between the microcontroller and the RFID board are as follows (except the power pins):

RFID Board	Microcontroller
UART_RX/IRQ_IN	RC6
UART_TX/IRQ/OUT	RC7
SSI_0	RA2
SSI_1	RE1

CR95HF Operational Modes

The CR95HF chip operates in two modes: Wait For Event (WFE) mode and Active mode. The WFE mode includes four low-power states: Power-up, Hibernate, Sleep, and Tag Detector. The chip cannot communicate with the external host (e.g. a microcontroller), while in one of these states, it can only be woken-up by the host. In the Active mode, the chip communicates actively with a tag or an external host.

Power-up mode: This mode is entered after power is applied to the chip.

Hibernate mode: In this mode, the chip consumes the lowest power, and it has to be woken up to communicate.

Sleep mode: The chip can be woken up from this state by the Timer, IRQ_IN pin, or the SPI_SS pin.

Tag Detector mode: The chip can be woken up from this state by the Timer, IRQ_IN pin, SPI_SS pin, or by tag detection.

CR95HF Startup Sequence

After applying power to the chip, the IRQ_IN pin should be raised after a minimum time of 10 ms. The IRQ_IN pin should stay high for a minimum of 100 µs. The chip then waits for a low pulse (minimum 10 µs) on the IRQ_IN pin. The IRQ_IN pin should then go high for a minimum of 10 ms before the type of communications interface to be used (SPI or UART) is selected and the device is ready to receive commands from the host.

To select the UART interface, both the SSI_0 and SSI_1 pins must be low.

UART Communication

The CR95HF default Baud rate is 57,600 bps, although it can be changed by a command if desired. The host sends commands to the CR95HF and waits for replies. A command consists of the following bytes. <CMD> and <LEN> are always 1 byte long, but <DATA> can be from 0 to 255 bytes long:

```
<CMD><LEN><DATA.....DATA>
```

Table 7.1: List of Commands.

Code	Command	Description
0x01	IDN	Request short information and revision of the CR95HF
0x02	PROTOCOL_SELECT	Select the required RF protocol and its parameters
0x04	SEND_RECV	Send data and receive tag response
0x07	IDLE	Switch CR95HF into the low-power WFE mode, specify the wake-up source, and wait for an event to exit to ready state
0x08	RD_REG	Read wake-up event register or the analog ARC_B register
0x09	WR_REG	Write to ARC_B, Timer window, or the AutoDetect filter enable register (for ISO/IEC 18092 tags)
0x0A	BAUD_RATE	Write the UART baud rate
0x55	ECHO	CR95HF returns echo response (0x55)

Where CMD is the command type, LEN is the length of the command, and DATA are the data bytes. If the LEN field is zero, no data will be sent or received.

The response from the CR95HF is in the following format:

```
<ResponseCode><LEN><Data....DATA>
```

A list of valid commands is given in Table 7.1 (see the CR95HF Data Sheet for further information and command examples). The use of some of these commands is given later in the programming section.

Passive RFID Tags

Passive RFID tags are available in many shapes, forms, frequencies, and capacities. The format of a very simple popular read-only passive tag, known as EM4100, is given here for reference (this is not the one used in this project). One form of the EM4100 card is shown in Figure 7.30, although they are also available in credit-card shape.

Figure 7.30: Sample EM4100 RFID Tag.

```
1   1   1   1   1   1   1   1   1          HEADER
V00 V01 V02 V03 P0          Version number (or customer ID)+even parity bit (P0)
V04 V05 V06 V07 P1          Version number (or customer ID)+even parity bit (P1)
D08 D09 D10 D11 P2  ┐
D12 D13 D14 D15 P3  │
D16 D17 D18 D19 P4  │
D20 D21 D22 D23 P5  ├   32 Data Bits (D08:D39) + even parity bits (P2:P9)
D24 D25 D26 D27 P6  │
D28 D29 D30 D31 P7  │
D32 D33 D34 D35 P8  │
D36 D37 D38 D39 P9  ┘
PC0 PC1 PC2 PC3 STP         Column parity bits (PC0:PC3) + stop bit (STP)
```

Figure 7.31: EM4100 Tag Data Format.

The EM4100 tag consists of a 64-bit Read-Only Memory. This means that the tag is configured during the manufacturing process and the data on the tag cannot be changed.

The format of the EM4100 tag is as follows:

- The first 9 bits is all 1 s, and this is the header field.
- Next, we have 11 groups of 5 bits of data. In the first 10 groups, the first 4 bits are the data bits, while the last bit is the even parity bit. In the last group, the first 4 bits are the column parity bits, while the last bit is the stop bit. The first 8 bits are the version number (or customer ID).

The format of the data is shown in Figure 7.31.

An example tag string with its decoding is given below:

```
111111111        Header
00000            0
11110            F
00000            0
00000            0
00011            1
00011            1
01010            5
01010            5
01100            6
10100            A
11000            (column parities and stop-bit)
```

The version number (or customer ID) = 0x0F.

Data string = 0x0011556A.

The tag used in this project is called the MIFARE MF1ICS50, manufactured by NXP. This is the card supplied by mikroElektronika for use with their RFID Click board.

This tag has the following specifications (see manufacturers' data sheet for further information):

- A 13.56-MHz operating frequency,
- A 106-kbps data transfer rate,
- Up to a 100-mm detection range,
- A 1-kbyte electrically erasable programmable read-only memory (EEPROM), organized in 16 sectors with four blocks of 16 bytes each (one block consists of 16 bytes).

The memory on the tag is organized as shown in Figure 7.32. Block 0, Sector 0, is also known as the Manufacturer Block, and the tag serial number is stored in the first 5 bytes of this block (4-byte serial number + Check byte) as shown in Figure 7.33.

Project PDL

The project PDL is given in Figure 7.34.

Sector	Block	Byte Number within a Block																Description
		0	1	2	3	4	5	6	7	8	9	10	11	12	13	14	15	
15	3	Key A						Access Bits				Key B						Sector Trailer 15
	2																	Data
	1																	Data
	0																	Data
14	3	Key A						Access Bits				Key B						Sector Trailer 14
	2																	Data
	1																	Data
	0																	Data
⋮	⋮																	
1	3	Key A						Access Bits				Key B						Sector Trailer 1
	2																	Data
	1																	Data
	0																	Data
0	3	Key A						Access Bits				Key B						Sector Trailer 0
	2																	Data
	1																	Data
	0																	Manufacturer Block

Figure 7.32: MF1ICS50 Tag Memory Organization.

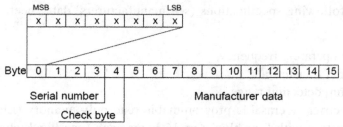

Figure 7.33: Serial Number is the First 4 Bytes.

Main Program:

BEGIN

      Define LCD to microcontroller connections

      Configure PORTA,B,C,E as digital

      ConfigureRA2, RC6, RE1 as outputs

      Initialize LCD

      Configure CR95HF to operate in UART mode

      CALL Initialize_CR95HF

      Initialize UART to 57600 Baud

      Display message "RFID" on LCD

      CALL Request_Short_Info to get CR95HF ID

      CALL Select _Protocol to select protocol 14443-A

      CALL RAQ to send RAQ and receive AQTA

      CALL UID to get tag serial number

END

BEGIN/Initialize_CR95HF

      Send startup initialization sequence to CR95HF chip

END/Initialzie_CR95HF

BEGIN/Request_Short_Info

      Send command 0x01 to CR95HF

      Get the ID and display on the LCD

END/Request_Short_Info

BEGIN/Select_Protocol

      Send command 0x02 to CR95hF to set ISO/IEC 14443-A protocol

END/Select_Protocol

BEGIN/RAQ

      Send command 0x04 to CR95HF with RAQ parameters

      CALL Get_Data to receive and display the data

END/RAQ

BEGIN/UID

      Send command 0x04 to CR95HF to read the serial number

      CALL Get_Data to receive and display the data

END/UID

BEGIN/Get_Data

      Get the data length

      Display the data on the LCD in hex format

END/Get_Data

Figure 7.34: Project PDL.

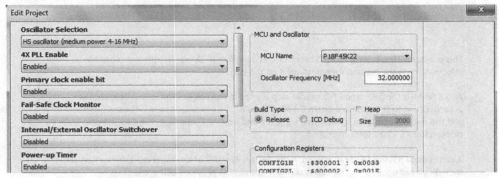

Figure 7.35: Setting the Microcontroller Clock Frequency to 32 MHz.

Project Program

mikroC Pro for PIC

The PIC18F45K22 microcontroller is operated from an 8-MHz crystal, but the PLL is enabled and the clock frequency is set to 32 MHz via the Project → Edit Project menu option for faster speed (Figure 7.35).

The mikroC Pro for PIC program is given in Figure 7.36 (MIKROC-RFID1.C). At the beginning of the program, the connections between the LCD and the microcontroller are defined, symbols IRQ_IN, SSI_0 and SSI_1 are assigned to port pins RC6, RA2 and RE1, respectively. Ports A, B, C, and E are then configured as digital, pins RA2, RC6, RE1 configured as outputs, and the LCD is initialized.

The program then configures the RFID reader to operate in the UART mode by clearing both SSI_0 and SSI_1 pins. Function Initialize_CR95HF is called to initialize the chip as explained earlier in the Startup Sequence. The UART is initialized to operate at 57,600 Baud, which is the default Baud rate of the RFID receiver. The LCD is cleared, and message RFID is displayed at the top row.

Function Request_Short_Info requests the ID of the CR95HF by sending command 1 with zero data length. Thus, the command sent to the chip is

```
0x01          Chip ID request command
0x00          Length of command
```

The chip returns result code and length of data. The result code must be 0x00 for success. The received result code is checked for validity, and if this is the case, ID data are received from the chip and displayed on the LCD as shown in Figure 7.37. The display consists of the following letters: "NFC FS2JAST2", which means read only memory (ROM) code revision 2. This text is followed by Cyclic Redundancy Check (CRC) bytes that are not ASCII and cannot be displayed on the LCD.

```
/*********************************************************************
                            RFID
                            ====

This project is about using an RFID receiver and a passive RFID tag. The project reads the
information on the tag and then displays it on the LCD.

An 13.56 MHz CR95HF type RFID reader chip is used in the design. The chip is operated in
accordance with the standard ISO/IEC 14443 Type A. The communication between the chip
and the microcontroller is established using a standard UART interface. The CR95HF chip is
clocked from a 27.12 MHz crystal.

A compatible passive RFID tag is used in the design. The UID of the tag is read, formatted, and
then displayed on an LCD. The LCD is connected as in the previous LCD projects.

An 8 MHz crystal is used for the PIC18F45K22 microcontroller in the project. The PLL is enabled
so that the effective microcontroller clock rate is X4. i.e. 32 MHz.

Author:  Dogan Ibrahim
Date:    October 2013
File:    MIKROC-RFID1.C
*********************************************************************/
// LCD module connections
sbit LCD_RS at LATB4_bit;
sbit LCD_EN at LATB5_bit;
sbit LCD_D4 at LATB0_bit;
sbit LCD_D5 at LATB1_bit;
sbit LCD_D6 at LATB2_bit;
sbit LCD_D7 at LATB3_bit;

sbit LCD_RS_Direction at TRISB4_bit;
sbit LCD_EN_Direction at TRISB5_bit;
sbit LCD_D4_Direction at TRISB0_bit;
sbit LCD_D5_Direction at TRISB1_bit;
sbit LCD_D6_Direction at TRISB2_bit;
sbit LCD_D7_Direction at TRISB3_bit;
// End LCD module connections

#define IRQ_IN LATC.RC6                        // IRQ_IN is also the URT TX pin
#define SSI_0 LATA.RA2                         // SSI_0 pin
#define SSI_1 LATE.RE1                         // SSI_1 pin

unsigned char Info[50];
unsigned char CMD, LEN;
//
// This function initializes the chip after power-up so that it is in Ready state. See the
// CR95HF Data sheet for the delays
//
void Initialize_CR95HF()
{
  Delay_Ms(10);                               // t4 delay time
```

Figure 7.36: mikroC Pro for the PIC Program.

```
    IRQ_IN = 1;                                    // Set IRQ_IN = 1
    Delay_us(100);                                 // t0 delay time
    IRQ_IN = 0;                                    // Lower IRQ_IN
    delay_us(500);                                 // t2 delay time (typical 250)
    IRQ_IN = 1;                                    // IRQ_IN high
    Delay_Ms(10);                                  // t3 delay time. The CR95HF is Ready now
}


//
// This function waits until the UART is ready and sends a bytes to it
//
void WriteUart(unsigned char c)
{
   while(Uart1_Tx_Idle() == 0);
   Uart1_Write(c);
}


//
// This function waits until the UART is ready and reads a byte from it
//
unsigned char ReadUart()
{
   unsigned char c;

   while(Uart1_Data_Ready() == 0);
   c = Uart1_Read();
   return c;
}


//
// This function requests short info (IDN) about the CR95HF chip. The received information
// is stored in character array called Info. Received short Info is displayed
//
void Request_Short_Info()
{
   unsigned char i, Cnt, flag, c;

   WriteUart(0x01);                               // Send IDN command
   WriteUart(0x00);                               // Send length
   c = ReadUart();
   if(c != 0x00)
   {
    Lcd_Out(1,1,"Error");                         // Error, abort
    while(1);
   }

   i = 0;
   flag = 0;
   Cnt = 0;
```

Figure 7.36
cont'd

```
        while(flag == 0)                                  // Read the data bytes
        {
          if(Uart1_Data_Ready() == 1)
          {
            Info[i] = Uart1_Read();
            if(i == 0)
            {
              LEN = Info[0];                              // Store the length
              Cnt = LEN + 1;
            }
            i++;
            if(i == Cnt)flag = 1;                         // All data read, terminate loop
          }
        }

        LCD_Cmd(_LCD_FIRST_ROW);                          // To first row of the LCD
        for(i=1; i < Cnt; i++)Lcd_Chr_Cp(Info[i]);        // Write short Info to LCD
        for(i=0; i < Cnt; i++)Info[i] = 0;                // Clear buffer
}

//
// This function selects protocol 14443 Type A
//
void Select_Protocol_14443()
{
   unsigned char c;

   WriteUart(0x02);                                       // Send Select Protocol command
   WriteUart(0x02);                                       // Send length
   WriteUart(0x02);                                       // Send protocol 14443 Type A
   WriteUart(0x00);
//
// Get result code (must be 0x00)
//
   c = ReadUart();                                        // Get result code
   LEN = ReadUart();                                      // get length

   if(c != 0x00)                                          // If error, abort
   {
     Lcd_Out(1,1,"Protocol Error");
     while(1);
   }
   else
   {
     Lcd_Out(1,1,"Protocol Selected");
   }
}

//
// This function reads the data length and then reads all the data and displays on the
// LCD in hexadecimal format
```

Figure 7.36
cont'd

```
//
void Get_Data()
{
    unsigned char i, Cnt, flag;
    unsigned char Txt[3];

    i = 0;
    flag = 0;
    Cnt = 0;

    while(flag == 0)                                    // Read data
    {
        if(Uart1_Data_Ready() == 1)                     // If UART has a character
        {
            Info[i] = Uart1_Read();                     // Get the character
            if(i == 0)
            {
                LEN = Info[0];                          // Get the data length
                Cnt = LEN + 1;
            }
            i++;
            if(i == Cnt)flag = 1;                       // If no more data, exit the loop
        }
    }

    LCD_Cmd(_LCD_FIRST_ROW);                            // Goto first row of LCD
    for(i=1; i < Cnt; i++)
    {
        ByteToHex(Info[i], Txt);                        // Convert to hex
        Lcd_Out_CP(Txt);                                // Display in hex
    }
}


//
// This function sends RAQ to the card and receives ATAQ response. The response is
// displayed on the LCD
//
void RAQ()
{
    unsigned char i, Cnt, flag, c;
    unsigned char Txt[3];

    WriteUart(0x04);                                    // Send IDN command
    WriteUart(0x02);                                    // Send length
    WriteUart(0x26);                                    // Send length
    WriteUart(0x07);                                    // Send length

    c = ReadUart();                                     // Read result code
    if(c != 0x80)
    {
```

Figure 7.36
cont'd

```
        Lcd_Out(1,1,"Read Error");
        while(1);
      }

      Get_Data();                                    // Get data and display it
    }


    //
    // This function receives the UID of the card and displays on the LCD
    //
    void UID()
    {
      unsigned char i, Cnt, flag, c;
      unsigned char Txt[3];

      WriteUart(0x04);                               // Send IDN command
      WriteUart(0x03);                               // Send length
      WriteUart(0x93);                               // Send length
      WriteUart(0x20);
      WriteUart(0x08);

      c = ReadUart();                                // Get result code
      if(c != 0x80)
      {
        c = ReadUart();                              // Read length
        while(1);
      }

      Get_Data();                                    // Get data and display it
    }


    void main()
    {
      ANSELA = 0;                                    // Configure PORTA as digital
      ANSELB = 0;                                    // Configure PORTB as digital
      ANSELC = 0;                                    // Configure PORTC as digital
      ANSELE = 0;                                    // Configure PORTE as digital
      TRISA.RA2 = 0;                                 // Configure RA2 as output
      IRQ_IN = 1;
      TRISC.RC6 = 0;                                 // Configure RC6 as output
      TRISE.RE1 = 0;                                 // Configure RE1 as output

      Lcd_Init();                                    // Initialize LCD
      Delay_ms(10);

      SSI_0 = 0;                                     // Configure CR95HF to use UART
      SSI_1 = 0;                                     // Configure CR95HF to se UART
      Initialize_CR95HF();                           // Ready mode after power-up

      UART1_Init(57600);                             // Initialize UART to 57600 Baud
```

Figure 7.36
cont'd

```
            Lcd_Cmd(_LCD_CURSOR_OFF);                // Disable cursor
            Lcd_Cmd(_LCD_CLEAR);                     // Clear LCD
            Lcd_Out(1,8,"RFID");                     // Display heading on LCD
            Delay_Ms(1000);                          // Wait 1 s for the display
            Lcd_Cmd(_LCD_CLEAR);
        //
        // Request short Info (IDN) about the CR95HF chip
        //
            Request_Short_Info();                    // Request short information
            Delay_Ms(2000);                          // Wait 2 s
            Lcd_Cmd(_LCD_CLEAR);
        //
        // Select the proocol used (14443 Type A)
        //
            Select_Protocol_14443();
            Delay_Ms(2000);                          // Wait 2 s
            Lcd_Cmd(_LCD_CLEAR);
        //
        // Send RAQ and get ATAQ
        //
            RAQ();
            Delay_ms(1000);
            Lcd_Cmd(_LCD_CLEAR);
        //
        // Read the Tag UID
        //
            UID();

            for(;;)                                  // Wait here forever
            {

            }
        }
```

Figure 7.36
cont'd

Figure 7.37: Displaying ID of the CR95HF Chip.

The next step is to select the RF communication protocol to be used. Function Select_Protocol_14443 is called for this purpose. In this example, ISO/IEC 14443-A protocol is used. The command to set a protocol is 0x02. The protocol code for the required protocol is 0x02, and the parameter is 0x00, which is the recommended setting to send and receive at 106 kbps (see the CR95HF data sheet). Thus, the command sent to the CR95HF is

```
0x02        Protocol select command
0x02        Command length is 2
0x02        Select ISO/IEC 14443-A
0x00        Command parameter
```

The chip returns the result code and length of data. The result code must be 0x00 if the required protocol is successfully selected. Selecting a protocol automatically turns ON the electromagnetic field around the RFID board so that tags can be detected.

The next step is to send an RAQ request command and receive the ATQA response from the chip. Command 0x04 (SEND_RECV) is used for this purpose. The command to send the RAQ is as follows:

```
0x04        SEND_RECV command
0x02        Command length 2
0x26        RAQ command
0x07        No of significant bits (RAQ is coded on 7 bits)
```

The tag must be near the receiver when this command is issued. The result code of 0x80 corresponds to success, and if this is the case, the command length and the data bytes (ATAQ) are returned. The card used by the author displayed the following data in hexadecimal format:

0400280000

This data decodes to the following (see ISO/IEC 14443-A specifications):

0400   - ATQA response

Twenty eight in binary is "0001 1000". This translates as

Bits 0: 3 = Number of significant bits in the first byte (here, 8),

Bit 4: 1 = Parity error, 0 = No parity error (here, no parity error),

Bit 5: 1 = CRC error, 0 = No CRC error (here this bit has no meaning since there is no CRC in AQTA response),

Bit 6: Not used,

Bit 7: 1 = Collision between multiple tags, 0 = No collision (here, no collision)

00 00   - Indexes to where collision detected (no collision here)

Figure 7.38: Data Returned by the Receiver.

In the next step, we send a command to read the manufacturers' block, which also contains the serial number. Command 0x04 (SEND_RECV) is used for this purpose. The command is as follows:

0x04	SEND_RECV command
0x03	Command length 3
0x93	Cascade Level 1 (CL1) command to get the serial number
0x20	Command tail
0x08	Command tail

The result code of 0x80 corresponds to success, and if this is the case, the command length and the data bytes are returned. The card used by the author displayed the following data in hexadecimal format (Figure 7.38):

   AD3D910706280000

The serial number is the first 5 bytes (40 bits), that is, AD3D910706, which corresponds to the binary number "1010 1101 0011 1101 1001 0001 0000 0111 0000 0110".

MPLAB XC8

It is left as an exercise for the reader to convert the program to compile under the MPLAB XC8 compiler.

Project 7.4—RFID Lock

In this project, we shall be using an RFID system to create a security lock. The lock will be based on a relay that will operate when the correct tag is placed near the reader.

The block diagram of the project is shown in Figure 7.39.

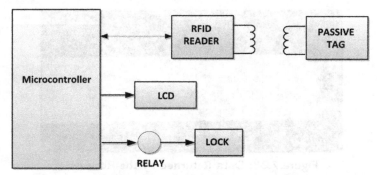

Figure 7.39: Block Diagram of the Project.

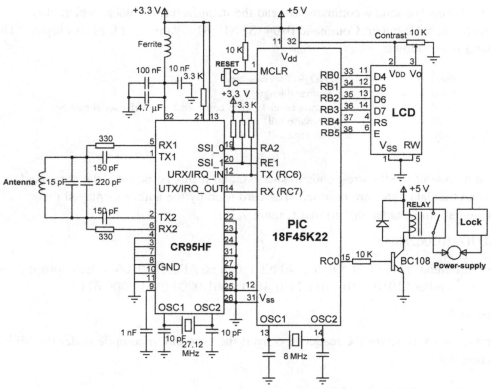

Figure 7.40: Circuit Diagram of the Project.

Project Hardware

The circuit diagram of the project is very similar to the one given in Figure 7.26, except that here additionally a relay and a lock mechanism are connected to the RC0 pin of the microcontroller via a transistor switch. Figure 7.40 shows the complete circuit diagram.

Project PDL

The project PDL is shown in Figure 7.41.

Project Program
mikroC Pro for PIC

The program listing of the project is given in Figure 7.42 (MIKROC-RFID2.C). In this project, it is assumed that the tag used in the previous project is used to open the lock. The program is similar to the one given in Figure 7.36. Additionally, at the beginning of the program, symbol RELAY is assigned to port RC0. Also, there is no need to read and display the ID of the CR95HF chip.

The program selects the required protocol (ISO/IEC 14443-A), displays the message READY on the LCD, and then enters an endless loop formed by a *while* statement. Inside this loop, the RAQ request is sent with no display of the AQAT. The serial number of the tag is read and compared with what was read in the earlier project (i.e. AD3D9107). If this is the matching tag, then the relay is energized by setting RELAY = 1, message "Opened" is sent to the LCD, and the program stays in this state for 5 s. After this time, the relay is deenergized, and message "Ready..." is displayed on the LCD to inform the user that the system is ready again.

Project 7.5—Complex SPI Bus Project

In Chapter 8, we have seen briefly how to use the SPI bus to generate various waveforms using a DAC. In this project, we will look at the operation of the SPI bus in greater detail as it is very important in the design of microcontroller-based systems. We will also develop a project to measure the temperature using an SPI-based temperature sensor. The ambient temperature will be measured and then displayed on an LCD.

The Master Synchronous Serial Port Module

The Master Synchronous Serial Port (MSSP) module is a serial interface module on PIC18F series of microcontrollers, used for communicating with other serial devices such as EEPROMs, display drivers, A/D converters, D/A converters, and SD cards. PIC18F45K22 microcontroller has two built-in MSSP modules.

The MSSP module can operate in one of two modes:

- SPI,
- Interintegrated Circuit (I^2C).

Main Program:

BEGIN

    Define LCD to microcontroller connections
    Configure PORTA,B,C,E as digital
    ConfigureRA2, RC6, RE1,RC0 as outputs
    Initialize LCD
    Configure CR95HF to operate in UART mode
    CALL Initialize_CR95HF
    Initialize UART to 57600 Baud
    Display message "RFID LOCK" on LCD
    CALL Select _Protocol to select protocol 14443-A
    Display "READY" on LCD
    DO FOREVER
        CALL RAQ to send RAQ and receive AQTA
        CALL UID to get tag serial number
        IF the tag matches the serial number
            Energize the relay
            Display "Opened" on LCD
            Wait for 5 seconds
            De-energize the relay
        ELSE
            Display "Ready..." on LCD to try again
        ENDIF
    ENDDO
END

BEGIN/Initialize_CR95HF
    Send startup initialization sequence to CR95HF chip
END/Initialzie_CR95HF

BEGIN/Select_Protocol
    Send command 0x02 to CR95hF to set ISO/IEC 14443-A protocol
END/Select_Protocol

BEGIN/RAQ
    Send command 0x04 to CR95HF with RAQ parameters
    CALL Get_Data to receive and display the data
END/RAQ

BEGIN/UID
    Send command 0x04 to CR95HF to read the serial number
    CALL Get_Data to receive and display the data
END/UID

BEGIN/Get_Data
    Get the data length
    Display the data on the LCD in hex format
END/Get_Data

Figure 7.41: Project PDL.

```
/*******************************************************************************
                              RFID LOCK
                              ========

This project is about using an RFID receiver and a passive RFID tag to operate a relay to open a lock.

An 13.56 MHz CR95HF type RFID reader chip is used in the design, operated in accordance with the
standard ISO/IEC 14443 Type A. The communication between the chip and the microcontroller is
established using a standard UART interface. The CR95HF chip is clocked from a 27.12 MHz crystal.

A compatible passive RFID tag is used in the design. The UID of the tag is read and if it is an
acceptable tag then the relay is operated to open the lock. The relay operates for 5 s to
open the lock and then stops. The program then waits for other activations.

An LCD is used to display various messages about the operation of the lock. The relay is connected
To RC0 pin of the microcontroller via a transistor switch. The lock is assumed to operate when the
Relay is energized.

An 8 MHz crystal is used for the PIC18F45K22 microcontroller in the project. The
PLL is enabled so that the effective microcontroller clock rate is X4. i.e. 32 MHz.

Author:  Dogan Ibrahim
Date:    September 2013
File:    MIKROC-RFID2.C
*******************************************************************************/
// LCD module connections
sbit LCD_RS at LATB4_bit;
sbit LCD_EN at LATB5_bit;
sbit LCD_D4 at LATB0_bit;
sbit LCD_D5 at LATB1_bit;
sbit LCD_D6 at LATB2_bit;
sbit LCD_D7 at LATB3_bit;

sbit LCD_RS_Direction at TRISB4_bit;
sbit LCD_EN_Direction at TRISB5_bit;
sbit LCD_D4_Direction at TRISB0_bit;
sbit LCD_D5_Direction at TRISB1_bit;
sbit LCD_D6_Direction at TRISB2_bit;
sbit LCD_D7_Direction at TRISB3_bit;
// End LCD module connections

#define IRQ_IN LATC.RC6              // IRQ_IN is also the URT TX pin
#define SSI_0 LATA.RA2              // SSI_0 pin
#define SSI_1 LATE.RE1             // SSI_1 pin
#define RELAY LATC.RC0             // Relay

unsigned char Info[50];
unsigned char CMD, LEN, ErrorFlag, ID[30];
//
// This function initializes the chip after power-up so that it is in Ready state
// See the CR95HF Data sheet for the delays
```

Figure 7.42: mikroC Pro for PIC Program.

```
//
void Initialize_CR95HF()
{
  Delay_Ms(10);                              // t4 delay time
  IRQ_IN = 1;                                // Set IRQ_IN = 1
  Delay_us(100);                             // t0 delay time
  IRQ_IN = 0;                                // Lower IRQ_IN
  delay_us(500);                             // t2 delay time (typical 250)
  IRQ_IN = 1;                                // IRQ_IN high
  Delay_Ms(10);                              // t3 delay time. The CR95HF is Ready now
}


//
// This function waits until the UART is ready and sends a byte to it
//
void WriteUart(unsigned char c)
{
  while(Uart1_Tx_Idle() == 0);
  Uart1_Write(c);
}


//
// This function waits until the UART is ready and reads a byte from it
//
unsigned char ReadUart()
{
  unsigned char c;

  while(Uart1_Data_Ready() == 0);
  c = Uart1_Read();
  return c;
}


//
// This function selects protocol 14443 Type A
//
void Select_Protocol_14443()
{
  unsigned char c;

  WriteUart(0x02);                           // Send Select Protocol command
  WriteUart(0x02);                           // Send length
  WriteUart(0x02);                           // Send protocol 14443 Type A
  WriteUart(0x00);
  //
  // Get result code (must be 0x00)
  //
  c = ReadUart();                            // Get result code
  LEN = ReadUart();                          // get length
```

Figure 7.42
cont'd

```
            if(c != 0x00)                                 // If error, abort
            {
              Lcd_Out(1,1,"Protocol Error");              // Error. Reset the device
              while(1);
            }
            else Lcd_Out(1,1,"Protocol Set  ");
            Delay_Ms(1000);
}


//
// This function reads the data length and then reads all the data and displays
// on the LCD in hexadecimal format
//
void Get_Data()
{
    unsigned char i, Cnt, flag;
    unsigned char Txt[3];

    i = 0;
    flag = 0;
    Cnt = 0;

    while(flag == 0)                                      // Read data
    {
      if(Uart1_Data_Ready() == 1)                         // If UART has a character
      {
        Info[i] = Uart1_Read();                           // Get the character
        if(i == 0)
        {
          LEN = Info[0];                                  // Get the data length
          Cnt = LEN + 1;
        }
        i++;
        if(i == Cnt)flag = 1;                             // If no more data, exit the loop
      }
    }

    ID[0] = 0x0;
    for(i=1; i < Cnt; i++)
    {
      ByteToHex(Info[i], Txt);                            // Convert to hex
      strcat(ID, Txt);                                    // Append to ID
    }
}


//
// This function sends RAQ to the card and receives ATAQ response. The response
// is displayed on the LCD
//
```

Figure 7.42
cont'd

```
void RAQ()
{
  unsigned char i, Cnt, flag, c;
  unsigned char Txt[3];

  WriteUart(0x04);                    // Send IDN command
  WriteUart(0x02);                    // Send length
  WriteUart(0x26);                    // Send length
  WriteUart(0x07);                    // Send length

  c = ReadUart();                     // Read result code
  ErrorFlag = 0;                      // Assume no error

  if(c != 0x80)                       // If error
  {
   c = ReadUart();                    // Read length
   ErrorFlag = 1;                     // Set error flag
   return;
  }

  Get_Data();                         // Get data
}

//
// This function receives the UID of the card and displays on the LCD
//
void UID()
{
  unsigned char i, Cnt, flag, c;
  unsigned char Txt[3];

  WriteUart(0x04);                    // Send IDN command
  WriteUart(0x03);                    // Send length
  WriteUart(0x93);                    // Send length
  WriteUart(0x20);
  WriteUart(0x08);

  c = ReadUart();                     // Get result code
  if(c != 0x80)
  {
   ErrorFlag = 1;                     // error flag
   c = ReadUart();                    // Read length
   return;
  }

  Get_Data();                         // Get data and display it
}


void main()
{
```

Figure 7.42
cont'd

```
ANSELA = 0;                                    // Configure PORTA as digital
ANSELB = 0;                                    // Configure PORTB as digital
ANSELC = 0;                                    // Configure PORTC as digital
ANSELE = 0;                                    // Configure PORTE as digital
TRISA.RA2 = 0;                                 // Configure RA2 as output
IRQ_IN = 1;
TRISC.RC6 = 0;                                 // Configure RC6 as output
TRISE.RE1 = 0;                                 // Configure RE1 as output
TRISC.RC0 = 0;                                 // Configure RC0 as output
RELAY = 0;                                     // Relay OFF to start with

Lcd_Init();                                    // Initialize LCD
Delay_ms(10);

SSI_0 = 0;                                     // Configure CR95HF to use UART
SSI_1 = 0;                                     // Configure CR95HF to se UART
Initialize_CR95HF();                           // Ready mode after power-up

UART1_Init(57600);                             // Initialize UART to 57600 Baud

Lcd_Cmd(_LCD_CURSOR_OFF);                      // Disable cursor
Lcd_Cmd(_LCD_CLEAR);                           // Clear LCD
Lcd_Out(1,1,"RFID LOCK");                      // Display heading on LCD
Delay_Ms(2000);                                // Wait 1 s for the display
Lcd_Cmd(_LCD_CLEAR);
//
// Select the proocol used (14443 Type A)
//
Select_Protocol_14443();
Delay_Ms(2000);                                // Wait 2 s
Lcd_Cmd(_LCD_CLEAR);
Lcd_Out(1,1,"READY");                          // Display "READY" message

//
// This is the tag detection and relay energization part of the program
//
while(1)
{
   RAQ();                                      // Send RAQ and get ATAQ
   UID();                                      // get serial number

   if(ErrorFlag == 0)                          // If no errors in RAQ and UID
   {
      if(ID[0] == 'A' && ID[1] == 'D' && ID[2] == '3' && ID[3] == 'D' &&        // If tag matches ?
        ID[4] == '9' && ID[5] == '1' && ID[6] == '0' && ID[7] == '7')
      {
         Lcd_Cmd(_LCD_CLEAR);
         Lcd_Out(1,1,"Opened");                // Display "Opened" message
         RELAY = 1;                            // Energize the relay
         Delay_Ms(5000);                       // Wait for 5 s
         RELAY = 0;                            // de-energize the relay
```

Figure 7.42
cont'd

```
                    Lcd_Cmd(_LCD_CLEAR);                        // Clear LCD
                }
            }
            else
            {
                Lcd_Out(1,1,"Ready...");                        // Error detected. Ready to try again
            }
            Delay_Ms(500);                                     // Wait before re-try
        }
    }
```

Figure 7.42
cont'd

Both SPI and I^{2}C are serial bus protocol standards. The SPI protocol was initially developed and proposed by Motorola for use in microprocessor and microcontroller-based interface applications. In a system that uses the SPI bus, one device acts as the master and other devices act as slaves. The master initiates a communication and also provides clock pulses to the slave devices.

The I^{2}C is a two-wire bus and was initially developed by Philips for use in low-speed microprocessor-based communication applications.

In this project, we shall be looking at the operation of the MSSP module in the SPI mode.

MSSP in the SPI Mode

The SPI mode allows 8 bits of data to be synchronously transmitted and received simultaneously. In the master mode, the device uses three signals, and in the slave mode, a fourth signal is used. In this project, we shall be looking at how to use the MSSP module in the master mode since in most microcontroller applications the microcontroller is the master device.

In the master mode, the following pins of the PIC18F45K22 microcontroller are used for SPI1 interface:

- Serial data out (SDO1)   -Pin RC5
- Serial data in (SDI1)   -Pin RC4
- Serial clock (SCK1)   -Pin RC3

And for SPI2 interface:

- Serial data out (SDO2)   -Pin RD4
- Serial data in (SDI2)   -Pin RD1
- Serial clock (SCK2)   -Pin RD0

Figure 7.43 shows the block diagram of the PIC18F45K22 microcontroller MSSP module when operating in the SPI mode.

SPI Mode Registers

The MSSP module has three registers when operating in the SPI master mode, 'x' is 1 or 2 and corresponds to the SPI module used:

* MSSP control register (SSPxCON1),
* MSSP status register (SSPxSTAT),
* MSSP receive/transmit buffer register (SSPxBUF),
* MSSP shift register (SSPxSR, not directly accessible).

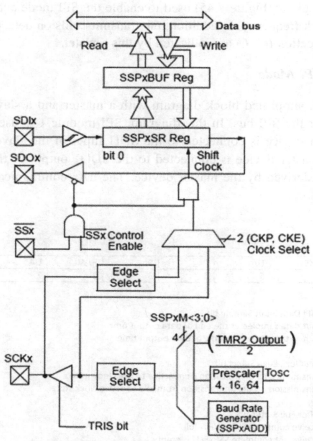

Figure 7.43: MSSP Module in the SPI Mode.

SSPxSTAT

This is the status register with the lower 6 bits read only and the upper two bits read/write. Figure 7.44 shows the bit definitions of this register. Only bits 0, 6, and 7 are related to operation in the SPI mode. Bit 7 (SMP) allows the user to select the input data sample time. When SMP = 0, input data are sampled at the middle of data output time, and when SMP = 1, the sampling is done at the end. Bit 6 (CKE) allows the user to select the transmit clock edge. When CKE = 0, transmit occurs on transition from the idle to active clock state, and when CKE = 1, transmit occurs on transition from the active to the idle clock state. Bit 0 (BF) is the buffer full status bit. When BF = 1, receive is complete (i.e. SSPxBUF is full), and when BF = 0, receive is not complete (i.e. SSPxBUF is empty).

SSPxCON1

This is the control register (Figure 7.45) used to enable the SPI mode and to set the clock polarity and the clock frequency. In addition, the transmit collision detection (bit 7), and receive overflow detection (bit 6) are indicated by this register.

Operation in the SPI Mode

Figure 7.46 shows a simplified block diagram with a master and a slave device communicating over the SPI bus. In this diagram, SPI module 1 is used. The SDO1 output of the master device is connected to the SDI1 input of the slave device, and the SDI1 input of the master device is connected to the SDO1 output of the slave device. The clock SCK1 is derived by the master device. The data communication is as follows:

SMP	CKE	D/A	P	S	R/W	UA	BF
7	6	5	4	3	2	1	0

Bit 7   **SMP**: SPI Data Input Sample bit
1 = Input data sampled at the end of data output time
0 = Input data sample at middle of data output time

Bit 6   **CKE**: SPI Clock Edge Select bit
1 = Transmit occurs on transition from active to idle clock state
0 = Transmission occurs on transition from idle to active clock state

Bit 0   **BF**: Buffer Full Status bit
1 = Receive complete SSPxBUF is full
0 = Receive not complete SSPxBUF is empty

Figure 7.44: SSPxSTAT Register Bit Configuration. Only the Bits when Operating in the SPI Master Mode are Described.

WCOL	SSPxOV	SSPxEN	CKP	SSPxM			
7	6	5	4	3	2	1	0

Bit 7   **WCOL**: Write Collision detect bit
1 = Collision detected
0 = No collision

Bit 6   **SSPxOV**: Receive Overflow Indicator bit
1 = Overflow occurred (can only occur in slave mode)
0 = No overflow

Bit 5   **SSPxEN**: Synchronous Serial Port Enable bit
1 = Enables serial port and configures SCKx,SDOx,SDIx as the source of port pins
0 = Disables serial port and configures these pins as I/O pins

Bit 4   **CKP**: Clock Polarity Select bit
1 = Idle state for clock is a high level
0 = Idle state for clock is a low level

Bit 3-0   **SSPxM**: Synchronous Serial Port Mode Select bits
0000 = SPI Master mode, clock = F_{osc}/4
0001 = SPI Master mode, clock = F_{osc}/16
0010 = SPI Master mode, clock = F_{osc}/64
0011 = SPI Master mode, clock = TMR2 output/2
0100 = SPI Master mode, clock = SCKx pin, SSx pin control enabled
0101 = SPI Master mode, clock = SCKx pin, SSx pin control disabled
1010 = SPI Master mode, clock = F_{osc}/4*(SSPxADD+1)

Figure 7.45: SSPxCON1 Register Bit Configuration. Only the Bits when Operating in the SPI Master Mode are Described.

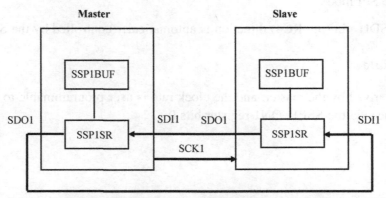

Figure 7.46: A Master and a Slave Device on the SPI Bus.

Sending Data to the Slave

To send data from the master to the slave, the master writes the data byte into its SSP1BUF register. This byte is also written automatically into the SSP1SR register of the master. As soon as a byte is written into the SSP1BUF register, eight clock pulses are sent out from the master SCK1 pin and at the same time the data bits are sent out from the master SSP1SR into the slave SSP1SR, that is, the contents of master and slave SSP1SR registers are swapped. At the end of this data transmission, the SSP1IF flag (PIR1 register) and the BF flag (SSP1STAT) will be set to show that the transmission is complete. Care should be taken not to write a new byte into SSP1BUF before the current byte is shifted out, otherwise an overflow error will occur (indicated by bit 7 of SSP1CON1).

Receiving Data From the Slave

To receive data from the slave device, the master has to write a "dummy" byte into its SSP1BUF register to start the clock pulses to be sent out from the master. The received data are then clocked into SS1PSR of the master, bit by bit. When the complete 8 bits are received, the byte is transferred to the SSP1BUF register and flags SSP1IF and BF are set. It is interesting to note that the received data are double buffered.

Configuration of MSSP for the SPI Master Mode

The following MSSP parameters for the master device must be set up before the SPI communication can take place successfully (SPI module 1 is assumed):

* Set data clock rate,
* Set clock edge mode,
* Clear bit 5 of TRISC (i.e. SDO1 = RC5 is output),
* Clear bit 3 of TRISC (i.e. SCK1 = RC3 is output),
* Enable the SPI mode.

Note that the SDI1 pin (pin RC4) direction is automatically controlled by the SPI module.

Data Clock Rate

The clock is derived by the master, and the clock rate is user programmable to one of the following values via the SSP1CON1 register bits 0−3:

* Fosc/4,
* Fosc/16,
* Fosc/64,
* Timer2 output/2.

Clock Edge Mode

The clock edge is user programmable via register SSP1CON1 (bit 4) and the user can either set the clock edge as *idle high* or *idle low*. In the idle high mode, the clock is high when the device is not transmitting, and in the idle low mode the clock is low when the device is not transmitting. Data can be transmitted either at the rising or the falling edge of the clock. The CKE bit of SSP1STAT (bit 6) is used to select the clock edge.

Enabling the SPI Mode

Bit 5 of SSP1CON1 must be set to enable the SPI mode. To reconfigure the SPI parameters, this bit must be cleared, SPI mode configured, and then the SSPEN bit set back to one.

An example is given below to demonstrate how to set the SPI parameters.

Example 7.1

It is required to operate the MSSP module 1 of a PIC18F45K22 microcontroller in the SPI mode. The data should be shifted on the rising edge of the clock and the SCK1 signal must be idle low. The required data rate is at least 1 Mbps. Assume that the microcontroller clock rate is 16 MHz and that the input data are to be sampled at the middle of data output time. What should be the settings of the MSSP registers?

Solution 7.1
Register SSP1CON1 should be set as follows
- Clear bits 6 and 7 of SSP1CON1 (i.e. no collision detect and no overflow);
- Clear bit 4 to 0 to select idle low for the clock;
- Set bits 0 through 3 to 0000 or 0001 to select the clock rate to Fosc/4 (i.e. 16/ 4 = 4 Mbps data rate), or Fosc/16 (i.e. 16/16 = 1 Mbps);
- Set bit 5 to enable the SPI mode.

Thus, register SSP1CON1 should be set to the following bit pattern:

00 1 0 0000 i.e. 0x20

Register SSP1STAT should be set as follows:

- Clear bit 7 to sample the input data at the middle of data output time;
- Clear bit 6—0 to transmit data on the rising edge (low to high) of the SCK1 clock;
- Bits 5—0 are not used in the SPI mode.

Thus, register SSP1STAT should be set to the following bit pattern:

0 0 0 00000 i.e. 0x00

Figure 7.47 shows the block diagram of the project. The temperature sensor TC72 is used in this project. The sensor is connected to the SPI bus pins of a PIC18F45K22 microcontroller. In addition, the microcontroller is connected to an LCD to display the temperature.

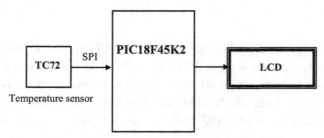

Figure 7.47: Block Diagram of the Project.

The specifications and operation of the TC72 temperature sensor are described below in detail.

TC72 Temperature Sensor

The TC72 is an SPI bus compatible digital temperature sensor IC that is capable of reading temperatures from −55 to +125 °C.

The device has the following features

- SPI bus compatible,
- A 10-bit resolution with 0.25 °C/bit,
- A ±2 °C accuracy from −40 to +85 °C,
- A 2.65- to 5.5-V operating voltage,
- A 250-µA typical operating current,
- A 1-µA shutdown operating current,
- Continuous and one-shot operating modes.

The pin configuration of TC72 is shown in Figure 7.48. The device is connected to an SPI bus via standard SPI bus pins SDI, SDO, and SCK. Pin CE is the chip-enable pin and is used to select a particular device in multiple TC72 applications. CE must be logic 1 for the device to be enabled. The device is disabled (output in tristate mode) when CE is logic 0.

1	NC	V_{DD}	8
2	CE	NC	7
3	SCK	SDI	6
4	GND	SDO	5

TC72

Figure 7.48: TC72 Pin Configuration.

The TC72 can operate either in the *one-shot* mode or in the *continuous* mode. In the one-shot mode, the temperature is read after a request is sent to read the temperature. In the continuous mode, the device measures the temperature approximately every 150 ms.

Temperature data are represented in 10-bit two's complement format with a resolution of 0.25 °C/bit. The converted data are available in two 8-bit registers. The most significant bit (MSB) register stores the decimal part of the temperature, whereas the least significant bit (LSB) register stores the fractional part. Only bits 6 and 7 of this register are used. The format of these registers is shown below:

MSB	S	2^6	2^5	2^4	2^3	2^2	2^1	2^0
LSB	2^{-1}	2^{-2}	0	0	0	0	0	0

Where S is the sign bit. An example is given below.

Example 7.2

The MSB and LSB settings of a TC72 are as follows:

> MSB: 00101011
> LSB: 10000000

Find the temperature read.

Solution 7.2

The temperature is found to be
$$MSB = 2^5 + 2^3 + 2^1 + 2^0 = 43,$$
$$LSB = 2^{-1} = 0.5$$

Thus, the temperature is 43.5 °C.

Table 7.2 shows sample temperature output data of the TC72 sensor.

Table 7.2: TC72 Temperature Output Data.

Temperature (°C)	Binary (MSB/LSB)	Hex
+125	0111 1101/0000 0000	7D00
+74.5	0100 1010/1000 0000	4A80
+25	0001 1001/0000 0000	1900
+1.5	0000 0001/1000 0000	0180
+0.5	0000 0000/1000 0000	0080
+0.25	0000 0000/0100 0000	0040
0	0000 0000/0000 0000	0000
−0.25	1111 1111/1100 0000	FFC0
−0.5	1111 1111/1000 0000	FF80
−13.25	1111 0010/1100 0000	F2C0
−25	1110 0111/0000 0000	E700
−55	1100 1001/0000 0000	C900

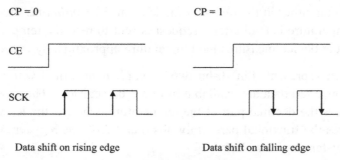

Figure 7.49: Serial Clock Polarity.

TC72 Read/Write Operations

The SDI input writes data into TC72's control register, while SDO outputs the temperature data from the device. The TC72 can operate using either the rising or the falling edge of the clock (SCK). The clock idle state is detected when the CE signal goes high. As shown in Figure 7.49, the clock polarity (CP) determines whether data are transmitted on the rising or the falling clock edge.

The maximum clock frequency (SCK) of TC72 is specified as 7.5 MHz. Data transfer consists of an address byte, followed by one or more data bytes. The most significant bit (A7) of the address byte determines whether a read or a write operation will occur. If $A7 = 0$, one or more read cycles will occur; otherwise, if $A7 = 1$, one or more write cycles will occur. The multibyte read operation will start by writing to the highest desired register and then reading from high to low addresses. For example, the temperature high byte address can be sent with $A7 = 0$ and then the result high byte, low byte, and the control register can be read as long as the CE pin is held active ($CE = 1$).

The procedure to read temperature from the device is as follows (assuming SPI module 1 is used):

- Configure the microcontroller SPI bus for the required clock rate and clock edge.
- Enable TC72 by setting $CE = 1$.
- Send temperature result high byte read address (0x02) to the TC72 (Table 7.3).

Table 7.3: TC72 Internal Registers.

Register	Read Address	Write Address	Bit7	Bit6	Bit5	Bit4	Bit3	Bit2	Bit1	Bit0
Control	0x00	0x80	0	0	0	OS	0	1	0	SHDN
LSB temperature	0x01	N/A	T1	T0	0	0	0	0	0	0
MSB temperature	0x02	N/A	T9	T8	T7	T6	T5	T4	T3	T2
Manufacturer ID	0x03	N/A	0	1	0	1	0	1	0	0

- Write a "dummy" byte into the SSP1BUF register to start eight pulses to be sent out from the SCK1 pin and then read the temperature result high byte.
- Write a "dummy" byte into the SSP1BUF register to start eight pulses to be sent out from the SCK1 pin and then read the temperature low byte.
- Set CE = 0 to disable the TC72 so that a new data transfer can begin.

Internal Registers of the TC72

As shown in Table 7.3, the TC72 has four internal registers: Control register, LSB temperature register, MSB temperature register, and the Manufacturer ID register.

Control Register

This is a read and write register used to select the mode of operation as shutdown, continuous, or one-shot. The address of this register is 0x00 when reading, and 0x80 when writing to the device. Table 7.4 shows how different modes are selected. At power-up, the shutdown bit (SHDN) is set to 1 so that the device is in the shutdown mode at startup and the device is used in this mode to minimize the power consumption.

A temperature conversion is initiated by a write operation to the Control register to select either the continuous mode or the one-shot mode. The temperature data will be available in the MSB and LSB registers after about 150 ms of the write operation. The one-shot mode performs a single temperature measurement after which time the device returns to the shutdown mode. In the continuous mode, new temperature data are available at 159-ms intervals.

LSB and MSB Registers

These are read-only registers that contain the 10-bit measured temperature data. The address of the MSB register is 0x02, and LSB register is 0x01.

Manufacturer ID

This is a read-only register with address 0x03. This register identifies the device as a temperature sensor, returning 0x054.

Table 7.4: Selecting the Mode of Operation.

Operating Mode	One Shot (OS)	Shutdown (SHDN)
Continuous	X	0
Shutdown	0	1
One-shot	1	1

Project Hardware

The circuit diagram of the project is shown in Figure 7.50. The TC72 temperature sensor is connected to the SPI bus pins of a PIC18F45K22 microcontroller, which is operated from a 4-MHz crystal. The CE pin of the TC72 is controlled from pin RC0 of the microcontroller. An LCD is connected to PORTB of the microcontroller as follows:

Microcontroller	LCD
RB0	D4
RB1	D5
RB2	D6
RB3	D7
RB4	E
RB5	R/S
RB6	RW

The connection between the TC72 and the microcontroller are as follows:

Microcontroller	TC72
RC0	CE
RC3	SCK
RC4	SDO
RC5	SDI

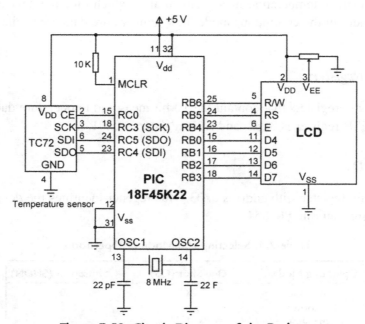

Figure 7.50: Circuit Diagram of the Project.

The microcontroller sends control commands to the TC72 sensor to initiate temperature conversions every second. The temperature data are then read and displayed on the LCD.

The Program

MPLAB XC8

The MPLAB XC8 program listing of the project is shown in Figure 7.51 (XC8-SPI.C). The program reads the temperature from the TC72 sensor and displays on the LCD every second. In this version of the program, only the positive temperatures and only the integer part are displayed.

The program consists of a number of functions. Some functions used in the program are

Init_LCD: This function initializes the LCD to 4-bit operation with 5×7 characters. The function also calls LCD_Clear to clear the LCD screen.

Init_SPI: This function initializes the microcontroller SPI bus to:

> Clock rate: Fosc/4 (i.e. 1 MHz)
> Clock Idle Low, Shift ta on clock falling edge
> Input data sample at end of data out

Send_To_TC72: This function loads a byte to SPI register SSP1BUF and then waits until the data are shifted out.

Read_Temperature: This function communicates with the TC72 sensor to read the temperature. The following operations are performed by this function:

1. Enable TC72 (CE = 1, for single byte write),
2. Send Address 0x80 (A7 = 1),
3. Clear BF flag,
4. Send One-Shot command (Control = 0001 0001),
5. Disable TC72 (CE = 0, end of single byte write),
6. Clear BF flag,
7. Wait at least 150 ms for the temperature to be available,
8. Enable TC72 (CE = 1, for multiple data transfer),
9. Send Read MSB command (Read address = 0x02),
10. Clear BF flag,
11. Send dummy output to start clock and read data (Send 0x00),
12. Read high temperature into variable MSB,
13. Send dummy output to start clock and read data (Send 0x00),
14. Read low temperature into variable LSB,
15. Disable TC72 data transfer (CE = 0),
16. Copy high result into variable "result".

```
/*******************************************************************
                SPI BUS BASED DIGITAL THERMOMETER
                =================================
```

In this project a TC72 type SPI bus based temperature sensor IC is used. The IC is
connected to the SPI bus pins of a PIC18F45K22 type microcontroller (i.e. to pins
RC3=SCK, RC4=SDI, and RC5=SDO) and the microcontroller is operated from an
8 MHz crystal.

In addition PORT B pins of the microcontroller are connected to a standard LCD.

The microcontroller reads the temperature every second and displays on the LCD as a
positive number (fractional part of the temperature, or negative temperatures are not
displayed in this version of the program).

An example display is:

        23

```
Author:   Dogan Ibrahim
Date:     October 2013
File:     XC8-SPI.C
*******************************************************************/
#include <xc.h>
#include <string.h>
#include <plib/usart.h>
#include <plib/xlcd.h>
#include <plib/spi.h>
#include <stdlib.h>

#pragma config MCLRE = EXTMCLR, WDTEN = OFF, FOSC = HSHP
#define _XTAL_FREQ 8000000

#define CE PORTCbits.RC0
#define Ready SSPSTATbits.BF

unsigned char LSB, MSB;
int result;

//
// This function creates seconds delay. The argument specifies the delay time in seconds
//
void Delay_Seconds(unsigned char s)
{
    unsigned char i,j;

    for(j = 0; j < s; j++)
    {
        for(i = 0; i <  100; i++)__delay_ms(10);
    }
}
```

Figure 7.51: MPLAB XC8 Program.

```
//
// This function creates milliseconds delay. The argument specifies the delay time in ms
//
void Delay_Ms(unsigned int ms)
{
   unsigned int i;

   for(i = 0; i <  ms; i++)__delay_ms(1);
}



//
// This function creates 18 cycles delay for the xlcd library
//
void DelayFor18TCY( void )
{
Nop(); Nop(); Nop(); Nop();
Nop(); Nop(); Nop(); Nop();
Nop(); Nop(); Nop(); Nop();
Nop(); Nop();
return;
}



//
// This function creates 15 ms delay for the xlcd library
//
void DelayPORXLCD( void )
{
__delay_ms(15);
   return;
}



//
// This function creates 5 ms delay for the xlcd library
//
void DelayXLCD( void )
{
   __delay_ms(5);
   return;
}

//
// This function clears the screen
//
void LCD_Clear()
{
   while(BusyXLCD());
   WriteCmdXLCD(0x01);
```

Figure 7.51
cont'd

```
        }


        //
        // Initialize the LCD, clear and home the cursor
        //
        void Init_LCD(void)
        {
                OpenXLCD(FOUR_BIT & LINE_5X7);                    // 8 bit, 5x7 character
                LCD_Clear();                                     // Clear LCD
        }


        //
        // Initialize the SPI bus
        //
        void Init_SPI(void)
        {
                OpenSPI(SPI_FOSC_4, MODE_01, SMPEND);            // SPI clk = 2MHz
        }


        //
        // This function sends a control byte to the TC72 and waits until the
        // transfer is complete
        //
        void Send_To_TC72(unsigned char cmd)
        {
                SSPBUF = cmd;                                    // Send control to TC72
                while(!Ready);                                   // Wait until data is shifted out
        }



        //
        // This function reads the temperature from the TC72 sensor
        //
        // Temperature data is read as follows:
        //
        // 1.  Enable TC72 (CE=1, for single byte write)
        // 2.  Send Address 0x80 (A7=1)
        // 3.  Clear BF flag
        // 4.  Send One-Shot command (Control = 0001 0001)
        // 5.  Disable TC72 (CE=0, end of single byte write)
        // 6.  Clear BF flag
        // 7.  Wait at leat 150ms for temperature to be available
        // 8.  Enable TC72 (CE=1, for multiple data transfer)
        // 9.  Send Read MSB command (Read address=0x02)
        // 10.  Clear BF flag
        // 11.  Send dummy output to start cloak and read data (Send 0x00)
        // 12. Read high temperature into variable MSB
```

Figure 7.51
cont'd

```
// 13. Send dummy output to start clock and read data (Send 0x00)
// 14. Read low temperature into variable LSB
// 15. Disable TC72 data transfer (CE=0)
// 16. Copy high result into variable "result"
void Read_Temperature(void)
{
        char dummy;

        CE = 1;                              // Enable TC72
        Send_To_TC72(0x80);                  // Send control write with A7=1
        dummy = SSPBUF;                      // Clear BF flag
        Send_To_TC72(0x11);                  // Set for one-shot operation
        CE = 0;                              // Disable TC72
        dummy = SSPBUF;                      // Clear BF flag
        Delay_Ms(200);                       // Wait 200 ms for conversion
        CE =1;                               // Enable TC72
        Send_To_TC72(0x02);                  // Read MSB temperature address
        dummy = SSPBUF;                      // Clear BF flag
        Send_To_TC72(0x00);                  // Read temperature high byte
        MSB = SSPBUF;                        // save temperature and clear BF
        Send_To_TC72(0x00);                  // Read temperature low byte
        LSB = SSPBUF;                        // Save temperature and clear BF
        CE = 0;                              // Disable TC72
        result = MSB;
}



//
// This function formats the temperature for displaying on the LCD.
// We have to convert to a string to display on the LCD.
//
// Only the positive MSB is displayed in this version of the program
//
void Format_Temperature(char *tmp)
{
        itoa(tmp,result,10);                 // Convert integer to ASCII
}



//
// This function clears the LCD, homes the cursor and then displays the
// temperature on the LCD
//
void Display_Temperature(char *d)
{
        LCD_Clear();                         // Clear LCD and home cursor
        putsXLCD(d);
}
```

Figure 7.51
cont'd

```
void main(void)
{
    char msg[] = "Temperature...";
    char tmp[3];

    ANSELB = 0;                              // Configure PORTB as digital
    ANSELC = 0;                              // Configure PORTC as digital

    TRISCbits.RC0 = 0;                       // Configure RC0 (CE) as output
    TRISB = 0;                               // Configure PORT B as outputs

    Delay_Seconds(1);
//
// Initialize the LCD
//
    Init_LCD();
    while(BusyXLCD());                       // Wait if the LCD is busy
    WriteCmdXLCD(DON);                       // Turn Display ON
    while(BusyXLCD());                       // Wait if the LCD is busy
    WriteCmdXLCD(0x06);                      // Move cursor right
    LCD_Clear();                             // Clear LCD
//
// Display a message on the LCD
//
    putsXLCD(msg);
    Delay_Seconds(2);
    LCD_Clear();
    Init_SPI();                              // Initialize the SPI bus
//
// Endless loop. Inside this loop read the TC72 temperature, display on the LCD,
// wait for 1 s and repeat the process
//
    for(;;)                                  // Endless loop
    {
        Read_Temperature();                  // Read the TC72 temperature
        Format_Temperature(tmp);             // Format the data for display
        Display_Temperature(tmp);            // Display the temperature
        Delay_Seconds(1);                    // Wait 1 s
    }
}
```

Figure 7.51
cont'd

Format_Temperature: This function converts the integer temperature into an ASCII string so that it can be displayed on the LCD.

Display_Temperature: This function calls to LCD_Clear to clear the LCD screen and home the cursor. The temperature is then displayed calling function putsXLCD.

Main Program: At the beginning of the main program, the port directions are configured, the LCD is initializes, the message "Temperature..." is sent to the LCD, and the microcontroller SPI bus is initialized. The program then enters an endless loop where the following functions are called inside this loop:

```
Read_Temperature();
Format_Temperature(tmp);
Display_Temperature(tmp);
One_Second_Delay();
```

Displaying Negative Temperatures

The program given in Figure 7.51 displays only the positive temperatures. Negative temperatures are stored in TC72 in two's complement format. If bit 8 of the MSB byte is set, the temperature is negative and two's complement should be taken to find the correct temperature. For example, if the MSB and LSB bytes are "1110 0111/1000 0000", the correct temperature is

1110 0111/1000 0000 → the complement is 0001 1000/0111 1111

adding "1" to find the two's complement gives: 0001 1000/1000 0000,

that is, the temperature is "−24.5 °C".

Similarly, if the MSB and LSB bytes are "1110 0111/0000 0000", the correct temperature is

1110 0111/0000 0000 → the complement is 0001 1000/1111 1111

adding "1" to find the two's complement gives: 0001 1001/0000 0000,

that is, the temperature is "−25 °C".

In the modified program, both negative and positive temperatures are displayed where the sign "−" is inserted in-front of negative temperatures. The temperature is displayed in integer format with no fractional part in this version of the program. The Format_Temperature function is modified such that if the temperature is negative the two's complement is taken, the sign bit is inserted, and then the value is shifted right by 8 digits and converted into an ASCII string for the display.

The new Format_Temperature function is shown below:

```
//
// Positive and negative temperatures are displayed in this version of the program
//
void Format_Temperature(char *tmp)
{
    if(result & 0x8000)           // If negative
```

```
    {
      result = ~result;        // Take complement
      result++;                // Take 2's complement
      result >>= 8;            // Get integer part
      *tmp++ = '-';            // Insert "-" sign
    }
    else
    {
      result >>= 8;            // Get integer part
    }
    itoa(tmp,result,10);       // Convert integer to ASCII
}
```

Displaying the Fractional Part

The program in Figure 7.51 does not display the fractional part of the temperature. The program can be modified to display the fractional part as well. In the new function, the LSB byte of the converted data is taken into consideration and the fractional part is displayed as ".00", ".25", ".50", or ".75". The two most significant bits of the LSB byte are shifted right by 6 bits. The fractional part then takes one of the following values:

Two Shifted LSB Bits	Fractional Part
00	.00
01	.25
10	.50
11	.75

Figure 7.52 shows the modified program (XC8-SPI2.C).

Project 7.6—Real-Time Clock Using an RTC Chip

In this project, we will design a clock using a real-time clock (RTC) chip. We will be using three push-button switches to set the clock initially: Mode, Up, and Down. Mode button will select the date and time field, Up and Down buttons will increment and decrement the selected field, respectively.

There are several RTC chips available. The one that we will be using in this project is the PCF8583 eight-pin DIL chip. The specifications of this chip are as follows:

- Clock, alarm, and timer functions;
- I^2C bus interface;
- A +2.5- to +6-V operation;
- A 32.768-kHz time base (requires an external 32.768-kHz crystal);
- Programmable alarm, timer, and interrupt functions;
- A 240 × 8 random access memory (RAM).

```
/******************************************************************
                SPI BUS BASED DIGITAL THERMOMETER
                =================================
```

In this project a TC72 type SPI bus based temperature sensor IC is used. The IC is connected to the SPI bus pins of a PIC18F45K22 type microcontroller (i.e. to pins RC3=SCK, RC4=SDI, and RC5=SDO) and the microcontroller is operated from an 8 MHz crystal.

In addition PORT B pins of the microcontroller are connected to a standard LCD.

The microcontroller reads the temperature every second and displays on the LCD

This version of the program displays the sign as well as the fractional part of the temperature. An example display is:

$$-23.75$$

```
Author:  Dogan Ibrahim
Date:    October 2013
File:    XC8-SPI2.C
*******************************************************************/
#include <xc.h>
#include <string.h>
#include <plib/usart.h>
#include <plib/xlcd.h>
#include <plib/spi.h>
#include <stdlib.h>

#pragma config MCLRE = EXTMCLR, WDTEN = OFF, FOSC = HSHP
#define _XTAL_FREQ 8000000

#define CE PORTCbits.RC0
#define Ready SSPSTATbits.BF

unsigned char LSB, MSB;
int result,int_part,fract_part;

//
// This function creates seconds delay. The argument specifies the delay time in seconds
//
void Delay_Seconds(unsigned char s)
{
   unsigned char i,j;

   for(j = 0; j < s; j++)
   {
      for(i = 0; i <  100; i++)__delay_ms(10);
   }
}
```

Figure 7.52: Modified MPLAB XC8 Program to Display Fractional Part as Well.

```
//
// This function creates milliseconds delay. The argument specifies the delay time in ms
//
void Delay_Ms(unsigned int ms)
{
   unsigned int i;

   for(i = 0; i <  ms; i++)__delay_ms(1);
}



//
// This function creates 18 cycles delay for the xlcd library
//
void DelayFor18TCY( void )
{
Nop(); Nop(); Nop(); Nop();
Nop(); Nop(); Nop(); Nop();
Nop(); Nop(); Nop(); Nop();
Nop(); Nop();
return;
}



//
// This function creates 15 ms delay for the xlcd library
//
void DelayPORXLCD( void )
{
   __delay_ms(15);
   return;
}



//
// This function creates 5 ms delay for the xlcd library
//
void DelayXLCD( void )
{
   __delay_ms(5);
   return;
}

//
// This function clears the screen
//
void LCD_Clear()
{
   while(BusyXLCD());
   WriteCmdXLCD(0x01);
}
```

Figure 7.52
cont'd

```
//
// Initialize the LCD, clear and home the cursor
//
void Init_LCD(void)
{
        OpenXLCD(FOUR_BIT & LINE_5X7);                  // 8 bit, 5x7 character
        LCD_Clear();                                    // Clear LCD
}


//
// Initialize the SPI bus
//
void Init_SPI(void)
{
        OpenSPI(SPI_FOSC_4, MODE_01, SMPEND);           // SPI clk = 2MHz
}


//
// This function sends a control byte to the TC72 and waits until the
// transfer is complete
//
void Send_To_TC72(unsigned char cmd)
{
        SSPBUF = cmd;                                   // Send control to TC72
        while(!Ready);                                  // Wait until data is shifted out
}




//
// This function reads the temperature from the TC72 sensor
//
// Temperature data is read as follows:
//
// 1.  Enable TC72 (CE=1, for single byte write)
// 2.  Send Address 0x80 (A7=1)
// 3.  Clear BF flag
// 4.  Send One-Shot command (Control = 0001 0001)
// 5.  Disable TC72 (CE=0, end of single byte write)
// 6.  Clear BF flag
// 7.  Wait at leat 150ms for temperature to be available
// 8.  Enable TC72 (CE=1, for multiple data transfer)
// 9.  Send Read MSB command (Read address=0x02)
// 10. Clear BF flag
// 11. Send dummy output to start cloak and read data (Send 0x00)
// 12. Read high temperature into variable MSB
// 13. Send dummy output to start clock and read data (Send 0x00)
```

Figure 7.52
cont'd

```
// 14. Read low temperature into variable LSB
// 15. Disable TC72 data transfer (CE=0)
// 16. Copy high result into variable "result"
void Read_Temperature(void)
{
        char dummy;

        CE = 1;                                 // Enable TC72
        Send_To_TC72(0x80);                     // Send control write with A7=1
        dummy = SSPBUF;                         // Clear BF flag
        Send_To_TC72(0x11);                     // Set for one-shot operation
        CE = 0;                                 // Disable TC72
        dummy = SSPBUF;                         // Clear BF flag
        Delay_Ms(200);                          // Wait 200 ms for conversion
        CE =1;                                  // Enable TC72
        Send_To_TC72(0x02);                     // Read MSB temperature address
        dummy = SSPBUF;                         // Clear BF flag
        Send_To_TC72(0x00);                     // Read temperature high byte
        MSB = SSPBUF;                           // save temperature and clear BF
        Send_To_TC72(0x00);                     // Read temperature low byte
        LSB = SSPBUF;                           // Save temperature and clear BF
        CE = 0;                                 // Disable TC72
        result = MSB*256+LSB;                   // The complete temperature
}


//
// Positive and negative temperatures are displayed as well as the fractional part
//
void Format_Temperature(char *tmp)
{
    if(result & 0x8000)                         // If negative
       {
                result = ~result;               // Take complement
                result++;                       // Take 2's complement
                int_part = result >> 8;         // Get integer part
                *tmp++ = '-';                    // Insert "-" sign
       }
    else
       {
                int_part = result >> 8;         // Get integer part
       }

        itoa(tmp,result,10);                    // Convert integer to ASCII
//
// Now find the fractional part. First we must find the end of the string "tmp"
// and then append the fractional part to it
//
        while(*tmp != '\0')tmp++;               // find end of string "tmp"
//
```

Figure 7.52
cont'd

```
// Now add the fractional part as ".00", ".25", ".50", or ".75"
//
          fract_part = result &0x00C0;                // fractional part
          fract_part = fract_part >> 6;               // fract is between 0-3
          switch(fract_part)
          {
                case 1:                                 // Fractional part = 0.25
                          *tmp++ = '.';                 // decimal point
                          *tmp++ = '2';                 // "2"
                          *tmp++ = '5';                 // "5"
                          break;
                case 2:                                 // Fractional part = 0.50
                          *tmp++ = '.';                 // decimal point
                          *tmp++ = '5';                 // "5"
                          *tmp++ = '0';                 // "0"

          break;
                case 3:                                 // Fractional part = 0.75
                          *tmp++ = '.';                 // decimal point
                          *tmp++ = '7';                 // "7"
                          *tmp++ = '5';                 // "5"
                          break;
                case 0:                                 //  Fractional part = 0.00
                          *tmp++ = '.';                 // decimal point
                          *tmp++ = '0';                 // "0"
                          *tmp++ = '0';                 // "0"
                          break;
          }
          *tmp++ = '\0';                                // Null terminator
}


//
// This function clears the LCD and then displays the temperature on the LCD
//
void Display_Temperature(char *d)
{
          LCD_Clear();                                  // Clear LCD and home cursor
          putsXLCD(d);
}


void main(void)
{
          char msg[] = "Temperature...";
          char tmp[8];

          ANSELB = 0;                                   // Configure PORTB as digital
          ANSELC = 0;                                   // Configure PORTC as digital

          TRISCbits.RC0 = 0;                            // Configure RC0 (CE) as output
```

Figure 7.52
cont'd

```
            TRISB = 0;                                  // Configure PORT B as outputs

            Delay_Seconds(1);
    //
    // Initialize the LCD
    //
            Init_LCD();
            while(BusyXLCD());                          // Wait if the LCD is busy
            WriteCmdXLCD(DON);                          // Turn Display ON
            while(BusyXLCD());                          // Wait if the LCD is busy
            WriteCmdXLCD(0x06);                         // Move cursor right
            LCD_Clear();                                // Clear LCD
    //
    // Display a message on the LCD
    //
            putsXLCD(msg);
            Delay_Seconds(2);
            LCD_Clear();
            Init_SPI();                                 // Initialize the SPI bus
    //
    // Endless loop. Inside this loop read the TC72 temperature, display on the LCD,
    // wait for 1 s and repeat the pro      cess
    //
        for(;;)                                         // Endless loop
        {
            Read_Temperature();                         // Read the TC72 temperature
            Format_Temperature(tmp);                    // Format the data for display
            Display_Temperature(tmp);                   // Display the temperature
            Delay_Seconds(1);                           // Wait 1 s
        }
    }
```

Figure 7.52
cont'd

The PCF8583 operates as a slave I^2C device with devices addresses 0xA1 or 0xA3 for reading, and 0xA0 or 0xA2 for writing.

Figure 7.53 shows the block diagram of the project.

Before going into the details of the design, it is worthwhile to review the basic principles of the I^2C bus communications protocol. I^2C is a bidirectional two-line communication between a master and one or more slave devices. The two lines are named SDA (serial data) and SCL (serial clock). Both lines must be pulled up to the supply voltage using suitable resistors. Figure 7.54 shows a typical system configuration with one master and three slaves communicating over the I^2C bus.

Most high-level language compilers provide libraries for I^2C communication. We can also easily develop our own I^2C library. Although the available libraries can easily be used, it is worthwhile to look at the basic operating principles of the bus.

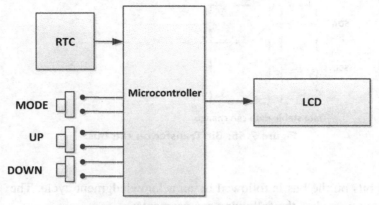

Figure 7.53: Block Diagram of the Project.

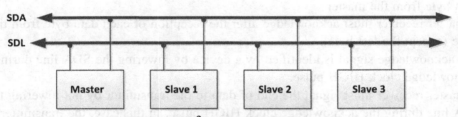

Figure 7.54: I²C System Configuration.

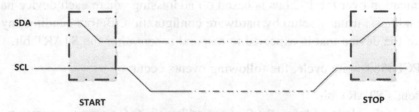

Figure 7.55: Start and STOP Bit Conditions.

The I²C bus must not be busy before data can be sent over the bus. Data are sent serially, and synchronized with the clock. Both SDA and SCL lines are HIGH when the bus is not busy. The START bit is identified by the HIGH-to-LOW transition of the SDA line while the SCL is HIGH. Similarly, a LOW-to-HIGH transition of the SDA line while the SCL is HIGH is identified as the STOP bit. Figure 7.55 shows both the START and STOP bit conditions.

One bit of data is transferred during each clock pulse. Data on the bus must be stable when SCL is HIGH; otherwise, the data will be interpreted as a control signal. Data can change when the SCL line is LOW. Figure 7.56 shows how bit transfer takes place on the bus.

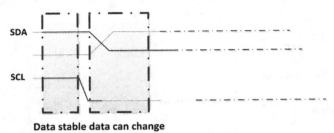

Data stable data can change

Figure 7.56: Bit Transfer on the Bus.

Each byte of 8 bits on the bus is followed by an acknowledgment cycle. The acknowledgement cycle has the following requirements:

- An addressed slave device must generate an acknowledgement after the reception of each byte from the master.
- A master receiver must acknowledge after the reception of each data byte from the slave (except the last byte).
- The acknowledge signal is identified by a device by lowering the SDA line during the acknowledge clock HIGH pulse.
- A master receiver must signal the end of data to the transmitter by not lowering the SDA line during the acknowledge clock HIGH pulse. In this case, the transmitter leaves the SCL line HIGH so that the master can generate the STOP bit.

The communication over the I^2C bus is based on addressing where each device has a unique 8-bit address, usually setup by hardware configuration. Before sending any data, the address of the device that is expected to respond is sent after the START bit.

During the PCF8583 write cycle, the following events occur:

- Master sends START bit.
- Master sends the slave address. Bit 0 of the address is 0 for a write operation.
- Slave sends an acknowledgement bit.
- Master sends the register address to specify the slave register to be accessed.
- Slave sends an acknowledgement bit.
- Slave sends the required data.
- Slave sends an acknowledgement bit.
- Master sends a STOP bit.

During the PCF8583 read cycle, the following events occur (8-bit address mode):

- Master sends START bit.
- Master sends the slave address. Bit 0 of the address is 1 for a read operation.
- Slave sends an acknowledgement bit.

- Slave sends data.
- Master sends acknowledgement bit.
- Slave sends the last byte.
- Master does not send the acknowledgement bit.
- Master sends the STOP bit.

In some read applications, as in the RTC clock project, it is sometimes necessary to send a 16-bit address to the slave device in the form of the actual device address, followed by the address of the register to be accessed. In such applications, the PCF8583 read cycle is as follows:

- Master sends START bit.
- Master sends the slave address in the write mode. Bit 0 of the address is 0 for a write operation.
- Slave sends an acknowledgement bit.
- Master sends the register address.
- Slave sends the acknowledgement bit.
- Master sends repeated START bit.
- Master sends the slave address in the read mode. Bit 0 of the address is 1 for a read operation.
- Slave sends the acknowledgement bit.
- Slave sends data.
- Master sends the acknowledgement bit.
- Slave sends the last byte.
- Master does not send the acknowledgement bit.
- Master sends the STOP bit.

When in the clock mode, the operation of the PCF8583 RTC chip is configured with seven registers. Figure 7.57 shows all the registers of the chip. The remaining registers are used to configure the timer and alarm functions as we shall see in the next project.

The first register is the control/status register. This register by default is loaded by 0x00 after reset. The important bit here as far as the clock operation is concerned is bit 7. This bit stops and restarts the internal clock counter. The normal state of this bit is 0, but it must be set to 1 to stop the counter during loading the current date and time information to the chip.

The date and time information is stored in the BCD format, the upper nibble holding the 10 s and the lower nibble holding the 1 s. For example, number 25 is stored in binary pattern as "0010 0101". The data should be converted into the correct format before being displayed on the LCD or before new data are loaded into the registers.

It is important to know the format of the registers during programming. Figures 7.58—7.60 give the format of the hour register, year—date register, and the weekday—month register, respectively. Note that the format of the year—day register is different from the others. The year is stored in 2 bits, having values 0—3. Thus, for example, to display year 2013, we have to provide the first three digits (201x) and read the last digit from the clock chip.

		Bit7	Bit 4 Bit 3	Bit 0
	0	Control/status		
Hundreths of a second	1	1/10 s	1/100 s	
Seconds	2	10 s	1 s	
Minutes	3	10 min	1 min	
Hours	4	10 h	1 h	
Year/day	5	10 day	1 day	
Weekday/month	6	10 month	1 month	
Timer	7	10 day	1 day	
Alarm control	8	Alarm control		
Hundreths of a second	9	1/10 s	1/100 s	
Alarm seconds	10	10 s	1 s	
Alarm minutes	11	10 min	1 min	
Alarm hours	12	10 h	1 h	
Alarm date	13	10 day	1 day	
Alarm month	14	10 month	1 month	
Alarm timer	15			
Free RAM	16	RAM (240 Bytes)		

Figure 7.57: PCF8583 Registers.

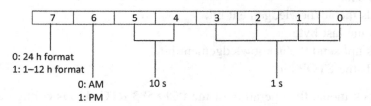

Figure 7.58: Hours Register.

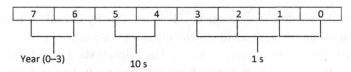

Figure 7.59: Year—Date Register.

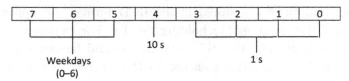

Figure 7.60: Weekday—Month Register.

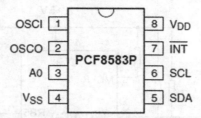

Figure 7.61: PCF8583 Pin Layout.

Project Hardware

Figure 7.61 shows the pin layout of the PCF8583 chip. The pin descriptions are as follows:

OSC1	oscillator or event pulse input
OSC0	oscillator input/output
A0	address input
V_{SS}, V_{DD}	power lines
SDA	data pin
SCL	clock pin
INT	interrupt output pin (active LOW, open drain)

If pin A0 is connected LOW the device responds to addresses 0xA0 and 0xA1 for writing and reading, respectively; otherwise, it responds to 0xA2 and 0xA3 for writing and reading, respectively.

The circuit diagram of the project is shown in Figure 7.62. The SDA and SCL pins of the PCF8583 are connected to microcontroller pins SDA1 (RC4) and SCL1 (RC3), respectively. Pin A0 is connected to the ground to select slave addresses 0xA0 and 0xA1. An LCD is connected to PORTB to display the clock. Three push-button switches connected to PORTD pins are used to set the initial date and time:

MODE (RD0)	this button is used to select the date or time field that will be set
UP (RD1)	this button increments the selected field
DOWN (RD2)	this button decrements the selected field

A 32.768-kHz crystal is used to provide timing pulses to the chip. A button shaped 3-V lithium battery (CR2032) is used to power the RTC chip so that it keeps the time even after the microcontroller power is turned off.

If you are using the mikroElektronika RTC board with the EasyPIC V7 development board, set the following jumpers on the RTC board and plug-in the board to PORTC:

Set switch 3 ON (Connect SDA line)
Set switch 4 ON (Connect SCL line)

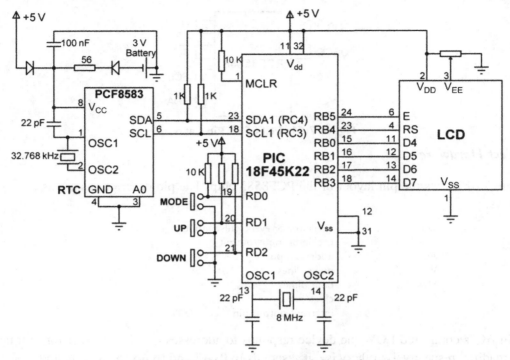

Figure 7.62: Circuit Diagram of the Project.

In addition, set the following jumper on the EasyPIC V7 development board so that when a button is pressed the button state goes from logic HIGH to logic LOW:

Set J17 to GND
PORTD RD0 switch to pull-up
PORTD RD1 switch to pull-up
PORTD RD2 switch to pull-up

Figure 7.63 shows the mikroElektronika RTC board.

Figure 7.63: mikroElektronika RTC Board.

Project PDL

The project PDL is shown in Figure 7.64.

Project Program

mikroC Pro for PIC

The mikroC Pro for the PIC compiler I^2C library supports the following functions
('x' refers to the I^2C module used in microcontrollers with more than one module):

I2Cx_Init	Initialize the I^2C library. The I^2C clock rate must be entered as an argument
I2Cx_Start	Sends START bit on the bus
I2C_Repeated_Start	Sends repeated START bits
I2C_Is_Idle	Returns 1 if the bus is free, otherwise returns 0
I2Cx_Rd	Reads one byte from the slave. If the argument is 1 an acknowledgement is sent, otherwise acknowledgement is not sent.
I2Cx_Wr	Sends a byte to the slave device. Returns 0 if there are no errors
I2Cx_Stop	Sends STOP bit on the bus

The mikroC Pro for PIC program is given in Figure 7.65 (MIKROC-I2C.C). Before
looking at the software in detail let us assume that we wish to set the date and time to 10-
09-2013 08:10:15. The steps are given below:

- Reset the microcontroller while pressing the MODE button.
- The LCD should display:

DAY:
31

- Keep pressing the UP button until 10 is displayed.
- Press MODE button to change the field to month:

MONTH:
12

- Keep pressing the DOWN button until 9 is displayed.
- Press MODE button to change the field to year:

YEAR (201x):
6

- Keep pressing the UP button until 3 is displayed (only the last digit of the year is entered).
- Press MODE button to change the field to hour:

HOUR:
23

Main Program:
BEGIN
        Define LCD – microcontroller interface
        Assign symbols MODE, UP, DOWN to port pins
        Configure PORTB, PORTC, PORTD as digital
        Configure RD0, RD1, RD2 as inputs
        Initialize LCD
        Initialize I^{2}C bus
        IF SETUP mode
                CALL Set_Date_Time to read new date and time values
                CALL SET_RTC to load the new values into clock chip
        ENDIF
        DO FOREVER
                CALL Read_Date_Time to read date and time from the clock chip
                CALL Convert_Date_Time to convert into displayable format
                CALL Display_Date_Time to display the date and time on the LCD
        ENDDO
END

BEGIN/Set_Date_Time
        Display maximum field value
        WHILE MODE button not pressed
                IF UP button pressed
                        Increment value
                        IF value > maximum
                                Set value = minimum
                        ENDIF
                ENDIF
                IF DOWN button pressed
                        Decrement value
                        IF value < minimum
                                Set value = maximum
                        ENDIF
                ENDIF
        WEND
        Return new value to the calling program
END/Set_Date_Time

BEGIN/SET_RTC
        Convert date and time into BCD
        Load date and time to the clock chip
END/SET_RTC

BEGIN/Read_Date_Time
        Get date and time from the clock chip
END/Read_Date_Time

BEGIN/Convert_Date_Time

Figure 7.64: Project PDL.

Convert date and time into ASCII for display
END/Convert_Date_Time

BEGIN/Display_Date_Time
Display converted date on row 1
Display converted time on row 2
END/Display_Date_Time

Figure 7.64
cont'd

- Keep pressing the UP button until 8 is displayed.
- Press the MODE button to change the field to minutes:

```
MINUTES:
59
```

- Keep pressing the UP button until 10 is displayed.
- Press the MODE button to change the field to seconds:

```
SECONDS:
59
```

- Keep pressing the UP button until 15 is displayed.
- Press the MODE button to terminate the setup. The clock should start working from the set date and time.

At the beginning of the program, the connections between the LCD and the microcontroller are defined. Symbols MODE, UP, DOWN and SETUP are assigned to port bits, PORTB, PORTC, PORTD are configured as digital with RD0, RD1, and RD2 pins configured as inputs. Then, the LCD and the I^2C modules are initialized. If the MODE button is pressed (MODE = 0) during the reset, the program enters the SETUP phase. Here, the new data and time values are read via the UP/DOWN/MODE buttons as described earlier, using function Set_Date_Time. This function displays the field to be modified at the first row of the LCD. The user changes the displayed value by pressing the UP or DOWN buttons. When the required value is displayed, the MODE button is pressed to move to the next field. This process continues until the last field value is selected and then the program calls function SET_RTC to load the new date and time values into the registers of PCF8583.

The rest of the program is executed in an endless loop. Here, the date and time are read from the PCF8583 using function Read_Date_Time. Function Convert_Date_Time is called to convert these values into a form that can be displayed on the LCD. Finally, function Display_Date_Time is called to display the date and time.

```
/*******************************************************************************
                            REAL TIME CLOCK
                            ===============
```

This project is about designing an accurate real time clock (RTC) using the RTC chip PCF8583

The PCF8583 chip is connected to the I2C pins (modul e 1) of a PIC18F45K22 microcontroller. The connections between the PCF8583 and the microcontroller are as follows (the SDA and SCL lines are pulled high with resistors):

```
PCF8583         Microcontroller
=======         ==============
  SDA           SDA1 (RC4)
  SCL           SCL1 (RC3)
```

Clock timing to the PCF8583 is provided with a 32.768 kHz crystal. 3 Push-button switches are used to set the clock initially:

```
    MODE (RD0):   Selects the date or time field to be set
    UP (RD1):     Increments the value
    DOWN (RD2):   Decrements the value
```

AN LCD is connected to PORTB of the microcontroller to help in setting the clock and also for displaying the clock data in real time.

The microcontroller is operated with an 8 MHz crystal with the PLL disabled (i.e. the actual running clock frequency is 8 MHz).

The software has 2 phases: SETUP and RUNNING. The SETUP phase is entered if the MODE button is pressed during the startup. In this phase the clock is set to current date and time. The RUNNING phase is entered if the MODE button is not pressed during the startup and this is the normal running state of the clock where the date and time are displayed on the LCD in the following format:

```
    dd-mm-yyyy
    hh:mm:ss
```

```
Author:  Dogan Ibrahim
Date:    October 2013
File:    MIKROC-I2C.C
*******************************************************************************/
// LCD module connections
sbit LCD_RS at LATB4_bit;
sbit LCD_EN at LATB5_bit;
sbit LCD_D4 at LATB0_bit;
sbit LCD_D5 at LATB1_bit;
sbit LCD_D6 at LATB2_bit;
sbit LCD_D7 at LATB3_bit;

sbit LCD_RS_Direction at TRISB4_bit;
sbit LCD_EN_Direction at TRISB5_bit;
```

Figure 7.65: mikroC Pro for PIC Program.

```
sbit LCD_D4_Direction at TRISB0_bit;
sbit LCD_D5_Direction at TRISB1_bit;
sbit LCD_D6_Direction at TRISB2_bit;
sbit LCD_D7_Direction at TRISB3_bit;
// End LCD module connections

#define MODE PORTD.RD0                                  // MODE button
#define UP PORTD.RD1                                    // UP button
#define DOWN PORTD.RD2                                  // DOWN button
#define SETUP 0

unsigned char seconds, minutes, hours, day, month, year;
unsigned char newday, newmonth, newyear, newhour, newminutes, newseconds;
//
// Ths function reads the Date and Time from the RTC chip
//
void Read_Date_Time()
{
  I2C1_Start();                                         // Send START bit to RTC chip
  I2C1_Wr(0xA0);                                        // Address the RTC chip
  I2C1_Wr(0x02);                                        // Start from address 2 (seconds)
  I2C1_Repeated_Start();                                // Issue repeated START bit
  I2C1_Wr(0xA1);                                        // Address the RTC chip for reading
  seconds = I2C1_Rd(1);                                 // Read seconds, send ack
  minutes = I2C1_Rd(1);                                 // Read minutes, send ack
  hours = I2C1_Rd(1);                                   // Read hours, send ack
  day = I2C1_Rd(1);                                     // Read year/day, send ack
  month = I2C1_Rd(0);                                   // Read month, no ack (last byte)
  I2C1_Stop();                                          // Send STOP bit to RTC chip

}


//
// This function converts the date-time into correct format for displaying on the LCD. The
// numbers in RTC memory are in BCD form and are converted as follows:
//   "extract upper byte, shift right 4 bits, multiply by 10, add lower byte".
// For example, number 25 in RTC memory is stored as 37. i.e bit pattern: "0010 0101".
// After the conversion we obtain the required number 25.
//
Convert_Date_Time()
{
  seconds = ((seconds & 0xF0) >> 4)*10 + (seconds & 0x0F);
  minutes = ((minutes & 0xF0) >> 4)*10 + (minutes & 0x0F);
  hours = ((hours & 0xF0) >> 4)*10 + (hours & 0x0F);
  month = ((month & 0x10) >> 4)*10 + (month & 0x0F);;
  year = (day & 0xC0) >> 6;
  day = ((day & 0x30) >> 4)*10 + (day & 0x0F);
}

//
```

Figure 7.65
cont'd

```
// Display the date and time on the LCD
//
void Display_Date_Time()
{
//
// Write day, month, year as: dd=mm=xxxy
//
    Lcd_Chr(1, 1, (day / 10) + '0');
    Lcd_Chr(1, 2, (day % 10) + '0');
    Lcd_Chr(1, 4, (month / 10) + '0');
    Lcd_Chr(1, 5, (month % 10) + '0');
    Lcd_Chr(1 , 10, year + '0');
//
// Write hour, minutes, seconds as: hh:mm:ss
//
    Lcd_Chr(2, 1, (hours / 10) + '0');
    Lcd_Chr(2, 2, (hours % 10) + '0');
    Lcd_Chr(2, 4, (minutes / 10) + '0');
    Lcd_Chr(2, 5, (minutes % 10) + '0');
    Lcd_Chr(2, 7, (seconds / 10) + '0');
    Lcd_Chr(2, 8, (seconds % 10) + '0');
}

//
// This function gets the date and Time from the user via the 3 buttons. New values
// of Date and Time are returned to the calling program. Initially the maximum values
// are shown and these can be changed using the UP and DOWN buttons.
//
unsigned char Set_Date_Time(unsigned char *str, unsigned char min, unsigned char max)
{
    unsigned char c, Txt[4];

    ByteToStr(max, Txt);                      // Convert max value to string in Txt
    Lcd_Cmd(_LCD_CLEAR);
    Lcd_Out(1,1,str);                         // Display field name (e.g. "DAY:")
    Lcd_Out(2,1,Txt);                         // Display max value to start with
    c = max;

    while(MODE == 1)                          // While MODE button not pressed
    {
      if(UP == 0)                             // If UP button pressed (increment)
      {
        Delay_Ms(10);
        while(UP == 0);                       // Wait until UP button is released
        c++;                                  // Increment value
        if(c > max)c = min;                   // If greater than max rollover to min
      }
      if(DOWN == 0)                           // If DOWN button is pressed
      {
        Delay_Ms(10);
        while(DOWN == 0);                     // Wait until DOWN button is released
```

Figure 7.65
cont'd

```
          c--;                                    // Decrement value
          if(c < min || c == 255)c = max;         // If less than min, rollover to max
        }
        ByteToStr(c, Txt);                        // Convert selected value to string
        Lcd_Out(2,1,Txt);                         // Display selected value on LCD
      }
      Delay_Ms(10);
      while(MODE == 0);                           // Wait until MODE button is released
      return c;                                   // return number to calling program
}


//
// This function sets the RTC with the new Date and Time values. The number to be sent to
// the RTC chip is divided by 10, shifted left 4 digits, and the remainder is added to it. Thus,
// for example if the number is decimal 25, it is converted into bit pattern "0010 0101" and
// stored at the appropriate RTC memory
//
void Set_RTC()
{
//
// Convert Date and Time into a format compatible with the RTC chip
//
   seconds = ((newseconds / 10) << 4) + (newseconds % 10);
   minutes = ((newminutes / 10) << 4) + (newminutes % 10);
   hours = ((newhour / 10) << 4) + (newhour % 10);
   month = ((newmonth / 10) << 4) + (newmonth % 10);
   day = (newyear << 6) + ((newday / 10) << 4) + (newday % 10);

   I2C1_Start();                                  // Send START bit to RTC chip
   I2C1_Wr(0xA0);                                 // Address the RTC chip
   I2C1_Wr(0x00);                                 // Start from address 0
   I2C1_Wr(0x80);                                 // Pause RTC counter
   I2C1_Wr(0x00);                                 // Write to hundreths memory location
   I2C1_Wr(seconds);                              // Write to seconds memory location
   I2C1_Wr(minutes);                              // Write to minutes memory location
   I2C1_Wr(hours);                                // Write to hours memory location;
   I2C1_Wr(day);                                  // Write to year/day memory location
   I2C1_Wr(month);                                // Write to month memory location
   I2C1_Stop();                                   // Send STOP bit

   I2C1_Start();                                  // Send START bit to RTC chip
   I2C1_Wr(0xA0);                                 // Address the RTC chip
   I2C1_Wr(0);                                    // Start from address 0 (Configuration reg)
   I2C1_Wr(0);                                    // Write 0 to Conf reg to start counter
   I2C1_Stop();                                   // Send STOP bit
}



void main()
```

Figure 7.65
cont'd

```
{
    ANSELB = 0;                              // Configure PORTB as digital
    ANSELC = 0;                              // Configure PORTC as digital
    ANSELD = 0;                              // Configure PORTD as digital
    TRISD.RD0 = 1;                           // Configure RD0 (MODE)as input
    TRISD.RD1 = 1;                           // Configure RD1 (UP) as input
    TRISD.RD2 = 1;                           // Configure RD2 (DOWN) as input

    Lcd_Init();                              // Initialize LCD
    Lcd_Cmd(_LCD_CURSOR_OFF);                // Disable cursor
    Lcd_Cmd(_LCD_CLEAR);                     // Clear LCD

    I2C1_Init(100000);                       // Initialize I2C module
//
// If the MODE button is pressed on startup we must get into SETUP phase
//
    if(MODE == SETUP)                        // If SETUP mode
    {
        while(MODE == SETUP);                // Wait until MODE button is released
        newday = Set_Date_Time("DAY:",1,31);       // Get current day
        newmonth = Set_Date_Time("MONTH:", 1,12);  // Get current month
        newyear = Set_Date_Time("YEAR (201x):", 3,6);  // Get current year (201x)
        newhour = Set_Date_Time("HOUR:", 0, 23);   // Get current hour
        newminutes = Set_Date_Time("MINUTES:", 0, 59);  // Get current minutes
        newseconds = Set_Date_Time("SECONDS:", 0, 59);  // get current seconds
        //
        // We have got all the new Date and Time values. Now set the RTC with these values
        //
        Set_RTC();
    }


//
// Read the Date and Time from the RTC chip and display on the LCD in the following format:
// Row 1: dd-mm-yyyy
// Row 2: hh:mm:ss
//
    Lcd_Out(1, 1, "dd-mm-2013");
    Lcd_Out(2, 1, "hh:mm:ss");
    while(1)
    {
        Read_Date_Time();                    // Read Date and Time from RTC chip
        Convert_Date_Time();                 // Convert into a form to display
        Display_Date_Time();                 // Display Date and Time on the LCD
    }
}
```

Figure 7.65
cont'd

It is important to notice that the PCF8583 RTC chip stores the date and time values in the BCD format. Thus, for example, if the current minutes is 28, it is stored internally as "00101000" (which is decimal 40). Function Convert_Date_Time converts the BCD minutes into decimal using the following code:

$$minutes = ((minutes \ \& \ 0xF0) >> 4)*10 + (minutes \ \& \ 0x0F);$$

The upper nibble ("0010") is shifted right by four digits, which becomes decimal 2. It is then multiplied by 10 to give 20. The lower nibble (8) is then added to give the required value 28.

Function Display_Date_Time divides the given number by 10 to extract the 10s digit. Then, character '0' is added to convert it into ASCII, and it is then displayed using the LCD_Chr function. The 1s digit is also determined by finding the remainder of the division by 10. Again the value is converted into ASCII and displayed on the LCD. The following code is used:

```
Lcd_Chr(2, 4, (minutes/10) + '0');
Lcd_Chr(2, 5, (minutes % 10) + '0');
```

Before loading the clock registers, the opposite process is done, that is, the number is converted into BCD. As an example, the following code is used to convert the minutes:

```
minutes = ((newminutes/10) << 4) + (newminutes % 10);
```

The decimal number is divided by 10 to find the 10s digit. This number is shifted left by four digits so that it occupies the upper nibble position. Then the 1s digit is determined by finding the remainder, and added to the number as the second BCD digit.

MPLAB XC8

The MPLAB XC8 compiler supports the following I^{2}C functions (header file <plib/ i2C.h> must be included at the beginning of the program):

AckI2C	Generate acknowledgement
CloseI2C	disable MSSP module
DataRdyI2C	Check if data is available in the I^{2}C buffer
getcI2C	Read a byte from the I^{2}C bus
getsI2C	Read a string from the I^{2}C bus (in master mode)
IdleI2C	Loop until I^{2}C bus is idle
NotAckI2C	Generate not acknowledgement condition
OpenI2C	Configure the I^{2}C module
putcI2C	Write a byte to the I^{2}C bus
RestartI2C	Generate a restart condition
StartI2C	Generate a start condition
StopI2C	Generate a stop condition

The program is basically similar to the mikroC Pro for PIC version, but here the initialization routines and the LCD functions are different. Also, the E, RS, and RW pins of the LCD are connected to port pins RB4, RB5, and RB6, respectively.

The MPLAB XC8 version of the program is left as an exercise for the reader.

Project 7.7—Real-Time Alarm Clock

This project is an extension to Project 7.5. In this project, we set up a daily alarm using the PCF8583 RTC chip. The time of the alarm can be set as in the previous project. An LED is used to indicate the alarm condition (we could have also used a buzzer) and the LED turns ON when alarm occurs. The alarm condition stays until a button is pressed to stop the alarm. The alarm occurs daily at the same time every day.

Figure 7.66 shows the block diagram of the project. The functions of the buttons are as follows:

MODE: Used to enter the clock setup mode. Keep this button pressed while resetting the microcontroller. This button is also used to move between the fields while setting the clock or the alarm time.

UP: Used during clock or alarm setup. Pressing the button increments the value in the selected field.

DOWN: Used during clock or alarm setup. Pressing the button decrements the value in the selected field.

ALARM SETUP: Used to enter the alarm setup mode. Keep this button pressed while resetting the microcontroller.

STOP ALARM: Pressing this button stops the present alarm condition (turns OFF the LED). The alarm will occur at the same time every day.

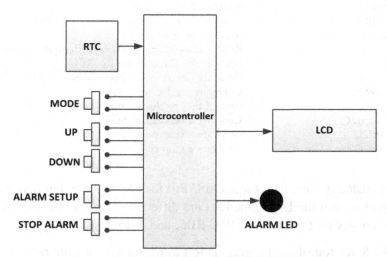

Figure 7.66: Block Diagram of the Project.

Bit 2 of the control and status register (at address 0x00) must be set for the alarm functions to be enabled. When an alarm occurs the INT pin of the PCF8583 goes from logic 1 to logic 0 to indicate the alarm condition. The INT bit can be cleared by clearing bit 0 of the control and status register.

Alarm functions are configured via register 0x08. Figure 7.67 shows the bit definitions of this register. To set daily alarms, the following bits must be set:

Bit 7: Lower INT pin when alarm occurs.
Bit 4: Set daily alarms.

When daily alarms are set, the day, month, and year fields are ignored. An alarm is generated when the contents of the alarm registers match the involved counter registers.

New date and time are loaded into the chip using the MODE, UP, and DOWN buttons as described in the previous project. New daily alarm time is loaded into the chip using the ALARM SETUP, MODE, UP, and DOWN buttons. The steps are given below as an example to set the daily alarm to occur every day at 10:00:00:

- Reset the microcontroller while pressing the ALARM SETUP button.
- The LCD should display:

```
ALRM HOUR:
23
```

- Keep pressing the UP button until 10 is displayed.
- Press MODE button to change the field to minutes:

```
ALRM MINS:
59
```

- Keep pressing the UP button until 0 is displayed.
- Press MODE button to change the field to seconds:

```
ALRM SECS:
59
```

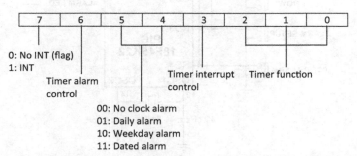

Figure 7.67: Alarm Control Register (Address 0x08).

- Keep pressing the UP button until 0 is displayed.
- Press MODE button to return to the clock mode. The daily alarm time will be set to 10:00:00.

Project Hardware

The circuit diagram of the project is shown in Figure 7.68. This circuit is similar to the one given in Figure 7.62, but here additional buttons and an LED are used for the alarm part of the project. Also, the alarm output pin (INT) of the PCF8583 is connected to the RC2 pin of the microcontroller. Note that this pin is active LOW, that is, it is normally HIGH and goes LOW when an alarm occurs.

Project PDL

The project PDL is shown in Figure 7.68.

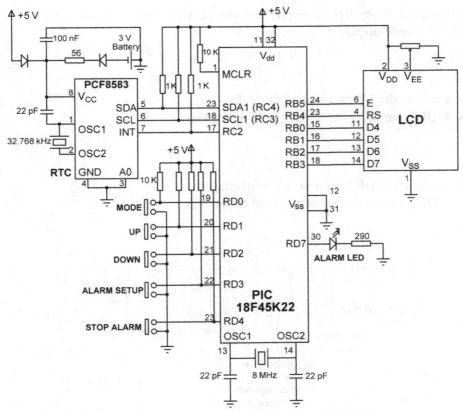

Figure 7.68: Circuit Diagram of the Project.

Project Program

mikroC Pro for PIC

The mikroC Pro for the PIC program listing is given in Figure 7.69 (MIKEOC-I2C2.C). The major part of the program is the same as the clock setting and display program given in the previous project. At the beginning of the program, the connections between the LCD and the microcontroller are defined. Symbols MODE, UP, DOWN, ALARM SETUP, STOP ALARM, and SETUP are assigned to port bits, PORTB, PORTC, PORTD are configured as digital with RD0:RD4 pins configured as inputs. Then, the LCD and the I^2C modules are initialized. If the MODE button is pressed (MODE = 0) during the reset, the program enters the clock setup phase.

If the ALARM SETUP button is pressed during the reset, the program enters the alarm setup phase. Here, the alarm hours, minutes, and seconds are read via the UP/DOWN/ MODE buttons as described earlier, using function Set_Date_Time function. The, Set_RTC_Alarm function is called to load the PCF8583 registers for the daily alarm so that the alarm occurs every day exactly at the selected time. Here, the alarm register is loaded with 0x90, which enables daily alarms and selects the INT pin as the alarm output. While displaying the current date and time, the program continuously checks the state of the INT pin (this could also be configured as an external interrupt, but the external interrupt pins are used for the LCD) and an alarm condition is said to occur if this pin goes LOW. The program turns ON the ALARM LED to indicate the alarm condition. The present alarm condition can be cleared by pressing the STOP ALARM button. Pressing this button calls function Reset_Alarm_Flag, which clears the timer flag (located at bit 0 of the control and status register) to set the INT pin back to HIGH to stop the alarm condition.

Project 7.8—SD Card Projects—Write Text To a File

In this and the next few projects, we will be using SD cards as storage devices. But before going into the details of these projects, we should take a look at the basic principles and operation of SD card memory devices.

SD cards are commonly used in many electronic devices where a large amount of nonvolatile data storage is required. Some application areas are as follows:

- Digital cameras,
- Camcorders,
- Printers,
- Laptop computers,
- GPS receivers,

```
/*******************************************************************************
                          REAL TIME ALARM CLOCK
                          ====================
```

This project is about designing an accurate real time alarm clock using the PCF8583 RTC chip

The PCF8583 chip is connected to the I2C pins (modul 1) of a PIC18F45K22 microcontroller.
The connections between the PCF8583 and the microcontroller are as follows The SDA, SCL
and INT lines are pulled high with resistors):

```
PCF8583  Microcontroller
=======  ===============
  SDA     SDA1 (RC4)
  SCL     SCL1 (RC3)
  INT     RC2
```

Clock timing to the PCF8583 is provided with a 32.768 kHz crystal. In this project both the
clock and daily alarms can be set. When an alarm occurs, an LED connected to pin RD7 is
turned ON.

```
MODE (RD0):        Enter clock setup mode. Also change fields during setup
UP (RD1) :         Increments the value
DOWN (RD2):        Decrements the value
ALARM SETUP (RD3): Enter Alarm setup mode
STOP ALARM (RD4):  Stop present alarm condition
```

AN LCD is connected to PORTB of the microcontroller to help in setting the clock and alarm
time and also for displaying the clock data in real time.

The microcontroller operated with an 8 MHz crystal with the PLL disabled (i.e. the actual
running clock frequency is 8 MHz).

```
Author:  Dogan Ibrahim
Date:    October 2013
File:    MIKROC-I2C2.C
*******************************************************************************/
// LCD module connections
sbit LCD_RS at LATB4_bit;
sbit LCD_EN at LATB5_bit;
sbit LCD_D4 at LATB0_bit;
sbit LCD_D5 at LATB1_bit;
sbit LCD_D6 at LATB2_bit;
sbit LCD_D7 at LATB3_bit;

sbit LCD_RS_Direction at TRISB4_bit;
sbit LCD_EN_Direction at TRISB5_bit;
sbit LCD_D4_Direction at TRISB0_bit;
sbit LCD_D5_Direction at TRISB1_bit;
sbit LCD_D6_Direction at TRISB2_bit;
sbit LCD_D7_Direction at TRISB3_bit;
// End LCD module connections
```

Figure 7.69: mikroC Pro for PIC Program.

```
#define MODE PORTD.RD0                  // MODE button
#define UP PORTD.RD1                    // UP button
#define DOWN PORTD.RD2                  // DOWN button
#define ALARM_SETUP PORTD.RD3           // ALARM SETUP button
#define STOP_ALARM PORTD.RD4            // Alarm stop button
#define ALARM_INT PORTC.RC2             // RTC clock alarm INT pin
#define ALARM_LED PORTD.RD7             // LED connected to RD7
#define SETUP 0


unsigned char seconds, minutes, hours, day, month, year;
unsigned char newday, newmonth, newyear, newhour, newminutes, newseconds;
//
// Ths function reads the Date and Time from the RTC chip
//
void Read_Date_Time()
{
  I2C1_Start();                         // Send START bit to RTC chip
  I2C1_Wr(0xA0);                        // Address the RTC chip
  I2C1_Wr(0x02);                        // Start from address 2 (seconds)
  I2C1_Repeated_Start();                // Issue repeated START bit
  I2C1_Wr(0xA1);                        // Address the RTC chip for reading
  seconds = I2C1_Rd(1);                 // Read seconds, send ack
  minutes = I2C1_Rd(1);                 // Read minutes, send ack
  hours = I2C1_Rd(1);                   // Read hours, send ack
  day = I2C1_Rd(1);                     // Read year/day, send ack
  month = I2C1_Rd(0);                   // Read month, no ack (last byte to read)
  I2C1_Stop();                          // Send STOP bit to RTC chip

}


//
// This function convert the date-time into correct format for displaying on the LCD. The numbers
// in RTC memory are in BCD form and are converted as follows:
//    "extract upper byte, shift right 4 bits, multiply by 10, add lower byte".
// For example, number 25 in RTC memory is stored as 37. i.e bit pattern: "0010 0101". After the
// conversion we obtain the required number 25.
//
Convert_Date_Time()
{
  seconds = ((seconds & 0xF0) >> 4)*10 + (seconds & 0x0F);
  minutes = ((minutes & 0xF0) >> 4)*10 + (minutes & 0x0F);
  hours =  ((hours & 0xF0) >> 4)*10 + (hours & 0x0F);
  month = ((month & 0x10) >> 4)*10 + (month & 0x0F);;
  year = (day & 0xC0) >> 6;
  day = ((day & 0x30) >> 4)*10 + (day & 0x0F);
}

//
```

Figure 7.69
cont'd

```
// Display the date and time on the LCD
//
void Display_Date_Time()
{
//
// Write day, month, year as: dd=mm=xxxy
//
    Lcd_Chr(1, 1, (day / 10) + '0');
    Lcd_Chr(1, 2, (day % 10) + '0');
    Lcd_Chr(1, 4, (month / 10) + '0');
    Lcd_Chr(1, 5, (month % 10) + '0');
    Lcd_Chr(1 , 10, year + '0');
//
// Write hour, minutes, seconds as: hh:mm:ss
//
    Lcd_Chr(2, 1, (hours / 10) + '0');
    Lcd_Chr(2, 2, (hours % 10) + '0');
    Lcd_Chr(2, 4, (minutes / 10) + '0');
    Lcd_Chr(2, 5, (minutes % 10) + '0');
    Lcd_Chr(2, 7, (seconds / 10) + '0');
    Lcd_Chr(2, 8, (seconds % 10) + '0');
}


//
// This function gets the date and Time from the user via the 3 buttons. New values of Date
// and Time are returned to the calling program. Initially the maximum values are shown
// and these can be changed using the UP and DOWN buttons.
//
unsigned char Set_Date_Time(unsigned char *str, unsigned char min, unsigned char max)
{
    unsigned char c, Txt[4];

    ByteToStr(max, Txt);                    // Convert maximum value to string in Txt
    Lcd_Cmd(_LCD_CLEAR);
    Lcd_Out(1,1,str);                       // Display field name (e.g. "DAY:")
    Lcd_Out(2,1,Txt);                       // Display maximum value to start with
    c = max;

    while(MODE == 1)                        // While MODE button is not pressed
    {
      if(UP == 0)                           // If UP button is pressed (increment)
      {
        Delay_Ms(10);
        while(UP == 0);                     // Wait until UP button is released
        c++;                                // Increment value
        if(c > max)c = min;                 // If greater than max, rollover to min
      }
      if(DOWN == 0)                         // If DOWN button is pressed
      {
        Delay_Ms(10);
        while(DOWN == 0);                   // Wait until DOWN button is released
```

Figure 7.69
cont'd

```
      c--;                                // Decrement value
      if(c < min || c == 255)c = max;     // If less than min, rollower to max
    }
    ByteToStr(c, Txt);                    // Convert selected value to string
    Lcd_Out(2,1,Txt);                     // Display selected value on LCD
  }
  Delay_Ms(10);
  while(MODE == 0);                       // Wait until MODE button is released
  return c;                               // Return selected number to calling program
}


//
// This function sets the RTC with the new Date and Time values. The number to be sent to
// the RTC chip is divided by 10, shifted left 4 digits, and the remainder is added to it. Thus
// for example if the number is decimal 25, it is converted into bit pattern  "0010 0101" and
// stored at the appropriate RTC memory
//
void Set_RTC()
{
//
// Convert Date and Time into a format compatible with the RTC chip
//
  seconds = ((newseconds / 10) << 4) + (newseconds % 10);
  minutes = ((newminutes / 10) << 4) + (newminutes % 10);
  hours = ((newhour / 10) << 4) + (newhour % 10);
  month = ((newmonth / 10) << 4) + (newmonth % 10);
  day = (newyear << 6) + ((newday / 10) << 4) + (newday % 10);

  I2C1_Start();                           // Send START bit to RTC chip
  I2C1_Wr(0xA0);                          // Address the RTC chip
  I2C1_Wr(0x00);                          // Start from address 0
  I2C1_Wr(0x80);                          // Pause RTC counter
  I2C1_Wr(0x00);                          // Write to hundreths memory location
  I2C1_Wr(seconds);                       // Write to seconds memory location
  I2C1_Wr(minutes);                       // Write to minutes memory location
  I2C1_Wr(hours);                         // Write to hours memory location;
  I2C1_Wr(day);                           // Write to year/day memory location
  I2C1_Wr(month);                         // Write to month memory location
  I2C1_Stop();                            // Send STOP bit

  I2C1_Start();                           // Send START bit to RTC chip
  I2C1_Wr(0xA0);                          // Address the RTC chip
  I2C1_Wr(0);                             // Start from address 0 (Configuration reg)
  I2C1_Wr(0);                             // Write 0 to Configuration reg to start
counter
  I2C1_Stop();                            // Send STOP bit
}

//
// This function sets the RTC alarm. The alarm is configured to occur every day
```

Figure 7.69
cont'd

```
// at the set time. The BUZZER sounds when the alarm occurs
//
void Set_RTC_Alarm()
{
//
// Convert Alarm Time into a format compatible with the RTC chip. For daily Alarm
// setup the Date fields (day, month, year) are ignored
//
  seconds = ((newseconds / 10) << 4) + (newseconds % 10);
  minutes = ((newminutes / 10) << 4) + (newminutes % 10);
  hours = ((newhour / 10) << 4) + (newhour % 10);

  I2C1_Start();                         // Send START bit to RTC chip
  I2C1_Wr(0xA0);                        // Address the RTC chip
  I2C1_Wr(0x00);                        // Start from address 0
  I2C1_Wr(0x04);                        // Enable Alarm Control register
  I2C1_Stop();

  I2C1_Start();
  I2C1_Wr(0xA0);                        // Address the RTC chip
  I2C1_Wr(0x08);                        // Start from address 8
  I2C1_Wr(0x90);                        // Enable Daily alarms, enable INT output
  I2C1_Wr(0x00);                        // Write to hundreths memory location
  I2C1_Wr(seconds);                     // Write to seconds memory location
  I2C1_Wr(minutes);                     // Write to minutes memory location
  I2C1_Wr(hours);                       // Write to hours memory location;
  I2C1_Stop();                          // Send STOP bit
}


//
// This function stops the alarm by clearing the timer register of the
// control and status register
//
void Reset_Alarm_Flag()
{
  I2C1_Start();                         // Send START bit to RTC chip
  I2C1_Wr(0xA0);                        // Address the RTC chip
  I2C1_Wr(0x00);                        // Start from address 0
  I2C1_Wr(0x04);                        // Reset Alarm flag
  I2C1_Stop();                          // Send STOP bit
}


void main()
{
  ANSELB = 0;                           // Configure PORTB as digital
  ANSELC = 0;                           // Configure PORTC as digital
  ANSELD = 0;                           // Configure PORTD as digital
  TRISC.RC2 = 1;                        // Alarm INT input
  TRISD.RD0 = 1;                        // Configure RD0 (MODE)as input
  TRISD.RD1 = 1;                        // Configure RD1 (UP) as input
```

Figure 7.69
cont'd

```
    TRISD.RD2 = 1;                                    // Configure RD2 (DOWN) as input
    TRISD.RD3 = 1;                                    // Configure RD3 (Alarm setup)
    TRISD.RD4 = 1;                                    // Configure RD4 as input (Alarm STOP)
    TRISD.RD7 = 0;                                    // Alarm LED is output

    ALARM_LED = 0;                                    // Alarm LEDOFF to start with
    Lcd_Init();                                       // Initialize LCD
    Lcd_Cmd(_LCD_CURSOR_OFF);                         // Disable cursor
    Lcd_Cmd(_LCD_CLEAR);                              // Clear LCD


    I2C1_Init(100000);                                // Initialize I2C module
//
// If the MODE button is pressed on startup we must get into SETUP phase
//
    if(MODE == SETUP)                                 // If CLOCK SETUP mode
    {
        while(MODE == SETUP);                         // Wait until MODE button is released
        newday = Set_Date_Time("DAY:",1,31);          // Get current day
        newmonth = Set_Date_Time("MONTH:", 1,12);         // Get current month
        newyear = Set_Date_Time("YEAR (201x):", 3,6);     // Get current year (201x)
        newhour = Set_Date_Time("HOUR:", 0, 23);          // Get current hour
        newminutes = Set_Date_Time("MINUTES:", 0, 59);    // Get current minutes
        newseconds = Set_Date_Time("SECONDS:", 0, 59);    // Get current seconds
        //
        // We have got all the new Date and Time values. Now set the RTC with these values
        //
        Set_RTC();
    }
    if(ALARM_SETUP == 0)                              // If ALARM SETUP mode
    {
        while(ALARM_SETUP = 0);                       // Wait until ALARM button is released
        newhour = Set_Date_Time("ALRM HOUR:", 0, 23);
        newminutes = Set_Date_Time("ALRM MINS:", 0, 59);
        newseconds = Set_Date_Time("ALRM SECS:", 0, 59);
        //
        // We have got the Daily Alarm time. Now set the RTC clock to generate alarm
        // at this time every day. Sound the BUZZER when alarm is generated.
        //
        Set_RTC_Alarm();
    }


//
// Read the Date and Time from the RTC chip and display on the LCD in the following format:
// Row 1: dd-mm-yyyy
// Row 2: hh:mm:ss
//
    Lcd_Out(1, 1, "dd-mm-2013");
    Lcd_Out(2, 1, "hh:mm:ss");
    while(1)
```

Figure 7.69

cont'd

```
{
    Read_Date_Time();                    // Read Date and Time from RTC chip
    Convert_Date_Time();                 // Convert into a form to display
    Display_Date_Time();                 // Display Date and Time on the LCD

    if(ALARM_INT == 0)                   // If alarm occurred (active LOW)
    {
        ALARM_LED = 1;                   // Turn ON ALARM LED
        if(STOP_ALARM == 0)              // if STOP ALARM button pressed
        {
            ALARM_LED = 0;               // Turn OFF ALARM LED
            Reset_Alarm_Flag();          // Reset alarm flag back to 1
        }
    }
}
}
```

Figure 7.69
cont'd

- Electronic games,
- Personal digital assistants (PDAs),
- Mobile phones,
- Embedded electronic systems.

Figure 7.70 shows the picture of a typical SD card.

Figure 7.70: A Typical SD Card.

Table 7.5: Different Size SD Card Specifications.

	Standard SD	miniSD	mikroSD
Dimensions	32 × 24 × 2.1 mm	21.5 × 20 × 1.4 mm	15 × 11 × 1 mm
Card weight	2.0 g	0.8 g	0.25 g
Operating voltage	2.7–3.6 V	2.7–3.6 V	2.7–3.6 V
Write protect	Yes	No	No
Pins	9	11	8
Interface	SD or SPI	SD or SPI	SD or SPI
Current consumption	100 mA (Write)	100 mA (Write)	100 mA (Write)

The SD card is a flash memory storage device designed to provide a high capacity, nonvolatile, and rewritable storage in small size. The memory capacity of the SD cards is increasing all the time. Currently, they are available in capacities from several gigabytes to >128 GB. SD cards are available in three sizes: *standard SD card*, *miniSD card*, and the *microSD card*. Table 7.5 lists the main specifications of different size cards.

SD card specifications are maintained by the *SD Card Association*, which has >600 members. MiniSD and microSD cards are electrically compatible with the standard SD cards, and they can be inserted in special adapters and used as standard SD cards in standard card slots.

SD card speeds are measured in three different ways: in kilobytes per second (kB/s), in megabytes per second (MB/s), or in an "x" rating similar to that of CD-ROMS where "x" is the speed corresponding to 150 kB/s. Thus, the various "x" based speeds are as follows:

- 4x: 600 kB/s,
- 16x: 2.4 MB/s,
- 40x: 6.0 MB/s,
- 66x: 10 MB/s.

As far as the memory capacity is concerned, we can divide SD cards into three families: Standard-Capacity (SDSC), High-Capacity (SDHC), and eXtended-Capacity (SDXC). SDSC are the older cards with capacities 1–2 GB. SDHC have capacities 4–32 GB, and SDXC cards have capacities >32–128 GB. The SD and SDHC families are available in all three sizes, but the SDXC family is not available in the mini size.

In the projects in this book, we shall be using the standard SD cards only. The use of the smaller size SD cards is virtually the same and is not described here any further.

SD cards can be interfaced to microcontrollers using two different protocols: SD card protocol and the SPI protocol. The SPI protocol is the most commonly used protocol and is the one used in the projects in this book. SPI bus is currently used by microcontroller interface circuits to talk to a variety of devices such as

- Memory devices (SD cards),
- Sensors,
- RTCs,
- Communications devices,
- Displays.

The advantages of the SPI bus are as follows:

- Simple communication protocol,
- Full duplex communication,
- Very simple hardware interface.

In addition, the disadvantages of the SPI bus are

- Requires four pins,
- No hardware flow control,
- No slave acknowledgement.

It is important to realize that there are no SPI standards governed by any international committee. As a result of this, there are several versions of the SPI bus implementation. In some applications, the MOSI and MISO lines are combined into a single data line, thus reducing the line requirements into three. Some implementations have two clocks, one to capture (or display) data and another to clock it into the device. Also, in some implementations, the chip select line may be active-high rather than active low.

The standard SD card has nine pins with the pin layout shown in Figure 7.71. Depending on the interface protocol used, pins have different functions. Table 7.6 gives the function of each pin in both the SD mode and the SPI mode of operation.

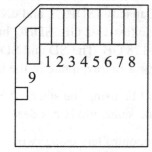

Figure 7.71: Standard SD Card Pin Layout.

Table 7.6: Standard SD Card Pin Definitions.

Pin	Name	SD Description	SPI Description
1	CD/DAT3/CS	Data line 3	Chip select
2	CMD/Datain	Command/response	Host to card command and data
3	V_{SS}	Supply ground	Supply ground
4	V_{DD}	Supply voltage	Supply voltage
5	CLK	Clock	Clock
6	V_{SS2}	Supply voltage ground	Supply voltage ground
7	DAT0	Data line 0	Card to host data and status
8	DAT1	Data line 1	Reserved
9	DAT2	Data line 2	Reserved

Operation of the SD Card in the SPI Mode

When the SD card is operated in the SPI mode only seven pins are used:

- Two power supply ground (pins 3 and 6),
- Power supply (pin 4),
- Chip select (pin 1),
- Data out (pin 7),
- Data in (pin 2),
- CLK (pin 5).

Three pins are used for the power supply, leaving four pins for the SPI mode of operation:

- Chip select (pin 1),
- Data out (pin 7),
- Data in (pin 2),
- CLK (pin 5).

At power-up, the SD card defaults to the SD bus protocol. The card is switched to the SPI mode if the CS signal is asserted during the reception of the reset command. When the card is in the SPI mode, it only responds to SPI commands. The host may reset a card by switching the power supply off and on again.

Most high-level language compilers normally provide a library of commands for initializing, reading, and writing to SD cards. In general, it is not necessary to know the internal structure of an SD card before it can be used since the available library functions can easily be used. It is however important to have some knowledge about the internal structure of an SD card so that it can be used efficiently. In this section, we shall be looking briefly at the internal architecture and the operation of SD cards.

An SD card has a set of registers that provide information about the status of the card. When the card is operated in the SPI mode, these registers are as follows:

- Card Identification Register (CID),
- Card Specific Data Register (CSD),
- SD Configuration Register (SCR),
- Operation Control Register (OCR).

The CID consists of 16 bytes, and it contains the manufacturer ID, product name, product revision, card serial number, manufacturer date code, and a checksum byte.

The CSD consists of 16 bytes, and it contains card-specific data such as the card data transfer rate, read/write block lengths, read/write currents, erase sector size, file format, write protection flags, and checksum.

The SCR is 8 bytes long, and it contains information about the SD card's special features capabilities such as the security support, and data bus widths supported.

The OCR is only 4 bytes long, and it stores the V_{DD} voltage profile of the card. The OCR shows the voltage range in which the card data can be accessed.

All SD card SPI commands are 6 bytes long with the MSB transmitted first. The first byte is known as the "command" byte, and the remaining 5 bytes are "command arguments". Bit 6 of the command byte is set to "1" and the MSB bit is always "0". With the remaining 6 bits, we have 64 possible commands, named CMD0 to CMD63. Some of the important commands are

- CMD0   GO_IDLE_STATE (Resets the SD card),
- CMD1   SEND_OP_COND (Initializes the card),
- CMD9   SEND_CSD (Get CSD data),
- CMD10   SEND_CID (Get CID data),
- CMD16   SET_BLOCKLEN (Selects a block length in bytes),
- CMD17   READ_SINGLE_BLOCK (Reads a block of data),
- CMD24   WRITE_BLOCK (Writes a block of data),
- CMD32   ERASE_WR_BLK_START_ADDR (Sets the address of the first write block to be erased),
- CMD33   ERASE_WR_BLK_END_ADDR (Sets the address of the last write block to be erased),
- CMD38   ERASE (Erases all previously selected blocks).

In response to a command, the card sends a status byte known as R1. The MSB bit of this byte is always "0" and the other bits indicate various error conditions.

Reading Data

The SD card in the SPI mode supports single block and multiple block read operations. The host should set the block length, and after a valid read command, the card responds with a response token, followed by a data block and a CRC check. The block length can be between 1 and 512 bytes. The starting address can be any valid address range of the card.

In multiple block read operations, the card sends data blocks with each block having its own CRC check attached to the end of the block.

Writing Data

The SD card in the SPI mode supports single or multiple block write operations. After receiving a valid write command from the host, the card will respond with a response token and will wait to receive a data block. A 1 byte "start block" token is added to the beginning of every data block. After receiving the data block, the card responds with a "data response" token and the card will be programmed as long as the data block has been received with no errors.

In multiple write operations, the host sends the data blocks one after the other, each preceded with a "start block" token. The card sends a response byte after receiving each data block.

A card can be inserted and removed from the bus without any damage. This is because all data transfer operations are protected by CRC codes, and any bit changes as a result of inserting or removing a card can easily be detected. SD cards operate with a typical supply voltage of 2.7 V. The maximum allowed power supply voltage is 3.6 V. If the card is to be operated from a standard 5.0-V supply, a voltage regulator should be used to drop the voltage to 2.7 V.

The use of an SD card requires the card to be inserted into a special card holder with external contacts (Figure 7.72). Connections can then be made easily to the required card pins.

Figure 7.72: SD Card Holder.

Project Description

This project is about using the card filing system. In this project, a file called
"MYFILE55.TXT" is created on an SD card. Text "This is MYFILE.TXT" is written to
the file initially. The text "This is the added data…" is appended to the file. The file can
be opened on a PC and its contents can be verified.

Project Hardware

Figure 7.73 shows the circuit diagram of the project. The SD card is inserted into a card
holder, and the holder is connected to PORTC of the microcontroller. The interface
between the SD card and the microcontroller ports is as follows:

SD Card Pin	Microcontroller Pin
CS	RC2
CLK	RC3
DO	RC3
DI	RC5

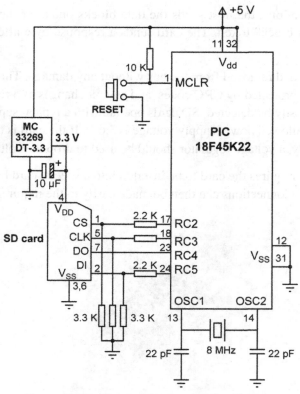

Figure 7.73: Circuit Diagram of the Project.

BEGIN
  Define CS port pin and direction
  Define file open argument definitions
  Configure PORTC as digital
  Initialize SPI bus
  Initialize Mmc_FAT library
  Create new file (if it does not exist)
  Position the cursor at the beginning for writing
  Write text "This is MYFILE.TXT." to the file
  Write "This is the added data..." to the file
  Close the file
END

Figure 7.74: Project PDL.

According to the SD card specifications, when the card is operating with a supply voltage of $V_{DD} = 3.3$ V, the input–output pin voltage levels are as follows:

- Minimum produced output HIGH voltage, VOH = 2.475 V,
- Maximum produced output LOW voltage, VOL = 0.4125 V,
- Minimum required input HIGH voltage, VIH = 2.0625,
- Maximum input HIGH voltage, VIH = 3.6 V,
- Maximum required input LOW voltage, VIL = 0.825 V.

Although the output produced by the card (2.475 V) is sufficient to drive the input port of a PIC microcontroller, the logic HIGH output of the microcontroller (about 4.3 V) is too high for the SD card inputs (maximum 3.6 V). As a result of this, a potential divider is setup at the three inputs of the SD card using 2.2 and 3.3 K resistors. Thus, the maximum voltage at the inputs of the SD card is limited to about 2.5 V:

$$SD \text{ card input voltage} = 4.3 \text{ V} \times 3.3 \text{ K}/(2.2 \text{ K} + 3.3 \text{ K}) = 2.48 \text{ V}$$

The microcontroller is powered from a 5-V supply, which is obtained using a 7805 type 5-V regulator with a 9-V input. The 2.7- to −3.6-V supply required by the SD card is obtained using a MC33269DT-3.3 type regulator with a 3.3-V output, and is driven from the 5-V input voltage.

Project PDL

The project PDL is shown in Figure 7.74.

Project Program
mikroC Pro for PIC

The program listing of the project is given in Figure 7.75 (MIKROC-SD1.C). mikroC Pro for PIC language provides an extensive set of library functions to read and write data to

```
/*******************************************************************************
                      SD CARD PROJECT - WRITE TEXT TO A FILE
                      ===================================

In this project an SD card is connected to PORT C as follows:

    CS   RC2
    CLK  RC3
    DO   RC4
    DI   RC5


The program creates a new file called MYFILE55.TXT on the SD card and writes the text
"This is MYFILE.TXT." to the file. Then the string "This is the added data..." is appended to the file.

Author:  Dogan Ibrahim
Date:    October 2013
File:    MIKROC-SD1.C
*******************************************************************************/
// MMC module connections
sbit Mmc_Chip_Select at LATC2_bit;
sbit Mmc_Chip_Select_Direction at TRISC2_bit;
// MMC module connections

#define FILE_READ 0x01                              // read only
#define FILE_WRITE 0x02                             // write only
#define FILE_APPEND 0x04                            // append to file

char filename[] = "MYFILE55.TXT";
unsigned char txt[] = "This is the added data...";
unsigned short character;
unsigned long file_size,i;

void main()
{
    unsigned char fhandle;

    ANSELC = 0;                                     // Configure PORTC as digital
//
// Initialise the SPI bus
//
    SPI1_Init_Advanced(_SPI_MASTER_OSC_DIV64, _SPI_DATA_SAMPLE_MIDDLE,
                       _SPI_CLK_IDLE_LOW, _SPI_LOW_2_HIGH);
//
// Initialize the Mmc library. Wait until card detected
//
    while(Mmc_Fat_Init());
//
// Create new file (if it does not exist)
//
    fhandle = MMc_Fat_Open(&filename,FILE_WRITE,0x80);
//
// Clear the file, start at the beginning for writing
//
    Mmc_Fat_Rewrite();
//
// Write to the file, specifying the length of the text
```

Figure 7.75: mikroC Pro for PIC Program.

```
                //
                    Mmc_Fat_Write("This is MYFILE.TXT.",19);
                //
                // Add more data to the end...
                //
                    Mmc_Fat_Append();
                    Mmc_Fat_Write(txt,sizeof(txt));
                //
                // Now close the file (releases the handle)
                //
                    Mmc_Fat_Close();

                    while(1);                                    // Wait here forever
                }
```

Figure 7.75
cont'd

SD cards (and also MultiMedia Cards, MMC). Data can be written or read from a given sector of the card, or the file system on the card can be used for more sophisticated applications.

mikroC Pro for PIC compiler supports an SD card library (called the Mmc library) with many functions. Some commonly used functions for file handling are listed below:

- Mmc_Fat_Init   (Initialize the card)
- Mmc_Fat_QuickFormat   (Format the card)
- Mmc_Fat_Assign   (Assign the file we will be working with)
- Mmc_Fat_Reset   (Reset the file pointer. Opens the currently assigned file for reading)
- Mmc_Fat_Rewrite   (Reset the file pointer and clear assigned file. Opens the assigned file for writing)
- Mmc_Fat_Append   (Move file pointer to the end of assigned file so that new data can be appended to the file)
- Mmc_Fat_Read   (Read the byte at which file pointer points to)
- Mmc_Fat_Write   (Write a block of data to the assigned file)
- Mmc_Fat_Delete   (Delete a file)
- Mmc_Set_File_Date   (Write system timestamp to a file)
- Mmc_Fat_Get_File_Dat   (Read file timestamp)
- Mmc_Fat_Get_File_Size   (Get file size in bytes)
- Mmc_Fat_Tell   (Get cursor position in a file)
- Mmc_Fat_Seek   (Set cursor position in a file)
- Mmc_Fat_Rename   (Rename a file)
- Mmc_Fat_MakeDir   (Create a new directory)
- Mmc_Fat_Exists   (Returns information about a file's existence)
- Mmc_Fat_Activate   (Select a file when multiple files are used)

- Mmc_Fat_ReadN   (Read multiple bytes)
- Mmc_Fat_Open   (Open/create a file)
- Mmc_Fat_Close   (Close currently open file)
- Mmc_Fat_EOF   (Check if end of file is reached)

The SPI module has to be initialized through SPIx_Init_Advanced routine with the following parameters. Once the MMC/SD card is initialized, the SPI module can be operated at higher speeds:

- SPI Master,
- Primary prescaler 64,
- Data sampled in the middle of data output time,
- Clock idle low,
- Serial output data changes on transition form low to high edge.

In this project, a new file is created, text is written inside the file, and then the file is closed. At the beginning of the program, the program creates file MYFILE55.TXT by calling library function Mmc_Fat_Open with the arguments as the *filename* and the creation flag 0x80, which tells the function to create a new file if the file does not exist. The filename should be in "filename.extension" format, although it is also possible to specify an eight-digit filename and a three-digit extension with no "." in between as the "." will be inserted by the function. Other allowed values of the creation flag are given in Table 7.7. The Mmc_Fat_Open function returns a file handle. In multiple file operations, we can select the file we wish to operate on by specifying this handle in the Mmc_Fat_Activate function call. Note that the SD card must have been formatted in FAT16 before we can read or write to the card. Most new cards are already formatted, but we can also use the Mmc_Fat_QuickFormat function to format a card.

Function Mmc_Fat_Rewrite is called to clear the file and position the cursor to the beginning, ready for writing. Initial text is written to the file using function

Table 7.7: Mmc_Fat_Open File Creation Fags.

Flag	Description
0x01	Read only
0x02	Hidden
0x04	System
0x08	Volume label
0x10	Subdirectory
0x20	Archive
0x40	Device (internal use only, never found on disk)
0x80	File creation flag. If the file does not exist and this flag is set, a new file with specified name will be created.

Mmc_Fat_Write. Then, function Mmc_Fat_Append is called to append the second text to the file. Finally, function Mmc_Fat_Close is called to close the file and release the handle.

Note that one of the arguments to the Mmc_Fat_Open function is the file mode (FILE_READ, FILE_WRITE, or FILE_APPEND). The following definitions must be made at the beginning of the program before one of these arguments can be used:

```
#define FILE_READ 0x01    // read only
#define FILE_WRITE 0x02   // write only
#define FILE_APPEND 0x04  // append to file
```

MPLAB XC8

For the MPLAB XC8 version of the program, we shall be using the PICDEM PIC18 Explorer board (Chapter 5). A Microchip daughter SD card board (known as the PICtail Daughter Board for SD & MMC Cards, see Figure 7.76) is used as the SD card interface. This board directly plugs onto the PICDEM PIC18 Explorer board (Figure 7.77) and provides the SD card interface to the demonstration board (*note that there are minor design faults with the voltage level conversion circuitry on some of the PICtail Daughter Boards for SD & MMC Cards. You can get around these problems by providing a 3.3-V supply for the daughter board directly from the PICDEM Explorer board. Cut short the*

Figure 7.76: PICtail Daughter Board for SD & MMC Cards.

Figure 7.77: The Daughter Board Plugs onto the PICDEM Board.

power supply pin of the daughter board connector, and connect this pin to the +3.3-V test point on the PICDEM Explorer board).

The SD card daughter board has an on-board positive-regulated charge pump direct current (DC)/DC converter chip (MCP1253) used to convert the +5-V supply to +3.3 V required for the SD card. In addition, the board has buffers to provide correct voltages for the SD card inputs. Seven jumpers are provided on the board to select the SD card signal interface. The following jumpers should be selected:

Jumper		Description
JPI	Pin 1-2	SCK connected to RC3
JP2	Pin 1-2	SDI connected to RC4
JP3	Pin 1-2	SDO connected to RC5
JP4	Pin 2-3 (default)	Card detect to RB4 (not used)
JP5	Pin 2-3 (default)	Write protect to RA4 (not used)
JP6	Pin 2-3 (default)	CS connected to RB3
JP7	Pin 2-3 (default)	Shutdown (not used)

The default jumper positions are connected by circuit tracks on the board, and these tracks should be cut to change the jumper positions if different connections are desired. Signals "card detect", "write protect", and "shutdown" are not used in this book, and the jumper settings can be left as they are. We shall be programming the PIC18 Explorer board using the ICD 3 programmer/debugger as described in earlier chapters.

The MPLAB XC8 version of the SD card program is slightly more complex. This is because a number of included files will have to be extracted and loaded in their correct places. The steps in loading the required files are given below (see Microchip Inc "Application Note 1045 Implementing File I/O Functions Using Microchip's Memory Disk Drive File System Library", located in folder "microchip_solutions_v2013-06-15\Microchip\MDD File System\Documentation" for further information).

- Download and install the "microchip solutions" library from the Microchip Inc website. At the time of writing this book, this library had the name "microchip_solutions_v2013-06-15". The installation process should create a folder called "microchip_solutions_v2013-06-15" under the main C:\ folder.
- Create a new MPLAB XC8 project as before. Choose the processor type as PIC18F8722 since the PICDEM PIC18 Explorer board is shipped with this microcontroller. Create a new main C source file.
- Right click Header Files on the left top window of MPLAB X IDE and select Add Existing Item (Figure 7.78). Find folder "microchip_solutions_v2013-06-15" and select the following files from the subfolders (one at a time):
  C:\Microchip Solutions\Microchip\MDD File System\FSIO.c
  C:\Microchip Solutions\Microchip\MDD File System\SD-SPI.c
  C:\Microchip Solutions\Microchip\PIC18 salloc\salloc.c
  C:\Microchip Solutions\Microchip\Include\Compiler.h
  C:\Microchip Solutions\Microchip\Include\GenericTypeDefs.h

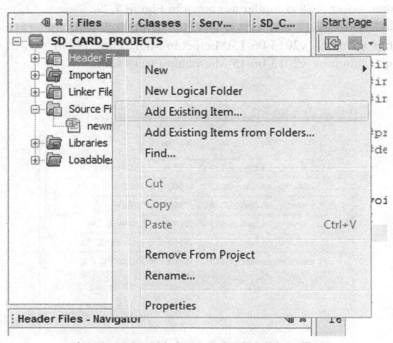

Figure 7.78: Add the Required Header Files.

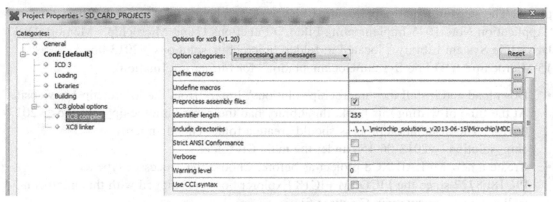

Figure 7.79: Select File Properties and XC8 Compiler.

C:\Microchip Solutions\MDD File System-SD Card\Pic18f\FSconfig.h
C:\Microchip Solutions\MDD File System-SD Card\Pic18f\HardwareProfile.h
C:\Microchip Solutions\Microchip\Include\MDD File System\FSDefs.h
C:\Microchip Solutions\Microchip\Include\MDD File System\SD-SPI.h
C:\Microchip Solutions\Microchip\Include\MDD File System\FSIO.h
C:\Microchip Solutions\Microchip\Include\PIC18 salloc\salloc.h

- Set the file paths in the compiler by: Select File → Project Properties in MPLAB X IDE and then select XC8 compiler as shown in Figure 7.79.
- Click Include Directories, and browse and include the following folders (Figure 7.80):
    microchip_solutions_v2013-06-15\Microchip\MDD File System
    microchip_solutions_v2013-06-15\Microchip\PIC18 salloc

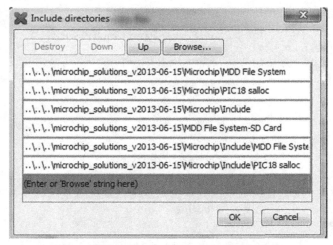

Figure 7.80: Set Folders to be Included.

```
// File: 18f8722.lkr
// Sample linker script for the PIC18F8722 processor

LIBPATH .

FILES c018i.o
FILES clib.lib
FILES p18f8722.lib

CODEPAGE   NAME=page     START=0x0        END=0x1FFFF
CODEPAGE   NAME=idlocs   START=0x200000   END=0x200007   PROTECTED
CODEPAGE   NAME=config   START=0x300000   END=0x30000D   PROTECTED
CODEPAGE   NAME=devid    START=0x3FFFFE   END=0x3FFFFF   PROTECTED
CODEPAGE   NAME=eedata   START=0xF00000   END=0xF003FF   PROTECTED

ACCESSBANK NAME=accessram START=0x0       END=0x5F
DATABANK   NAME=gpr0     START=0x60       END=0xFF
DATABANK   NAME=gpr1     START=0x100      END=0x1FF
DATABANK   NAME=gpr2     START=0x200      END=0x2FF
DATABANK   NAME=gpr3     START=0x300      END=0x3FF
DATABANK   NAME=gpr4     START=0x400      END=0x4FF
DATABANK   NAME=gpr5     START=0x500      END=0x5FF
DATABANK   NAME=gpr6     START=0x600      END=0x6FF
DATABANK   NAME=buffer1  START=0x700      END=0x8FF          PROTECTED
DATABANK   NAME=buffer2  START=0x900      END=0xAFF          PROTECTED
DATABANK   NAME=gpr7     START=0xB00      END=0xBFF
DATABANK   NAME=gpr8     START=0xC00      END=0xCFF
DATABANK   NAME=gpr9     START=0xD00      END=0xEFF
//DATABANK  NAME=gpr9     START=0xE00      END=0xEFF
//DATABANK  NAME=gpr10    START=0xF00      END=0xFFF
DATABANK   NAME=gpr11    START=0xF00      END=0xF5F
ACCESSBANK NAME=accesssfr START=0xF60      END=0xFFF   PROTECTED

SECTION    NAME=CONFIG              ROM=config
SECTION    NAME=_SRAM_ALLOC_HEAP             RAM=gpr7
SECTION    NAME=dataBuffer          RAM=buffer1
SECTION    NAME=FATBuffer           RAM=buffer2

STACK SIZE=0x200 RAM=gpr9
```

Figure 7.81: Modified Linker File.

microchip_solutions_v2013-06-15\Microchip\Include
microchip_solutions_v2013-06-15\MDD File System-SD Card
microchip_solutions_v2013-06-15\Microchip\Include\MDD File System
microchip_solutions_v2013-06-15\Microchip\Include\PIC18 salloc

Finally, before writing our program, we have to modify the compiler linker file to include a 512-byte section for the data read and write and also a 512-byte section for the FAT allocation. This is done by editing the linker file 18f8722.lkr in the folder and adding lines for a dataBuffer, and an FATBuffer. In addition, it is required to add a section named _SRAM_ALLOC_HEAP to the linker file. The modified linker file is shown in Figure 7.81 (check the last part of the file for the modifications).

Setting the Configuration Files

It is now necessary to customize some of the header files for our requirements. You should make the following modifications when using the PICDEM PIC18 Explorer Demonstration board with the PICtail SD Card Daughter board (you are recommended to make copies of the original files before modifying them in case you may want to return to them):

- Modify File C:\Microchip Solutions\MDD File System-SD Card\Pic18f\FSconfig.h and enable the following defines:

1. #define FS_MAX_FILES_OPEN 2
2. #define MEDIA_SECTOR_SIZE 512
3. #define ALLOW_FILESEARCH
   #define ALLOW_WRITES
   #define ALLOW_DIRS
   #define ALLOW_PGMFUNCTIONS
4. #define USERDEFINEDCLOCK
5. Make sure that the file object allocation is dynamic. i.e.
   #if 1

- Modify File C:\Microchip Solutions\MDD File System-SD Card\Pic18f\Hardwar-eProfile.h and set the following options (notice that the system clock is 10 MHz, but the configuration option OSC = HSPLL will be used in our projects to multiply the clock by a factor of four, and it should be set to 40 MHz):

1. Set clock rate to 10 MHz:
   #define GetSystemClock( ) 40000000
2. Enable SD-SPI interface.
   #define USE_SD_INTERFACE_WITH_SPI
3. Define SD card interface pins and SPI bus pins to be used:

```
        #define SD_CS              PORTBBits.RB3
        #define SD_CS_TRIS         TRISBBits.TRISB3
        #define SD_CD              PORTBBits.RB4
        #define SD_CD_TRIS         TRISBBits.TRISB4
        #define SD_WE              PORTABits.RA4
        #define SD_WE_TRIS         TRISABits.TRISA4
        #define SPICON1            SSP1CON1
        #define SPISTAT            SSP1STAT
        #define SPIBUF             SSP1BUF
        #define SPISTAT_RBF        SSP1STATbits.BF
        #define SPICON1bits        SSP1CON1bits
        #define SPISTATbits        SSP1STATbits
        #define SPICLOCK           TRISCbits.TRISC3
        #define SPIIN              TRISCbits.TRISC4
        #define SPIOUT             TRISCbits.TRISC5
        #define SPICLOCKLAT        LATCbits.LATC3
        #define SPIINLAT           LATCbits.LATC4
        #define SPIOUTLAT          LATCbits.LATC5
        #define SPICLOCKPORT       PORTCbits.RC3
        #define SPIINPORT          PORTCbits.RC4
        #define SPIOUTPORT         PORTCbits.RC5
```

MPLAB XC8 MDD Library Functions

Before writing our program, let us look at the MPLAB XC8 MDD library functions.

The MDD library provides a large number of "File and Disk Manipulation" functions that can be called and used from our programs. The functions can be collected into following groups:

- Initialize a card.
- Open/create/close/delete/locate/rename a file on the card.
- Read/write to an opened file.
- Create/delete/change/rename a directory on the card.
- Format a card.
- Set file creation and modification date and time.

Table 7.8—Table 7.13 give a summary of each function briefly.

Table 7.8: Initialize a Card Function.

Function	Description
FSInit	Initialize the card

Table 7.9: Open/Create/Close/Delete/Locate/Rename Functions.

Function	Description
FSfopen/FSfopenpgm	This function opens an existing file for reading, or appending at the end of the file, or creates a new file for writing.
FSfclose	Updates and closes a file. The file time-stamping information is also updated
FSRemove/FSremovepgm	Delete a file
FSrename	Change the name of a file
FindFirst/FindFirstpgm	Locate a file in the current directory that matches the specified name and attributes
FindNext	Locate the next file in the current directory that matches the name and attributes specified earlier

pgm versions are to be used with PIC18 microcontrollers where the arguments are specified in the ROM.

Table 7.10: Read/Write Functions.

Function	Description
FSfread	Reads data from an open file to a buffer
FSfwrite	Writes data from a buffer to an open file
FSftell	Return the current position in a file
FSfprintf	Write a formatted string to a file

Table 7.11: Create/Delete/Change/Rename Directory.

Function	Description
FSmkdir	Create a new subdirectory in the current working directory
FSrmdir	Delete the specified directory
FSchdir	Change the current working directory
FSrename	Change the name of a directory
FSgetcwd	Return the name of the current working directory

Table 7.12: Format a Card.

Function	Description
FSformat	Format a card

Table 7.13: File Time-Stamping Function.

Function	Description
SetClockVars	Set the date and time that will be applied to files when they are created or modified

Table 7.14: MDD Library Options (in File FSconfig.h).

Library Option	Description
ALLOW_WRITES	Enables write functions to write to the card
ALLOWS_DIRS	Enables directory functions (Writes must be enabled)
ALLOW_FORMATS	Enable card formatting function (Writes must be enabled)
ALLOW_FILESEARCH	Enables file and directory search
ALLOW_PGMFUNCTIONS	Enabled pgm functions for getting parameters from the ROM
ALLOW_FSFPRINTF	Enables Fsfprintf function (Writes must be enabled)
SUPPORT_FAT32	Enables FAT32 functionality

Library Options

A number of options are available in the MDD library. These options are enabled or disabled by uncommenting or commenting them, respectively, in include file FSconfig.h. The available options are given in Table 7.14.

Microcontroller Memory Usage

The MPLAB XC8 program memory and data memory usage with the MDD library functions is shown in Table 7.15. Note that 512 bytes of data are used for the data buffer

Table 7.15: MPLAB C18 Memory Usage with MDD Library.

Functions Included	Program Memory (bytes)	Data memory (bytes)
Read-only mode (basic)	11099	2121
File search enabled	+2098	+0
Write enabled	+7488	+0
Format enabled	+2314	+0
Directories enabled	+8380	+90
pgm functions enabled	+288	+0
FSfprintf enabled	+2758	+0
FAT32 support enabled	+407	+4

and an additional 512 bytes are used for the file allocation table buffer. The amount of required memory also depends on the number of files opened at a time. In Table 7.15, it is assumed that two files are opened. The first row shows the minimum memory requirements, and additional memory will be required when any of the subsequent row functionality is enabled.

Sequence of Function Calls

The sequence of the function calls to read or write data to a file, or to delete an existing file are given in this section.

Reading From an Existing File

The steps to open en existing file and read from it are as follows:

> Call FSInit to initialize the card and SPI bus
> Call FSfopen or FSfopenpgm to open the existing file in read mode
> Call FSfread to read data from the file
> Call FSfclose to close the file

The FSread function can be called as many times as required.

Writing to an Existing File

The steps to open an existing file and append data to it are as follows:

> Call FSInit to initialize the card and SPI bus
> Call FSfopen or FSfopenpgm to open the existing file in append mode
> Call FSwrite to write data to the file
> Call FSfclose to close the file

The FSwrite function can be called as many times as required.

Deleting an Existing File

The steps to delete an existing file are as follows:

Call FSInit to initialize the card and SPI bus
Call FSfopen or FSfopenpgm to open the existing file in **write** mode
Call FSremove or FSremovepgm to delete the file
Call FSfclose to close the file

The circuit diagram of the XC8 version of the project is shown in Figure 7.82.

The program listing for the MPLAB XC8 version of the project is shown in Figure 7.83 (XC8-SD.C).

Project 7.9—SD Card-Based Temperature Data Logger

In this project, the design of a temperature data logger system is described. The ambient temperature is read every 10 s, and 100 records are stored on an SD card, or the contents

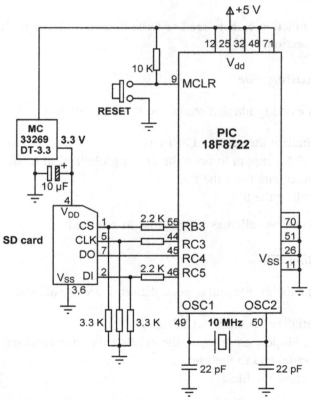

Figure 7.82: Circuit Diagram of the XC8 Version of the Project.

```
/*****************************************************************
              PROJECT TO WRITE SHORT TEXT TO AN SD CARD
              =========================================

In these projects a PIC18F8722 type microcontroller is used. The microcontroller
is operated with an 10 MHz crystal.

An SD card is connected to the microcontroller as follows:

SD card        microcontroller
CS                  RB3
CLK                 RC3
DO                  RC4
DI                  RC5

The program uses the Microchip MDD library functions to read and write to the
SD card.

In this version of the program an LED is connected to port RD0 and the LED is turned
ON when the program is terminated successfully.

Author:  Dogan Ibrahim
Date:    October 2013
File:    XC8-SD.C
*****************************************************************/
#include <p18f8722.h>
#include <FSIO.h>

#pragma config WDT = OFF, OSC = HSPLL,LVP = OFF
#pragma config MCLRE = ON,CCP2MX = PORTC, MODE = MC

#define LED PORTDbits.RD0
#define ON 1
#define OFF 0

/* ================ START OF MAIN PROGRAM ============== */
//
// Start of MAIN Program
//
void main(void)
{
        FSFILE *MyFile;
        unsigned char txt[]="This is a TEXT message";

        TRISD = 0;
        PORTD = 0;
//
// Initialize the SD card routines
//
        while(!FSInit());
//
// Create a new file called MESSAGE.TXT
//
        MyFile = FSfopenpgm("MESSAGE.TXT", "w+");
        if(MyFile == NULL)while(1);
//
// Write message to the file
//
        if(FSfwrite((void *)txt, 1, 22, MyFile) != 22)while(1);
//
// Close the file
//
        if(FSfclose(MyFile) != 0)while(1);
//
// Success. Turn ON the LED
//
        LED = ON;
while(1);
}
```

Figure 7.83: MPLAB XC8 Program.

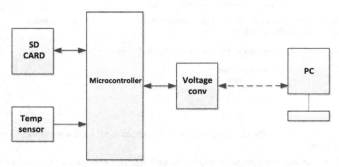

Figure 7.84: Block Diagram of the Project.

of an already saved file are sent to the PC. The program is menu based, and the user is given the options of (Figure 7.84)

- Send saved temperature readings on the SD card to a PC.
- Save temperature readings in a new file on SD card.
- Append the temperature readings to an existing file on SD card.

The block diagram of the project is shown in Figure 7.83.

Hardware Description

The circuit diagram of the project is shown in Figure 7.85. An SD card is connected to the microcontroller. Additionally, the UART pins (RX6 and RX7) are connected to an RS232

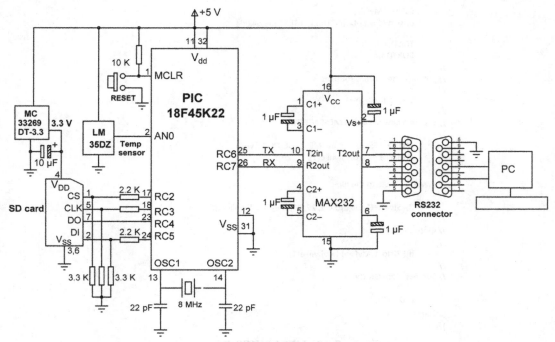

Figure 7.85: Circuit Diagram of the Project.

connector via a MAX232 voltage level translator chip. The temperature is sensed via the LM35DZ-type analog temperature sensor, connected to AN0 pin.

LM35DZ is a three-pin analog temperature sensor that can measure temperature with a 1 °C accuracy in the interval 0 to +100 °C. One pin of the device is connected to the supply (+5 V), the other pin to the ground and the third pin is the analog output. The output voltage of the sensor is directly proportional to the temperature, that is, Vo = 10 mV/°C. If, for example, the temperature is 10 °C, the output voltage will be 100 mV. Similarly, if the temperature is 35 °C, the output voltage of the sensor will be 350 mV.

Project Program

mikeoC Pro for PIC

When the program is started, the following menu will be displayed on the PC screen:

```
TEMPERATURE DATA LOGGER

1. Send temperature data to the PC
2. Save temperature data in a new file
3. Append temperature data to an existing file
   Choice ?
```

The user is then expected to choose the required option. At the end of an option, the program does not return to the menu and the system should be restarted to display the menu again.

The mikroC Pro for PIC program listing of the project is shown in Figure 7.86 (MIKROC-SD2.C). In this project, a file called "TEMPERTR.TXT" is created on the SD card to store the temperature readings.

The following functions are created at the beginning of the program, before the *main* program:

Newline: This function sends a carriage return and a line feed to the UART so that the cursor moves to the next line.

Get_Temperature: This function starts the A/D conversion and receives the converted data into a variable called *Vin*. The voltage corresponding to this value is then calculated in millivolts and divided by 10 to find the actual measured temperature in degrees Celsius. The decimal part of the temperature found is then converted into string form using function LongToStr. The leading spaces are removed from this string, and the resulting string is stored in character array *temperature*. Then the fractional parts of the measured temperature, a carriage return, and a line feed are added to this character array, which is later written onto the SD card.

```
/********************************************************************************
                    TEMPERATURE DATA LOGGER PROJECT
                    ================================
```

In this project an SD card is connected to PORTC as follows:

```
    CS    RC2
    CLK   RC3
    DO    RC4
    DI    RC5
```

In addition, a MAX232 type RS232 voltage level converter chip is connected to serial ports RC6 and RC7. Also, a LM35DZ type analog temperature sensor is connected to analog input AN0 of the microcontroller.

The program is menu based. The user is given options of either to send the saved temperature data to a PC, or to read and save new data on the SD card, or to read temperature data and append to the existing file. Temperature is read at every 10 s.

The temperature is stored in a file called "TEMPERTR.TXT"

```
Author: Dogan Ibrahim
Date:   September 2013
File:   MIKROC-SD2.C
********************************************************************************/
// MMC module connections
sbit Mmc_Chip_Select          at LATC2_bit;
sbit Mmc_Chip_Select_Direction at TRISC2_bit;
// End of MMC module connections

char filename[] = "TEMPERTR.TXT";
unsigned short character;
unsigned long file_size,i,rec_size;
unsigned char ch1,ch2,flag,ret_status,choice;
unsigned char temperature[10],txt[12];

//
// This function sends carriage-return and line-feed to USART
//
void Newline()
{
   Uart1_Write(0x0D);                           // Send carriage-return
   Uart1_Write(0x0A);                           // Send line-feed
}


//
// This function sends a space character to USART
//
void Space()
{
```

Figure 7.86: mikroC Pro for PIC Program.

```
      Uart1_Write(0x20);
}



//
// This function reads the temperature from analog input AN0
//
void Get_Temperature()
{
    unsigned long Vin, Vdec,Vfrac;
    unsigned char op[12];
    unsigned char i,j;

    Vin = Adc_Read(0);                          // Read from channel 0 (AN0)
    Vin = 488*Vin;                              // Scale up the result
    Vin = Vin /10;                              // Convert to temperature in C
    Vdec = Vin / 100;                           // Decimal part
    Vfrac = Vin % 100;                          // Fractional part
    LongToStr(Vdec,op);                         // Convert Vdec to string in "op"
//
// Remove leading blanks
//
    j=0;
    for(i=0;i<=11;i++)
    {
     if(op[i] != ' ')                           // If a blank
     {
        temperature[j]=op[i];
        j++;
     }
    }

    temperature[j] = '.';                       // Add "."
    ch1 = Vfrac / 10;                           // fractional part
    ch2 = Vfrac % 10;
    j++;
    temperature[j] = 48+ch1;                    // Add fractional part
    j++;
    temperature[j] = 48+ch2;
    j++;
    temperature[j] = 0x0D;                      // Add carriage-return
    j++;
    temperature[j] = 0x0A;                      // Add line-feed
    j++;
    temperature[j]='\0';
}

//
```

Figure 7.86
cont'd

```
// Start of MAIN program
//
void main()
{
  unsigned char i;

  ANSELC = 0;                                          // Configure PORTC as digital
  ANSELA = 1;                                          // Configure RA0 as analog
  TRISA = 0x1;                                         // RA0 (AN0) is input
//
// Configure the serial port
//
  Uart1_Init(2400);
//
// Initialise the SPI bus
//
  SPI1_Init_Advanced(_SPI_MASTER_OSC_DIV64, _SPI_DATA_SAMPLE_MIDDLE,
            _SPI_CLK_IDLE_LOW, _SPI_LOW_2_HIGH);
//
// Initialise the SD card FAT file system
//
  while(Mmc_Fat_Init());
//
// Display the MENU and get user choice
//
  while(1)
  {
  Newline();
  Newline();
  Uart1_Write_Text("TEMPERATURE DATA LOGGER");                  // Display heading on the PC
  Newline();
  Newline();
  Uart1_Write_Text("1. Send temperature data to the PC");       // Display opt 1 on the PC
  Newline();
  Uart1_Write_Text("2. Save temperature data in a new file");   // Display opt 2 on the PC
  Newline();
  Uart1_Write_Text("3. Append temperature data to an existing file"); // Display opt 3 on the PC
  Newline();
  Newline();
  Uart1_Write_Text("Choice ? ");                                // Get choice

//
// Read a character from the PC keyboard
//
  flag = 0;
  do {
  if (Uart1_Data_Ready() == 1)                                  // If data received
  {
    choice = Uart1_Read();                                      // Read the received data
    Uart1_Write(choice);                                        // Echo received data
    flag = 1;
```

Figure 7.86

cont'd

```
    }
  } while (!flag);
  Newline();
  Newline();
  rec_size = 0;

//
// Now process user choice
//
  switch(choice)
  {
     case '1':
         ret_status = Mmc_Fat_Assign(&filename,1);
         if(!ret_status)
         {
            Uart1_Write_Text("File does not exist...");
            Newline();
            Uart1_Write_Text("Try again...");
         }
         else
         {
          //
          // Read the data and send to UART
          //
          Uart1_Write_Text("Sending saved data to the PC...");
          Newline();
          Mmc_Fat_Reset(&file_size);
          for(i=0; i<file_size; i++)
          {
              Mmc_Fat_Read(&character);
              Uart1_Write(character);
          }
          Newline();
          Uart1_Write_Text("End of data...");
         }
         break;
     case '2':
         //
         // Start the A/D converter, get temperature readings every
         // 10 s, and then save in a NEW file
         //
         Uart1_Write_Text("Saving data in a NEW file...");
         Newline();
         Mmc_Fat_Assign(&filename,0x80);                                      // Assign the file
         Mmc_Fat_Rewrite();
         Mmc_Fat_Write("TEMPERATURE DATA - SAVED EVERY 10 SECONDS\r\n",43);
         //
         // Read the temperature from A/D converter, format and save
         //
         for(i = 0; i < 100; i++)
         {
```

Figure 7.86
cont'd

```
                Mmc_Fat_Append();
                Get_Temperature();
                Mmc_Fat_Write(temperature,9);
                rec_size++;
                LongToStr(rec_size,txt);
                Newline();
                Uart1_Write_Text("Saving record:");
                Uart1_Write_Text(txt);
                Delay_ms(10000);
              }
            break;
       case '3':
            //
            // Start the A/D converter, get temperature readings every
            // 10 s, and then APPEND to the existing file
            //
            Uart1_Write_Text("Appending data to the existing file...");
            Newline();
            ret_status = Mmc_Fat_Assign(&filename,1);                    // Assign the file
            if(!ret_status)
            {
               Uart1_Write_Text("File does not exist - can not append...");
               Newline();
               Uart1_Write_Text("Try again...");
               Newline();
            }
            else
            {
            //
            // Read the temperature from A/D converter, format and save
            //
               for(i = 0; i < 100; i++)
               {
                  Mmc_Fat_Append();
                  Get_Temperature();
                  Mmc_Fat_Write(temperature,9);
                  rec_size++;
                  LongToStr(rec_size,txt);
                  Newline();
                  Uart1_Write_Text("Appending new record:");
                  Uart1_Write_Text(txt);
                  Delay_ms(10000);
               }
            }
            break;
       default:
            Uart1_Write_Text("Wrong choice.Try again...");
     }
   }
 }
```

Figure 7.86
cont'd

The following operations are performed inside the main program:

- Initialize the UART to 2400 Baud.
- Initialize the SPI bus.
- Initialize the FAT file system.
- Display menu on the PC screen.
- Get a choice from the user (between 1 and 3).
- If the choice = 1, then open the saved temperature file, read the temperature records, and send them to the PC.
- If the choice = 2, then create a new temperature file, get new temperature readings every 10 s, and store 100 records in the file.
- If the choice = 3, then assign to the temperature file, get new temperature readings every 10 s, and append them to the existing temperature file. Hundred records are appended to the file.
- If the choice is not 1–3, display an error message on the screen.

The menu options are described below in more detail:

Option 1: The program attempts to open an existing temperature file with name TEMPERTR.TXT (notice here that Mmc_Fat_Assign function is used. We could have used the Mmc_Fat_Open function instead). If the file does not exist, the error messages: "**File does not exist...**" and "**Try again...**" are displayed on the screen. If on the other hand the temperature file already exists, then the message: "**Sending saved data to the PC...**" is displayed on the PC screen. **Mmc_Fat_Reset** function is called to set the file pointer to the beginning of the file and also to return the size of the file in bytes. Then a **for** loop is formed, temperature records are read from the card 1 byte at a time using function **Mmc_Fat_Read**, and these records are sent to the PC screen. At the end of the data, the message "**End of data...**" is sent to the PC screen.

Option 2: In this option, the message: "**Saving data in a NEW file...**"is sent to the PC screen, a new file is created, with the create flag set to 0x80. The message "**TEMPERATURE DATA − SAVED EVERY 10 SECONDS**" is written on the first line of the file using function **Mmc_Fat_Write**. Then, a **for** loop is formed, the SD card is set into file append mode by calling function **Mmc_Fat_Append**, and a new temperature reading is obtained by calling function **Get_Temperature**. The temperature is then written to the SD card. Also, the current record number is shown on the PC screen to indicate that the program is actually working. This process is repeated after a 10-s delay, until 100 records are written to the file. After this time, the main menu is displayed again.

Option 3: This option is very similar to Option 2. The only difference is that here a new file is not created, but the existing temperature file is opened in the append mode, and 100

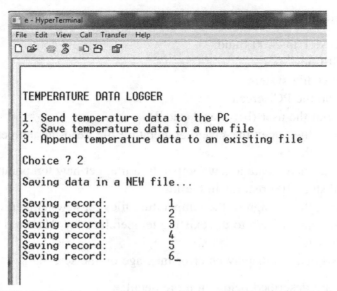

Figure 7.87: Saving Temperature Records on the SD Card with Option 2.

records are written to the file. If the file does not exist, then an error message is displayed on the PC screen.

Default: If the user entry is a number outside 1–3, then this option runs and displays the error message "**Wrong choice…Try again…**" on the PC screen.

The project can be tested by connecting the output of the microcontroller to the serial port of a PC (e.g. COM1) and then running a terminal emulation software (e.g. Hyperterm or mikroC Pro for PIC built-in terminal emulator—USART Terminal). Set communication parameters to 2400 baud, 8 data bits, 1 stop bit, no parity bit, and no flow control. Figure 7.87 shows a snapshot of the PC screen when Option 2 is selected to save temperature records in a new file. Note that the current record numbers are displayed on the screen as they are written to the SD card.

Figure 7.88 shows a screen snapshot where Option 1 is selected to read the temperature records from the SD card and display them on the PC screen.

Finally, Figure 7.89 shows a screen snapshot when option 3 is selected to append the temperature readings to the existing file.

MPLAB XC8

The MPLAB XC8 version of the program is left as an exercise to the reader.

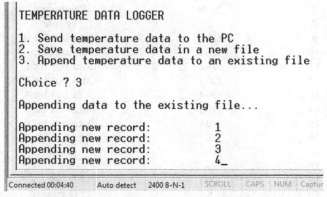

```
e - HyperTerminal
File  Edit  View  Call  Transfer  Help

1. Send temperature data to the PC
2. Save temperature data in a new file
3. Append temperature data to an existing file

Choice ? 1

Sending saved data to the PC...
TEMPERATURE DATA - SAVED EVERY 10 SECONDS
22.93
22.93
22.93
22.93
22.93
22.93
22.93
22.93
22.93
22.93
```

Figure 7.88: Displaying the Records on the PC Screen with Option 1.

Project 7.10—Using Graphics LCD—Displaying Various Shapes

Graphics LCD (GLCDs) are used in many consumer applications, such as mobile phones, MP3 players, GPS systems, games, and educational toys. Another important applications area of GLCDs is in industrial automation and control where various plant characteristics can easily be monitored or changed.

There are several GLCD screens and GLCD controllers in use currently. For small applications, the 128 × 64 pixel monochrome GLCD with the KS0107/8 controller is one of the most commonly used display. For larger display requirements and more complex projects, one can select the 240 × 128 pixel monochrome GLCD screen with the T6963 (or RA6963) controller. For color GLCD-based applications, thin film transistor (TFT)-type displays seem to be the best choice currently.

```
TEMPERATURE DATA LOGGER

1. Send temperature data to the PC
2. Save temperature data in a new file
3. Append temperature data to an existing file

Choice ? 3

Appending data to the existing file...

Appending new record:          1
Appending new record:          2
Appending new record:          3
Appending new record:          4_

Connected 00:04:40    Auto detect    2400 8-N-1    SCROLL  CAPS  NUM  Captur
```

Figure 7.89: Saving Temperature Records on an SD Card with Option 3.

In this project, we shall be looking at how the standard 128×64 GLCD can be interfaced and used in microcontroller-based projects. In this simple project, we shall see how to display various shapes on the GLCD.

The 128 × 64 Pixel GLCD

These GLCDs have dimensions of 7.8×7.0 cm and a thickness of 1.0 cm. The viewing area is 6.2×4.4 cm. The display consists of 128×64 pixels, organized as 128 pixels in the horizontal direction and 64 pixels in the vertical direction. The display operates with a +5-V supply, consumes typically 8-mA current, and comes with a built-in KS0108-type display controller. A backlight LED is provided for visibility in low ambient light conditions. This LED consumes about 360 mA when operated. Basically two controllers are used internally: one for segments 1–64, and the other one for segments 65–128.

The display is connected to the external world through a 20-pin SIL (Single-In-Line) type connector. Table 7.16 gives the pin numbers and corresponding pin names.

The description of each pin is as follows:

/CSA, /CSB: Chip select pins for the two controllers. The display is logically divided into two sections, and these signals control which half should be enabled at any time.

Table 7.16: 128 × 64 Pixel GLCD Pin Configuration.

Pin No	Pin Name	Function
1	\CSA or CS1	Chip select for controller 1
2	\CSB or CS2	Chip select for controller 2
3	V_{SS}	Ground
4	V_{DD}	+5 V
5	V0	Contrast adjustment
6	D/I	Register select
7	R/W	Read-write
8	E	Enable
9	DB0	Data bus bit 0
10	DB1	Data bus bit 1
11	DB2	Data bus bit 2
12	DB3	Data bus bit 3
13	DB4	Data bus bit 4
14	DB5	Data bus bit 5
15	DB6	Data bus bit 6
16	DB7	Data bus bit 7
17	RST	Reset
18	V_{EE}	Negative voltage
19	A	LED +4.2 V
20	K	LED ground

V$_{CC}$, GND: Power supply and ground pins.

V0: Contrast adjustment. A 10 K potentiometer should be used to adjust the contrast. The wiper arm should be connected to this pin, and the other two arms should be connected to V$_{EE}$ and the ground.

D/I: Register select pin. Logic HIGH is data mode, logic LOW is instruction mode.

R/W: Read–write pin. Logic HIGH is read, logic LOW is write.

E: Enable pin. Logic HIGH to LOW to enable.

DB0–DB7: Data bus pins.

RST: Reset pin. The display is reset if this pin is held LOW for at least 100 ns. During reset, the display is off, and no commands can be executed by the display controller.

V$_{EE}$: Negative voltage output pin for contrast adjustment.

A, K: Power supply and ground pins for the backlight. Pin K should be connected to the ground and pin A should be connected to a +5-V supply through a 10-Ω resistor.

Figure 7.90 shows the connection of the GLCD to a microcontroller with the contrast adjustment potentiometer and backlight LED connections shown as well.

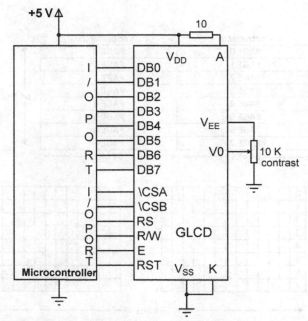

Figure 7.90: Connecting the 128 × 64 GLCD to a Microcontroller.

Operation of the GLCD

The internal operation of the GLCD and the KS0108 controller is very complex and beyond the scope of this book. Most high-level microcontroller compiler developers provide libraries for using these displays in their programming languages. In this project, only the basic information required before using the GLCD library is given.

Figure 7.91 shows the structure of the GLCD as far as programming the display is concerned. The 128 × 64 pixel display is logically split into two halves. There are two controllers: controller A controlling the left half of the display and controller B controlling the right half, where the two controllers are addressed independently. Each half of the display consists of 8 pages where each page is 8 bits high and 8 bytes (64 bits) wide. Thus, each half consists of 64 × 64 bits. Text is written to the pages of the display. Thus, a total of 16 characters across can be written for a given page on both halves of the display. Considering that there are 8 pages, a total of 128 characters can be written on the display.

The origin of the display is the top left hand corner (Figure 7.92). The X-direction extends toward the right, and Y-direction extends toward the bottom of the display. In the X-direction, the pixels range from 0 to 127, while in the Y-direction the pixels range from 0 to 63. Coordinate (127, 63) is at the bottom right-hand corner of the display.

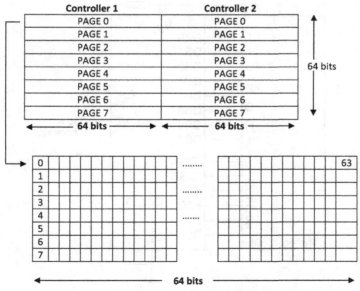

Figure 7.91: Structure of the GLCD.

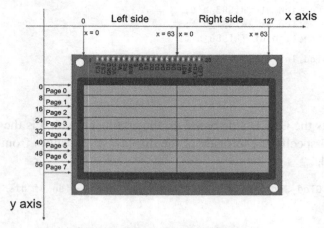

Figure 7.92: GLCD Coordinates.

mikroC Pro for PIC GLCD Library Functions

mikroC Pro for PIC language supports the 128 × 64 pixel GLCDs and provides a large library of functions for the development of GLCD-based projects. In actual fact there are libraries for several different types of GLCDs. In this section, we shall be looking at the commonly used library functions provided for the 128 × 64 GLCDs, working with the KS0108 controller.

Glcd_Init

This function initializes the GLCD module. The GLCD control and data lines can be configured by the user, but the eight data lines must be on a single port. Before this function is called, the interface between the GLCD and the microcontroller must be defined using *sbit* type statements of the following format. In the following example, it is assumed that the GLCD data lines are connected to PORTD, and in addition, the CS1, CS2, RS, RW, EN, and RST lines are connected to PORTB:

```
// GLCD pinout settings
char GLCD_DataPort at PORTD;

sbit GLCD_CS1 at RB0_bit;
sbit GLCD_CS2 at RB1_bit;
sbit GLCD_RS at RB2_bit;
sbit GLCD_RW at RB3_bit;
sbit GLCD_EN at RB4_bit;
sbit GLCD_RST at RB5_bit;

sbit GLCD_CS1_Direction at TRISB0_bit;
sbit GLCD_CS2_Direction at TRISB1_bit;
sbit GLCD_RS_Direction at TRISB2_bit;
```

```
sbit GLCD_RW_Direction at TRISB3_bit;
sbit GLCD_EN_Direction at TRISB4_bit;
sbit GLCD_RST_Direction at TRISB5_bit;

Example Call: Glcd_Init();
```

Glcd_Set_Side

This function selects the GLCD side based on the argument, which is the x coordinate. Values from 0 to 63 specify the left side of the display, while values from 64 to 127 specify the right side.

```
Example Call: Glcd_Set_Side(0); // Select left hand side of display
```

Glcd_Set_X

This function sets the x-axis position from the left border of the GLCD within the selected display side.

```
Example Call: Glcd_Set_X(10); // Set position to pixel 10
```

Glcd_Set_Page

This function selects a page of the GLCD. The argument to the function is the page number between 0 and 7.

```
Example Call: Glcd_Set_Page(2); // Select Page 2
```

Glcd_Write_Data

This function writes 1 byte to the current location on the GLCD memory and moves to the next location. The GLCD side and page number should be set before calling this function.

```
Example Call: Glcd_Write_Data(MyData);
```

Glcd_Fill

This function fills the GLCD memory with the specified byte pattern, where the pattern is passed as an argument to the function.

```
Example Call: Glcd_Fill(0); // Clears the screen
```

Glcd_Dot

This function draws a dot on the GLCD at coordinates x_pos, y_pos. The x and y coordinates and the color of the dot are passed as arguments. Valid x coordinates are

0—127, valid y coordinates are 0—63, and valid colors are 0—2, where 0 clears the dot, 1 places a dot, and 2 inverts the dot.

```
Example Call:  Glcd_Dot(0, 10, 1);  // Place a dot at x = 0, y = 10
```

Glcd_Line

This function draws a line on the GLCD. The arguments passed to the function are

```
x_start:    x coordinate of the line starting position (0 to 127)
y_start:    y coordinate of the line starting position (0 to 63)
x_end:      x coordinate of the line ending position (0 to 127)
y_end:      y coordinate of the line ending position (0 to 63)
colour:     The colour value between 0 and 2. 0 is white, 1 is black, and 2
            inverts each dot.
Example Call:  Glcd_Line(0, 0, 5, 10, 1);  // Draw a line from (0,0) to (5,10)
```

Glcd_V_Line

This function draws a vertical line on the GLCD. The arguments passed to the function are

```
y_start:    y coordinate of the line starting position (0 to 63)
y_end:      y coordinate of the line ending position (0 to 63)
x_pos:      x coordinate of the vertical line (0 to 127)
colour:     The colour value between 0 and 2. 0 is white, 1 is black, and 2
            inverts each dot.
Example Call:  Glcd_V_Line(4, 10, 5, 1); // Draw a line from (5,4) to (5,10)
```

Glcd_H_Line

This function draws a horizontal line on the GLCD. The arguments passed to the function are

```
x_start:    x coordinate of the line starting position (0 to 127)
x_end:      x coordinate of the line ending position (0 to 127)
y_pos:      y coordinate of the vertical line (0 to 63)
colour:     The colour value between 0 and 2. 0 is white, 1 is black, and 2
            inverts each dot.
Example Call:  Glcd_H_Line(15, 55, 25, 1); // Draw a line from (15,25) to (55,25)
```

Glcd_Rectangle

This function draws a rectangle on the GLCD. The arguments passed to the function are

```
x_upper_left:  x coordinate of the upper left corner of rectangle (0 to 127)
y_upper_left:  y coordinate of the upper left corner of rectangle (0 to 63)
```

```
x_bottom_right:    x coordinate of the lower right corner of rectangle (0 to 127)
y_bottom_right:    y coordinate of the lower right corner of rectangle (0 to 63)
colour:            The colour value between 0 and 2. 0 is white, 1 is black, and 2
                   inverts each dot.
Example Call:      Glcd_Rectangle(5, 5, 10, 10);  // Draw rectangle between (5,5)
                   and (10,10)
```

Glcd_Rectangle_Round_Edges

This function draws a rounded-edge rectangle on the GLCD. The arguments passed to the function are

```
x_upper_left:      x coordinate of the upper left corner of rectangle (0 to 127)
y_upper_left:      y coordinate of the upper left corner of rectangle (0 to 63)
x_bottom_right:    x coordinate of the lower right corner of rectangle (0 to 127)
y_bottom_right:    y coordinate of the lower right corner of rectangle (0 to 63)
round radius:      radius of the rounded edge
colour:            The colour value between 0 and 2. 0 is white, 1 is black, and 2
                   inverts each dot.
Example Call:      Glcd_Rectangle_Round_Edge(5, 5, 10, 10, 15, 1);
                   // Draw rectangle between (5,5) and (10,10) with edge radius 15
```

Glcd_Rectangle_Round_Edges_Fill

This function draws a filled rounded edge rectangle on the GLCD with color. The arguments passed to the function are

```
x_upper_left:      x coordinate of the upper left corner of rectangle (0 to 127)
y_upper_left:      y coordinate of the upper left corner of rectangle (0 to 63)
x_bottom_right:    x coordinate of the lower right corner of rectangle (0 to 127)
y_bottom_right:    y coordinate of the lower right corner of rectangle (0 to 63)
round radius:      radius of the rounded edge
colour:            colour of the rectangle border. The colour value is between 0 and
                   2. 0 is white, 1 is black, and 2 inverts each dot.
Example Call:      Glcd_Rectangle_Round_Edges_Fill(5, 5, 10, 10, 15, 1);
                   // Draw rectangle between (5,5) and (10,10) with edge radius 15
```

Glcd_Box

This function draws a box on the GLCD. The arguments passed to the function are:

```
x_upper_left:      x coordinate of the upper left corner of box (0 to 127)
y_upper_left:      y coordinate of the upper left corner of box (0 to 63)
x_bottom_right:    x coordinate of the lower right corner of box (0 to 127)
y_bottom_right:    y coordinate of the lower right corner of box (0 to 63)
colour:            colour of the box fill. The colour value is between 0 and 2. 0 is
                   white, 1 is black, 2 inverts each dot.
Example Call:      Glcd_Box(5, 15, 20, 30, 1);  // Draw box between (5,15) and (20,30)
```

Glcd_Circle

This function draws a circle on the GLCD. The arguments passed to the function are as follows:

```
x_center:       x coordinate of the circle center (0 to 127)
y_center:       y coordinate of the circle center (0 to 63)
radius:         radius of the circle
colour:         colour of the circle line. The colour value is between 0 and 2.
                0 is white, 1 is black, 2 inverts each dot.
Example Call:   Glcd_Circle(30, 30, 5, 1);
                // Draw circle with center at (30,30), and radius 5
```

Glcd_Circle_Fill

This function draws a filled circle on the GLCD. The arguments passed to the function are as follows:

```
x_center:       x coordinate of the circle center (0 to 127)
y_center:       y coordinate of the circle center (0 to 63)
radius:         radius of the circle
colour:         The colour value is between 0 and 2. 0 is white, 1 is black, 2
                inverts each dot.
Example Call:   Glcd_Circle_Fill(30, 30, 5, 1);
                // Draw a filled circle with center at (30,30), and radius 5
```

Glcd_Set_Font

This function sets the font that will be used with functions: Glcd_Write_Char and Glcd_Write_Text. The arguments passed to the function are

```
activeFont:     font to be set. Needs to be formatted as an array of char
aFontWidth:     width of the font characters in dots.
aFontHeight:    height of the font characters in dots.
aFontOffs:      number that represents difference between the mikroC Pro for PIC
                character set and regular ASCII set (e.g. if A is 65 in ASCII
                character, and A is 45 in the mikroC Pro for the PIC character set,
                aFontOffs is 20)
```

List of supported fonts are as follows:

- Font_Glcd_System3x5,
- Font_Glcd_System5x7,
- Font_Glcd_5x7,
- Font_Glcd_Character8x7.

```
Example Call:   Glcd_Set_Font(&MyFont, 5, 7, 32);
                //Use custom 5x7 font MyFont which starts with space character (32)
```

Glcd_Set_Font_Adv

This function sets the font that will be used with functions: Glcd_Write_Char_Adv and Glcd_Write_Text_Adv. The arguments passed to the function are

```
activeFont:          font to be set. Needs to be formatted as an array of char.
font_colour:         sets font colour.
font_orientation:    sets font orientation.
Example Call:        Glcd_Set_Font_Adv(&MyFont, 0, 0);
```

Glcd_Write_Char

This function displays a character on the GLCD. if no font is specified, then the default Font_Glcd_System5x7 font supplied with the library will be used. The arguments passed to the function are

```
chr:          character to be displayed
x_pos:        character starting position on x-axis (0 to 127- FontWidth)
page_num:     the number of the page on which the character will be displayed (0 to 7)
colour:       colour of the character between 0 and 2. 0 is white, 1 is black, 2
              inverts each dot
Example Call: Glcd_Write_Char('Z', 10, 2, 1);
              //Display character Z at x position 10, inside page 2
```

Glcd_Write_Char_Adv

This function displays a character on the GLCD at coordinates (x, y).

```
ch:           character to be displayed.
x:            character position on x-axis.
y:            character position on y-axis.
Example Call: Glcd_Write_Char_Adv('A', 20, 10,);  // Display A at (20,10)
```

Glcd_Write_Text

This function displays text on the GLCD. if no font is specified, then the default Font_Glcd_System5x7 font supplied with the library will be used. The arguments passed to the function are

```
text:         text to be displayed
x_pos:        text starting position on x-axis.
page_num:     the number of the page on which text will be displayed (0 to 7)
colour:       The colour parameter between 0 and 2. 0 is white, 1 is black and 2
              inverts each dot.
Example Call: Glcd_Write_Text("My Computer", 10, 3, 1);
              //Display "My Computer" at x position 10 in page 3
```

Glcd_Write_Text_Adv

This function displays text on the GLCD at coordinates (x, y). The arguments passed to the function are as follows:

```
text:          text to be displayed
x:             text position on x-axis.
y:             text position on y-axis.
Example Call:  Glcd_Write_Text_Adv("My Computer", 10, 10);
               //Display text "My Computer" at coordinates (10,10)
```

Glcd_Write_Const_Text_Adv

This function displays text on the GLCD, where the text is assumed to be located in the program memory of the microcontroller. The text is displayed at coordinates (x, y). The arguments passed to the function are

```
text:          text to be displayed
x:             text position on x-axis.
y:             text position on y-axis.

               const char Txt[ ] = "My Computer";
Example Call:  Glcd_Write_Text_Adv(Txt, 10, 10);
               //Display text "My Computer" at coordinates (10,10)
```

Glcd_Image

This function displays the bitmap image on the GLCD. The image to be displayed is passed as an argument to the function. The bitmap image array must be located in the program memory of the microcontroller. The GLCD Bitmap Editor of mikroC Pro for PIC compiler can be used to convert an image to a constant so that it can be displayed by this function.

```
Example Call:  Glcd_Image(MyImage);
```

Project Hardware

The circuit diagram of the project is shown in Figure 7.93. The GLCD is connected to PORTB and PORTD of the microcontroller. The microcontroller is operated from an 8-MHz crystal.

Project Program
mikroC Pro for PIC

The mikroC Pro for PIC program listing is given in Figure 7.94 (MIKROC-GLCD1.C). At the beginning of the program, the GLCD connections are

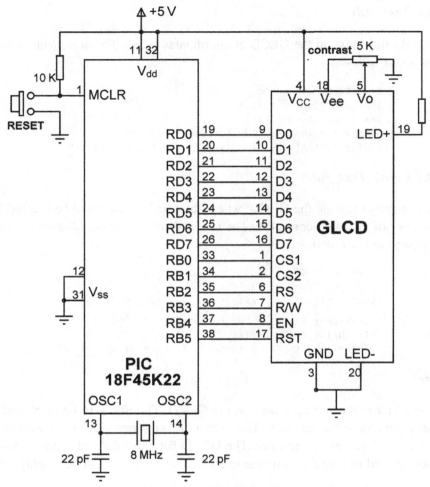

Figure 7.93: Circuit Diagram of the Project.

defined, and PORTB and PORTD are configured as digital. Then, the following shapes are drawn on the GLCD:

- A rectangle with rounded edges at coordinates (5, 5), (123, 59) and edge radius 10;
- A rectangle at coordinates (15, 15), (113, 49);
- A line from (50, 30) to (70, 30);
- A circle with center at (30, 30) and radius 10;
- A filled circle with the center at (50, 42) and radius 5;
- Text "Txt" at the x coordinate 80 and page 3;
- Text "LCD" at the x coordinate 80 and page 4;
- Text "micro" at coordinates (80, 38).

```
/*******************************************************************************
                    GLCD LIBRARY EXAMPLE
                    -----------------------------

This program uses some of the mikroC GLCD library functions to show how the functions should
be used in programs.

The program was loaded to a PIC18F45K22 microcontroller and operated with a 8 MHz crystal.
The EasyPIC 7 development board is used for this project

Author: Dogan Ibrahim
Date:    October, 2013
File:    MIKROC-GLCD1.C
*******************************************************************************/

// Glcd module connections
char GLCD_DataPort at PORTD;

sbit GLCD_CS1 at RB0_bit;
sbit GLCD_CS2 at RB1_bit;
sbit GLCD_RS  at RB2_bit;
sbit GLCD_RW  at RB3_bit;
sbit GLCD_EN  at RB4_bit;
sbit GLCD_RST at RB5_bit;

sbit GLCD_CS1_Direction at TRISB0_bit;
sbit GLCD_CS2_Direction at TRISB1_bit;
sbit GLCD_RS_Direction  at TRISB2_bit;
sbit GLCD_RW_Direction  at TRISB3_bit;
sbit GLCD_EN_Direction  at TRISB4_bit;
sbit GLCD_RST_Direction at TRISB5_bit;
// End Glcd module connections

void main()
{
    ANSELB = 0;                                              // Configure PORTB as digital
    ANSELD = 0;                                              // Configure PORTD as digital

    Glcd_Init();                                             // Initialize GLCD
    Glcd_Fill(0x00);                                         // Clear GLCD
    Glcd_rectangle_round_edges(5,5,123,59,10,1);             // Draw rectangle
    Glcd_Rectangle(15,15,113,49,1);                          // Draw rectangle
    Glcd_Line(50, 30, 70, 30, 1);                            // Draw line
    Glcd_Circle(30,30,10,1);                                 // Draw circle
    Glcd_Circle_Fill(50,42,5,1);                             // Draw filled circle
    Glcd_Set_Font(Font_Glcd_Character8x7, 8, 7, 32);         // Change Font

    Glcd_Write_Text("Txt", 80, 3, 2);                        // Write string "Txt"
    Glcd_Write_Text("LCD",80,4,1);                           // Write string "LCD"
    Glcd_Write_Text_Adv("micro",80,38);                      // Write string "micro"
}
```

Figure 7.94: mikroC Pro for PIC Program.

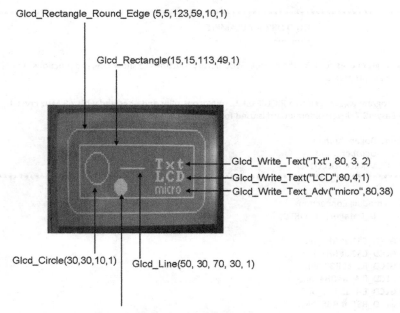

Glcd_Rectangle_Round_Edge (5,5,123,59,10,1)

Glcd_Rectangle(15,15,113,49,1)

Glcd_Write_Text("Txt", 80, 3, 2)
Glcd_Write_Text("LCD",80,4,1)
Glcd_Write_Text_Adv("micro",80,38)

Glcd_Circle(30,30,10,1)   Glcd_Line(50, 30, 70, 30, 1)

Figure 7.95: Shapes Drawn on the GLCD.

Figure 7.95 shows the shapes drawn on the GLCD.

Project 7.11—Barometer, Thermometer and Altimeter Display on a GLCD

This project is about using a sensor to read and display the pressure, temperature, and the altitude on a GLCD.

The project is based on using a MEMS sensor called LPS331AP. This is basically a pressure sensor, but it can also measure the ambient temperature. The altitude is calculated mathematically from the pressure measurements.

It is necessary to know the features and basic operation of this sensor before it can be used in projects. The features of this sensor are as follows:

- A 260—1260 mbar absolute pressure measurement;
- A 0 to +80 °C temperature measurement;
- Very low-power consumption (30 μA in the high-resolution mode);
- A 24-bit digital pressure output in millibars;
- A 16-bit digital temperature output in degrees Celsius;
- SPI and I^2C interfaces;
- A +1.71- to +3.6-V supply voltage.

The LPS331AP is a 16-pin device, having dimensions of $3 \times 3 \times 1$ mm. In this project, it is operated in the I^2C communications mode. When operated in this mode the following pins are used:

Pin	Description
1	VDD_IO (power supply)
4	SCL (I^2C clock)
5	GND
6	SDA (I^2C data)
7	SA0 (I^2C device address LSB)
8	CS (Set to 1 for I^2C mode)
9	INT2 (interrupt or data ready)
11	INT1 (interrupt or data ready)
12	GND
13	GND
14	VDD (power supply)
15	VCCA (analog power supply)
16	GND

When pin 7 is connected to the supply voltage, the device write and read addresses are 0xBA and 0xBB, respectively. Alternatively, when connected to the ground, the device write and read addresses are 0xB8 and 0xB9, respectively.

Figure 7.96 shows the output block diagram of the LPS331AP sensor. The device is controlled with 19 registers (Table 7.17). A 24-bit reference pressure output (registers REF_P_XL, REF_P_L, and REF_P_H) is subtracted from the measure sensor pressure to obtain the 24-bit output pressure via registers PRESS_POUT_XL_REH, PRESS_OUT_L, and PRESS_OUT_H. The measured pressure is compared with two threshold pressures preloaded into registers THS_P_LOW_REG and THS_P_HIGH_REG. If the measured pressure is higher than THS_P_HIGH_REG, then a High Pressure Interrupt (PH) is generated. Similarly, if the measured pressure is lower than THS_P_LOW_REG, then a Low Pressure Interrupt (PL) is generated.

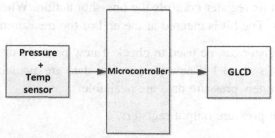

Figure 7.96: Block Diagram of the Project.

Table 7.17: LPS331AP Registers.

Name	Type	Address (Hex)
REF_P_XL	R/W	08
REF_P_L	R/W	09
REF_P_H	R/W	0A
WHO_AM_I	R	0F
RES_CONF	R/W	10
CTRL_REG1	R/W	20
CTRL_REG2	R/W	21
CTRL_REG3	R/W	22
INT_CFG_REG	R/W	23
INT_SOURCE_REG	R	24
THS_P_LOW_REG	R/W	25
THS_P_HIGH_REG	R/W	26
STATUS_REG	R	27
PRESS_POUT_XL_REH	R	28
PRESS_OUT_L	R	29
PRESS_OUT_H	R	2A
TEMP_OUT_L	R	2B
TEMP_OUT_H	R	2C
AMP_CTRL	R/W	30

The functions of some important registers are described below:

WHO_AM_I: This register is used to identify the device and returns 0xBB.

RES_CONF: This register is used to select the internal pressure and temperature averaging to be used in a measurement. Loading the recommended value 0x78 configures for 256 averages for the pressure and 128 averages for the temperature measurements.

CTRL_REG1: This register controls the active/power-down mode (bit 7), output data rates for pressure and temperature (bits 4–6), and output update control bit (bit 2). Loading 0x04 configures the device to enter power-down mode, one-shot pressure and temperature measurement, that is, a request must me done for a measurement.

CTRL_REG2: Bit 0 of this register controls the one-shot action. When the bit is set to 1, a new measurement starts. The bit is cleared at the end of the measurement.

STATUS_REG: This register can be used to check if new pressure or temperature data is available. Bit 0 (T_DA) is set to 1 if new temperature data are available. Similarly, bit 1 (P_DA) is set to 1 when new pressure data are available.

PRESS_OUTxxx: 24-bit pressure output registers.

TEMP_OUTxxx: 16-bit temperature output registers.

The block diagram of the project is shown in Figure 7.96. The pressure is displayed in integer format while the other two are displayed in fractional format:

```
P (mb) : nnnn
T(C)   : nn.n
A(ft)  : nnn.n
```

Project Hardware

The circuit diagram of the project is shown in Figure 7.97. The sensor is connected to the I²C pins of the microcontroller via pull-up resistors. The SA0 pin is connected to supply voltage so that the device write and read addresses are 0xBA and 0xBB, respectively. The CS pin is connected to the supply voltage to select I²C communication protocol. The GLCD is connected to PORTB and PORTD pins in the default configuration.

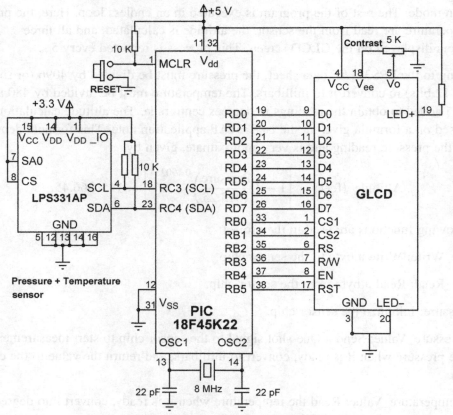

Figure 7.97: Circuit Diagram of the Project.

An 8-MHz crystal is used for timing, but the clock PLL is enabled so that the actual running clock frequency is 32 MHz.

Project PDL

The PDL of the project is shown in Figure 7.98.

Project Program

mikroC Pro for PIC

The mikroC Pro for PIC program is shown in Figure 7.99 (MIKRO-GLCD2.C). At the beginning of the program, the GLCD connections to the microcontroller are defined. Also, the register names and addresses of the LPS331AP chip are defined. Inside the main program, PORTB, PORTC, and PORTD are configured as digital. The GLCD module is initialized and the screen is cleared, the I^{2}C module is initialized (with 100 kHz clock rate).

The program then initializes the LPS331AP chip with full resolution and One-shot operation mode. The rest of the program is executed in an endless loop. Here, the pressure and temperature are read from the sensor, the altitude is calculated, and all three parameters displayed on the GLCD screen. This process is repeated every 5 s.

According to the LPS331AP data sheet, the pressure must be divided by 4096 (or shifted right by 12 bits) to convert it to millibars. The temperature must be divided by 480 and then 42.5 added to obtain the readings in degrees centigrade. The altitude calculation is done based on a formula given in the LPS331AP application note. This formula depends only on the pressure reading and is very approximate, given by

$$\text{Altitude (feet)} = \left[1 - \left(\frac{\text{Pressure}}{1013.25} \right)^{0.190284} \right] \times 145,366.45.$$

The following functions are used in the project:

Pressure_Write: Write a byte to the sensor chip.

Pressure_Read: Read a byte from the sensor chip.

Init_Pressure: Initialize the sensor chip.

Read_Pressure_Value: Send a One-shot signal to the sensor chip to start measurement. Read the pressure when it is ready, convert to millibars, and return the value to the calling program.

Read_Temperature_Value: Read the temperature when it is ready, convert into degrees Centigrade, and return to the calling program.

Main Program

BEGIN
    Define connections between LCD and microcontroller
    Define LPS331AP register addresses
    Configure PORTB, PORTC, PORTD as digital
    Initialize GLCD
    Initialize I²C module
    CALL Init_Pressure_Chip
    DO FOREVER
        CALL Read_Pressure_Value
        CALL Read_Temperature_Value
        CALL Read_Altimeter_Value
        Display pressure, temperature, and altitude
        Wait 5 seconds
        Clear GLCD
    ENDDO
END

BEGIN/Pressure_Write
    Write byte to sensor chip
END/Pressure_Write

BEGIN/Pressure_Read
    Read byte from sensor chip
    Return the byte to calling program
END/Pressure_Read

BEGIN/Init_Pressure_Chip
    Configure sensor resolution
    Configure sensor to one-shot mode
    Check chip identity
END/Init_Pressure_Chip

BEGIN/Read_Pressure_Value
    Start conversion
    Wait until reading is available
    Get a pressure reading
    Convert into millibars
    Return the pressure to calling program
END/Read_Pressure_Value

BEGIN/Read_Temperature_Value
    Wait until temperature is available
    Read temperature
    Convert into degrees Centigrade
    Return temperature to calling program
END/Read_Temperature_Value

BEGIN/Read_Altimeter_Value
    Convert pressure into altitude
    Return altitude to calling program
END/Read_Altimeter_Value

BEGIN/Display_PTA
    Display pressure
    Display temperature
    Display altitude
END/Dislay_PTA

Figure 7.98: Project PDL.

```
/*****************************************************************
            BAROMETER, THERMOMETER AND ALTIMETER DISPLAY
   ----------------------------------------------------------------
```

This program uses the LPS331AP MEMS pressure sensor chip. In addition to pressure, the chip also measures the temperature, and the altimeter reading can be obtained
from the pressure reading

The program displays the pressure, altitude, and temperature. The program works in One-shot mode. i.e. a new sample is requested and the sample is received and displayed. Then a new sample is requested and so on.

The clock PLL is enabled so that the actual clock frequency is 32 MHz (Enable the 4xPLL and set oscillator frequency to 32 MHz in Project -> Edit Project)

Author: Dogan Ibrahim
Date: October, 2013
File: MIKROC-GLCD2.C
***/

```c
// Glcd module connections
char GLCD_DataPort at PORTD;

sbit GLCD_CS1 at LATB0_bit;
sbit GLCD_CS2 at LATB1_bit;
sbit GLCD_RS  at LATB2_bit;
sbit GLCD_RW  at LATB3_bit;
sbit GLCD_EN  at LATB4_bit;
sbit GLCD_RST at LATB5_bit;

sbit GLCD_CS1_Direction at TRISB0_bit;
sbit GLCD_CS2_Direction at TRISB1_bit;
sbit GLCD_RS_Direction  at TRISB2_bit;
sbit GLCD_RW_Direction  at TRISB3_bit;
sbit GLCD_EN_Direction  at TRISB4_bit;
sbit GLCD_RST_Direction at TRISB5_bit;
// End Glcd module connections

// Define LPS331AP registers
#define REF_P_XL 0x08              // Reference pressure (LSB)
#define REF_P_L 0x09               // Reference pressure (middle)
#define REF_P_H 0x0A               // Reference pressure (MSB)
#define WHO_AM_I 0x0F              // Device identification
#define RES_CONF 0x10              // Pressure resolution
#define CTRL_REG1 0x20             // Control register 1
#define CTRL_REG2 0x21             // Control register 2
#define CTRL_REG3 0x22             // Control register 3
#define INT_CFG_REG 0x23           // Interrupt configuration register
#define INT_SOURCE_REG 0x24        // Interrupt source register
#define THS_P_LOW_REG 0x25         // Threshold LOW register
#define THS_P_HIGH_REg 0x26        // Threshold HIGH register
```

Figure 7.99: mikroC Pro for PIC Program.

```
#define STATUS_REG 0x27                          // Status register
#define PRESS_POUT_XL_REH 0x28                    // Pressure output register (LSB)
#define PRESS_OUT_L 0x29                          // Pressure output register (middle)
#define PRESS_OUT_H 0x2A                          // Pressure output register (MSB)
#define TEMP_OUT_L 0x2B                           // Temperature output register (LSB)
#define TEMP_OUT_H 0x2C                           // Temperature output register (MSB)
#define AMP_CTRL 0x30                             // Analog front end control register
#define Write_Addr 0xBA                           // Device write register
#define Read_Addr 0xBB                            // Device read register

void Pressure_Write(unsigned char address, unsigned char value)
{
    I2C1_Start();                                 // Send START bit
    I2C1_Wr(Write_Addr);                          // Send device address
    I2C1_Wr(address);                             // Send register address
    I2C1_Wr(value);                               // Send data
    I2C1_Stop();                                  // Send STOP bit
}

unsigned char Pressure_Read(unsigned char address)
{
    unsigned char c = 0;

    I2C1_Start();                                 // Send START bit
    I2C1_Wr(Write_Addr);                          // Send device address
    I2C1_Wr(address);                             // Send register address
    I2C1_Repeated_Start();                        // Send repeated START
    I2C1_Wr(Read_Addr);                           // Send device read address
    c = I2C1_Rd(0);                               // Read, send no ack
    I2C1_Stop();                                  // Send STOP bit
    return c;
}
//
// This function initializes the pressure chip
//
unsigned char Init_Pressure_Chip()
{
    unsigned char temp, flag = 0;

    Pressure_Write(RES_CONF, 0x78);              // Select pressure and temp resolution
    Pressure_Write(CTRL_REG1, 0x04);             // Configure One-shot mode
    Pressure_Write(CTRL_REG1, 0x84);             // Configure chip active mode
    temp = Pressure_Read(WHO_AM_I);              // Read chip identity
    if(temp != 0xBB)flag = 1;                    // Error if wrong chip identity
    return flag;
}

//
// This function reads and returns the pressure (24 bits). The reading is converted into
// millibars after dividing by 4019 (shifting right 12 bits)
//
```

Figure 7.99
cont'd

```
//
long int Read_Pressure_Value()
{
    long int outP;
    unsigned char stat, P_DA, PressureM, PressureL;

    Pressure_Write(CTRL_REG2, 1);                      // Send One-shot request
//
// Check if new pressure data is available, if so read it
//
    P_DA = 0;
    while(P_DA == 0)                                   // Wait until new pressure is available
    {
      stat = Pressure_Read(STATUS_REG);                // Read the status register
      P_DA = stat & 0x02;                              // Extract P_DA
    }

    OutP = Pressure_Read(PRESS_OUT_H);                 // Read high byte
    PressureM = Pressure_Read(PRESS_OUT_L);            // Read middle byte
    PressureL = Pressure_Read(PRESS_POUT_XL_REH);      // Read low byte

    OutP = (OutP << 8);                                // Move to middle byte position
    OutP = OutP | PressureM;                           // Add middle byte
    OutP = (OutP << 8);                                // Move to upper byte position
    OutP = (OutP | PressureL);                         // Add low byte
    OutP = (OutP >> 12);                               // Divide by 4096 (in mbars)
    return OutP;                                       // Return the pressure
}


//
// This function reads and returns the temperature (16 bits). The reading is converted into
// Degrees Centigrade after the following operation:
//
//   (Value / 480) + 42.5
//
//
float Read_Temperature_Value()
{
    int OutT;
    unsigned char stat, T_DA, TempL;
    float DegreesC;
//
// Wait until a new temperature data is available and if so get it
//
    T_DA = 0;
    while(T_DA == 0)                                   // Wait for new temperature
    {
      stat = Pressure_Read(STATUS_REG);                // Read status register
      T_DA = stat & 0x01;                              // extract T_DA bit
    }
```

Figure 7.99
cont'd

```
        OutT = Pressure_Read(TEMP_OUT_H);              // Read high byte
        TempL = Pressure_Read(TEMP_OUT_L);             // Read low byte

        OutT = OutT << 8;                              // Move to left byte
        OutT = OutT | TempL;                           // Add low byte
        DegreesC = ((OutT / 480.0) + 42.5);
        return DegreesC;                               // Return the temperature
}


//
// This function calculates the height (altimeter function) from the pressure
//
float Read_Altimeter_Value(long int Pressure)
{
    float Altitude_ft;

    Altitude_ft = Pressure/1013.25;
    Altitude_ft = pow(Altitude_ft, 0.190284);
    Altitude_ft = (1.0 - Altitude_ft)*145366.45;

    return Altitude_ft;
}


//
// This function displays the pressure, temperature and altitude on the GLCD
// in the following format:
//
//   P(mb): nnnn
//   T(C) : nn.n
//   A(ft): nnn.n
//
void Display_PTA(long int Pressure, float Temperature, float Altitude)
{
    unsigned char i, Txt[14];
    char *res;

    Glcd_Rectangle(5,5,120,55,1);                      // Draw rectangle
    Glcd_Write_Text("P(mb): ", 7,1,1);                 // Write string "Pressure (mb): "
    Glcd_Write_Text("T(C) : ", 7,3,1);
    Glcd_Write_Text("A(ft): ", 7,5,1);
// Display Pressure
    for(i=0; i<14; i++)Txt[i] = 0;
    LongWordToStr(Pressure, Txt);
    Ltrim(Txt);
    Glcd_Write_Text(Txt, 45,1,1);
// Display temperature
    for(i=0; i<14; i++)Txt[i] = 0;
    FloatToStr(Temperature, Txt);
    res = strrchr(Txt, '.');                           // Locate "."
    *(res + 2) = 0x0;                                  // Terminate after 1 digit
    Glcd_Write_text(Txt, 45, 3, 1);
```

Figure 7.99
cont'd

```
// Display altitude
   for(i=0; i<14; i++)Txt[i] = 0;
   FloatToStr(Altitude, Txt);
   res = strrchr(Txt, '.');
   *(res + 2) = 0x0;
   Glcd_Write_Text(Txt, 45, 5, 1);
}

//
// Display Error message
//
void Error()
{
   Glcd_Write_Text("Error...", 5,3,1);             // Wrong device ID
   while(1);
}


void main()
{
   unsigned char stat;
   long int P;
   float T, A;

   ANSELB = 0;                                     // Configure PORTB as digital
   ANSELC = 0;                                     // Configure PORTC as digital
   ANSELD = 0;                                     // Configure PORTD as digital

   Glcd_Init();                                    // Initialize GLCD
   Glcd_Fill(0x00);                                // Clear GLCD

   I2C1_Init(100000);                              // Initialize I2C
   Delay_Ms(10);

   stat = Init_Pressure_Chip();                    // Initialize pressure chip
   if(stat != 0)Error();
//
// Loop to read and display the pressure, temperature and altitude
//
   while(1)
   {
     P = Read_Pressure_Value();                    // Read pressure
     T = Read_Temperature_Value();                 // Read temperature
     A = Read_Altimeter_Value(P);                  // Read altitude
     Display_PTA(P, T, A);                         // Display Pressure, Temp, Altitude
     Delay_Ms(5000);                               // Wait 5 seconds
     Glcd_Fill(0x00);                              / Clear GLCD
   }
}
```

Figure 7.99
cont'd

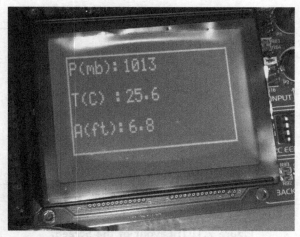

Figure 7.100: Sample Display from the Project.

Read_Altimeter_Value: Use the above formula to convert pressure into altitude.

Display_PTA: Display the pressure, temperature, and altitude on the GLCD.

Error: Display error message if the sensor chip is not identified.

Figure 7.100 shows a sample display.

Project 7.12—Plotting the Temperature Variation on the GLCD
Project Description

This project demonstrates how the ambient temperature can be measured and then plotted in real time on the GLCD. The temperature is measured every second using an LM35DZ-type analog sensor and is then plotted in real-time on the GLCD.

The X and Y axes are drawn on the GLCD, the axes ticks are displayed, and the Y axis is labeled as shown in Figure 7.101. The Y axis is the temperature, and the X axis is the time where every pixel corresponds to 1 s in real time.

Block Diagram

The block diagram of the project is shown in Figure 7.102.

Circuit Diagram

The circuit diagram of the project is as shown in Figure 7.103. The LM35DZ temperature sensor is connected to analog port RA0 (or AN0) of the microcontroller. The sensor

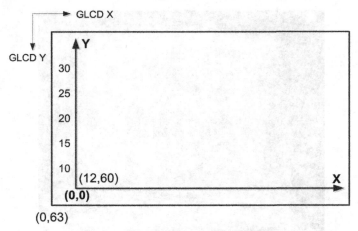

Figure 7.101: Layout of the Screen.

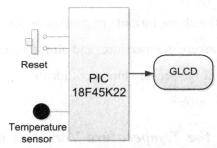

Figure 7.102: Block Diagram of the Project.

provides an output voltage directly proportional to the measured temperature. The output of the sensor is given by

Vo = 10 mV/°C.

PORT B and PORT D are connected to the GLCD as in the previous project.

Project PDL

The PDL of this project is given in Figure 7.104.

Project Program

mikroC Pro for PIC

The mikroC Pro for the PIC program is given in Figure 7.105 (MIKROC-GLCD3.C). The A/D converter on the PIC18F45K22 microcontroller is 10 bits wide. Thus, with a +5-V

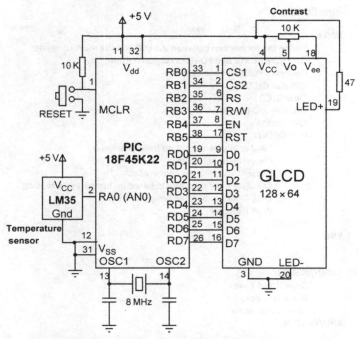

Figure 7.103: Circuit Diagram of the Project.

reference voltage the resolution will be 5000/1024 or 4.88 mV, which is accurate enough to measure the temperature to an accuracy of 0.5 °C.

The microcontroller is operated from an 8 MHz crystal. The PLL is disabled so that the actual running clock frequency is 8 MHz.

At the beginning of the program, the connections between the microcontroller and the GLCD are defined using *sbit* statements. The GLCD is connected to ports B and D of the microcontroller and thus both these ports are configured as digital output ports using ANSEL and TRIS statements. PORTA is configured as analog with pin RA0 (or AN0) being configured as an input.

The GLCD library is then initialized using the Glcd_Init function. This function must be called before calling to any other GLCD function. The GLCD screen is then cleared using the Glcd_Fill(0x0), which turns OFF all pixels of the GLCD.

The A/D converter is initialized by calling library function ADC_Init. The background of the display is drawn by calling function PlotAxis. This function draws the X and Y axes. The bottom left part of the screen with coordinates (12, 60) is taken as the (0, 0) coordinate of our display. Then, ticks are placed on both the X and the Y axes using

Main Program

BEGIN
    Define the connection between the LCD and the microcontroller
    Configure PORTB and PORTD as digital output
    Configure PORTA as analog input
    Initialise GLCD
    Clear GLCD
    Initialise A/D converter
    CALL PlotAxis
    DO FOREVER
        Read analog temperature from Channel 0
        Convert into millivolts
        Convert into Degrees centigrade
        Calculate the Y co-ordinate based on temperature reading
        CALL PlotXY to plot the temperature
        Wait 1 second
    ENDDO
END

BEGIN/PlotAxis
    Draw X and Y axes
    Draw axes ticks
    Draw Y axis labels
END/PlotAxis

BEGIN/PlotXY
    Draw a line to join previous and current temperature values
    Update the previous X and Y values with current values
END/PlotXY

Figure 7.104: PDL of the Project.

Glcd_Dot statements. The Y axis is labeled from 10 to 30 °C in steps of 5 °C using the Glcd_Write_Text_Adv statements.

The program then enters an endless loop formed by a *for* statement. Inside this loop, the analog temperature is converted into digital format and stored in variable *T* by calling function ADC_Get_Sample with the channel number specified as 0 (RA0 or AN0). This digital value is converted into millivolts by multiplying with 5000 and dividing by 1024. The actual temperature in degrees Celsius is calculated by dividing the voltage in millivolts by 10 (Vo = 10 mV/°C).

The graph is drawn using the GLCD function Glcd_Line. This function draws a line between the specified starting and ending X and Y coordinates. Variables old_x, old_y, new_x, and new_y are used to store the old and the new (current) X and Y values of the temperature, respectively. At the first iteration, the old and the current values are assumed to be the same, and this is identified by variable *flag* being cleared to 0. In all other

```
/********************************************************************************
                        TEMPERATURE PLOTTING ON GLCD
                        ==============================
```

This project shows how the temperature can be read from an analog temperature sensor and then plotted on a GLCD in real time.

In this project an LM35DZ type analog temperature sensor is used. This sensor has 3 pins: The ground, power supply (+5 V), and the output pin. The sensor gives an output voltage which is directly proportional to the measured temperature. i.e. V o = 10mV/C. Thus, for example at 15C the output voltage is 150 mV. Similarly, at 30C the output voltage is 300 mV and so on.

The temperature sensor is connected to analog input RA0 (or AN0) of a PIC18F45K22 type microcontroller. The microcontroller is operated from an 8 MHz crystal, with the PLL is disabled, so that the actual clock frequency is 8 MHz. The GLCD used in the project is based on KS0107/108 type controller with 128 x 64 pixels.

The program first draws the X and Y axes, axes ticks, and the Y axis labels. Then, the temperature is read from Channel 0 (RA0 or AN0), converted into digital, and then into Degrees C. The temperature is plotted in real-time every second. i.e. the horizontal axis is the time where each pixel corresponds to 1 s.

The GLCD is connected to PORTB and PORTD of the microcontroller as in the previous GLCD projects.

```
Author:  Dogan Ibrahim
Date:    October, 2013
File:    MIKROC-GLCD3.C
********************************************************************************/

unsigned char stp, old_x, old_y, new_x, new_y;

// Glcd module connections
char GLCD_DataPort at PORTD;

sbit GLCD_CS1 at RB0_bit;
sbit GLCD_CS2 at RB1_bit;
sbit GLCD_RS  at RB2_bit;
sbit GLCD_RW  at RB3_bit;
sbit GLCD_EN  at RB4_bit;
sbit GLCD_RST at RB5_bit;

sbit GLCD_CS1_Direction at TRISB0_bit;
sbit GLCD_CS2_Direction at TRISB1_bit;
sbit GLCD_RS_Direction  at TRISB2_bit;
sbit GLCD_RW_Direction  at TRISB3_bit;
sbit GLCD_EN_Direction  at TRISB4_bit;
sbit GLCD_RST_Direction at TRISB5_bit;
// End Glcd module connections
```

Figure 7.105: mikroC Pro for the PIC Program.

```
//
// This function plots the X and Y axis. The origin is set at screen co-ordinates (12,60).
// First the two axes are drawn. Then the axes ticks are displayed for both X and Y axis.
// Finally, the Y axis labels are displayed (i.e. the temperature labels)
//
void PlotAxis()
{
  unsigned char i;

  Glcd_Line(12, 0, 12, 60, 1);                      // Draw Y axis
  Glcd_Line(12, 60, 127, 60, 1);                    // Draw X axis
  for(i=12; i<127; i += 9)Glcd_Dot(i, 61, 1);       // Display x axis ticks
  for(i=0; i<60; i += 10)Glcd_Dot(11, i, 1);        // Display y axis ticks
  Glcd_Write_Text_Adv("30",0,5);                    // Y axis label
  Glcd_Write_Text_Adv("25",0,15);                   // Y axis label
  Glcd_Write_Text_Adv("20",0,25);                   // Y axis label
  Glcd_Write_Text_Adv("15",0,35);                   // Y axis label
  Glcd_Write_Text_Adv("10",0,45);                   // Y axis label
}


//
// This function plots the temperature in real-time. The temperature is plotted by joining
// the data points with straight lines. The X axis is the time where each pixel corresponds
// to one second. The Y axis is the temperature in Degrees C
//
void PlotXY(float Temperature)
{
  Glcd_Line(old_x,old_y,new_x,new_y,1);             // Draw temperature changes
  old_x = new_x;                                    // Update old points
  old_y = new_y;
}



//
// Start of main program
//
void main()
{
  unsigned int T;
  unsigned char flag = 0;
  float mV, C;

  ANSELA = 1;                                       // Configure PORTA as analog
  ANSELB = 0;                                       // Configure PORTB as digital
  ANSELD = 0;                                       // Configure PORTD as digital
  TRISA = 1;                                        // RA0 is input (analog)
  TRISB = 0;                                        // PORT B is output
  TRISD = 0;                                        // PORT D is output

  Glcd_Init();                                      // Initialise GLCD
  Glcd_Fill(0x0);                                   // Clear GLCD
```

Figure 7.105

cont'd

```
ADC_Init();                          // Initialise ADC
PlotAxis();                          // Plot X-Y axes and labels

for(;;)                              // DO FOREVER
{
  T = ADC_Get_Sample(0);             // Read temperature from channel 0
  mV = T*5000.0/1024.0;              // Temperature in mV
  C = mV /10.0;                      // Temperature in Degrees C

  if(flag == 0)                      // If first time
  {
    new_x = 12;                      // Start from x = 12
    old_x = new_x;
    new_y = -2*C+70;                 // New temperature value
    old_y = new_y;
    flag = 1;                        // Set so that not first time
  }
  else                               // Not first time
  {
    new_x++;                         // Inc x by 1 (1 second each pixel)
    new_y = -2*C+70;                 // New temperature value
  }
  PlotXY(C);                         // Plot the graph
  Delay_Ms(1000);                    // Wait 1 s
}
}
```

Figure 7.105
cont'd

iterations, variable *flag* is 1 and the *else* part of the *if* statement is executed. The X value is incremented by 1 to correspond to the next second and the new Y value is updated.

The Y coordinate (temperature) is calculated as follows:

The relationship between the Y axis ticks and the Y coordinates of data values can be derived from the table:

Y Axis Ticks Pixel Coordinates	Y Axis Data Coordinate (Degrees C)
10	30
20	25
30	20
40	15
50	10

The above relationship is linear and is in the form of a straight line $y = mx + C$, where m is the slope of the line and C is the point where the line crosses the Y axis. The equation of this line can be found from

$$y - y_1 = m(x - x_1),$$
$$\text{where } m = (y_2 - y_1)/(x_2 - x_1)$$

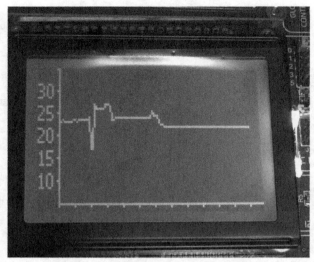

Figure 7.106: Sample Display of the Temperature.

by taking any two points on the line, we can easily find the equation. Considering the points

$$(x_1, y_1) = (30, 10) \text{ and } (x_2, y_2) = (10, 50)$$

The relationship is found to be

$$y = -2x + 70.$$

Therefore, given the temperature C in degrees Celsius, the y coordinate to be used for plotting can be calculated from

$$new\_y = -2^*C + 70.$$

After plotting a point, the *new_x* and *new_y* are copied to *old_x* and *old_y*, respectively, ready for the next sample to be plotted.

Figure 7.106 shows a sample display of the temperature in real time.

Project 7.13—Using the Ethernet—Web Browser-Based Control

The Ethernet has traditionally been implemented on PCs and laptops and has been used widely at homes, offices, and industries to access the worldwide Internet and companywide intranet networks. The Internet can nowadays be accessed using smaller handheld devices such as smart mobile phones, PDAs and, IPADs. Most of these devices are based on microcontrollers and use single-chip Ethernet-controller devices for connectivity. Such Ethernet controllers can easily be configured, programmed, and incorporated into embedded systems to provide the system with Ethernet connectivity with the outside world.

Ethernet Connectivity

Ethernet was originally invented by Xerox in 1972, and then developed jointly by Xerox, DEC, and Intel. It is a frame-based networking technology based on the standard IEEE 802.3. The physical medium of a typical Ethernet-based local area network (LAN) network uses coaxial cable, twisted pair wires, fiber optics, or can be in the form of Wireless LANs. Currently, the most common form of Ethernet is called 100Base-T, and it provides transmission speeds up to 100 Mbps. Slower Ethernet or 10Base-T is also commonly used in lower speed control and monitoring projects.

Devices on the Ethernet are all connected together, and the communication is based on the *Carrier Sense Multiple Access with Collision Detection* (CSMA/CD) protocol. Only one node transmits its data while all the other nodes listen to avoid collision. In the case of a possible collision, transmitting nodes wait for a random time and attempt to retransmit, hoping to avoid the collision. The maximum length of an Ethernet cable depends on the speed of transmission and the type of cable used.

As shown in Figure 7.107, an Ethernet packet consists of

*   Six-byte destination address,
*   Six-byte source address,
*   Two-byte data type,
*   Forty five to 1500 byte data,
*   Four-byte CRC.

In addition, when transmitted on the Ethernet medium, a 7-byte preamble field and Start-of-Frame (SOF) delimiter byte are appended to the beginning of the Ethernet packet.

Various network communication protocols are embedded inside Ethernet packets. For example, DECnet, IP, and ARP protocols all make use of the Ethernet as the communications protocol. In this article, we will be using a Web Browser command to establish communication between the PC and the microcontroller system. Web Browser is based on the transmission control protocol (TCP) and uses port 80. TCP is an advanced protocol requiring connection and providing guaranteed packet delivery with

DESTINATION ADDRESS	SOURCE ADDRESS	TYPE	DATA	CRC

Figure 7.107: Ethernet Packet Format.

retransmission if an error occurs. TCP packets are acknowledged to confirm the safe packet delivery.

Embedded Ethernet Controller Chips

There are many Ethernet controller chips in the market. Although these chips can be purchased as components, in most applications, it is easier and usually cheaper to use boards with incorporated Ethernet controller chips and network connection sockets (e.g. RJ45). The ENC28J6 is a standalone 28-pin Ethernet controller chip that meets the IEEE 802.3 specifications and is controlled using the SPI interface. This chip is used in the project given later in this section. This chip has the following basic features:

- Compatible with 10Base-T networks,
- Supports both half-duplex and full-duplex operation,
- Supports automatic polarity detection and correction,
- Automatic retransmit on collision,
- Eight-kilobyte transmit/receive buffer,
- Supports unicast, multicast, and broadcast addresses,
- Link and Activity LED interface,
- Differential signal interface to RJ45 connector.

Figure 7.108 shows the block diagram and connection of the ENC28J60 Ethernet controller chip to a microcontroller. Basically, the interface requires the SPI signals SI, SO, and SCK to be connected to the microcontroller. In addition, the CS pin can also be connected to a microcontroller I/O pin.

Some high-end PIC microcontrollers incorporate Ethernet controllers. For example, the PIC18F97J60 is a microcontroller that includes a 10Base-T Ethernet controller with 8-kbyte transmit/receive buffers. The advantage of using an Ethernet-based microcontroller chip is that in addition to the Ethernet functions, the chip provides analog and digital I/O ports and many other microcontroller features.

Embedded Ethernet Access Methods

In general, there are four access methods that can be used to establish the connectivity between a PC and an embedded Ethernet controller (see Application Note AN292, Silicon Labs):

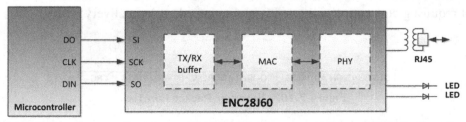

Figure 7.108: Connecting the ENC28J60 Ethernet Controller to a Microcontroller.

- Using a Web Browser on the PC,
- Using a HyperTerminal on the PC,
- Having the embedded system to send E-mail,
- Using a custom application based on developing software on both the PC and the embedded system,

Using a Web Browser on the PC

This is perhaps the easiest and the most reliable method of establishing connectivity with no software development on the PC. This method is based on HTTP, which has been in use since the 1990s as the most widely used protocol to transfer data on the internet. The aim of HTTP protocol is to allow the transfer of HTML files between a browser (usually a PC) and a Web Server where the data item is located. In this method, the PC is termed the *Client* and the microcontroller system is termed the *Server.* The client sends a request by entering the *url* of the server. Assuming that the server *url* is 192.168.10.15, then entering the following command on the PC will establish a link to the microcontroller system:

> http://192.168.10.15

The microcontroller system, for example, can then send an HTML page as a response to the client to display a menu with buttons. By clicking a button on the menu, a command (e.g. GET) will be sent to the server with the appropriate command tail. The server can decode this command tail and take appropriate actions.

Figure 7.109 shows the connectivity using a Web Browser interface.

Using a HyperTerminal

The HyperTerminal interface is also known as the *Telnet* interface. Here, the user connects to the microcontroller system by issuing Telnet commands and specifying the IP address. This kind of interface is usually an interactive interface and requires the connectivity to be initiated and terminated by the user on the PC.

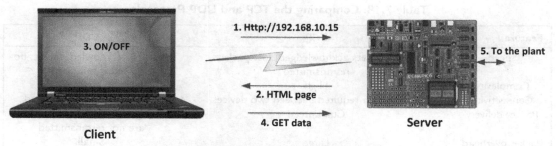

Figure 7.109: Web Browser Connectivity.

Figure 7.110: Connectivity using the E-mail Method.

Embedded System Sending E-mail

In this method, the microcontroller system sends its data using the E-mail protocols. The outgoing E-mail is handled by SMTP, while the incoming mail is handled by mail servers such as POP, IMAP, or HTTP. Using this method has disadvantages that an incoming mail may stay in the input buffers for a long time until it is discovered and read by the user. Figure 7.110 shows the connectivity using the E-mail method.

Using Custom Application

The development of custom applications for network connectivity provides a highly flexible interface. This method however has the greatest disadvantage that network software should be developed on both the PC and the microcontroller system. Two protocols are usually used for custom development: user datagram package (UDP) and TCP. The TCP protocol is used in applications where a guaranteed packet delivery is required with the delivery of each packet being acknowledged. Lost packets are retransmitted. The UDP protocol on the other hand is used where a high transmission speed is more important than the safe delivery of packets. There is no acknowledgement and no retransmission if a packet is lost.

Table 7.18 compares the TCP and the UDP protocols.

Table 7.18: Comparing the TCP and UDP Protocols.

Feature	TCP	UDP
Speed	Slow. Packets acknowledged, lost packets retransmitted	Fast. No acknowledgement, no retransmission
Complexity	High	Low
Connectivity	Connection required between two devices	Connection is not required
Packet delivery	Guaranteed	Not guaranteed. Lost packets are not retransmitted
Packet overhead	Large	Small

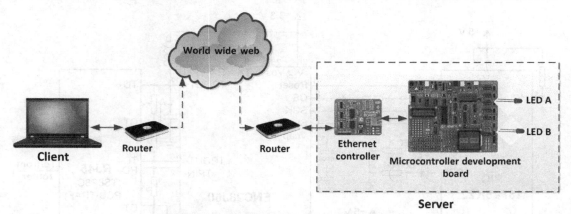

Figure 7.111: Block Diagram of the Project.

Example Ethernet-Based Embedded Control Project

This section describes the design of a simple microcontroller-based automation system using the Ethernet as the communication medium. In this project, a web browser-based communication is used where the PC is a client and the microcontroller system is the server. Figure 7.111 shows the block diagram of the project. The project hardware is in two parts, connected using a network hub or a switch (or a crossed network cable for local testing): the Ethernet controller (or the Server) and the PC (or the Client).

The Ethernet controller consists of a microcontroller and an ENC28J60 Ethernet controller chip. Two LEDs (LED A and LED B) are connected to the microcontroller RD0 and RD1 output pins to simulate two lamps. These LEDs are toggled under the control of a standard Web Browser command initiated on the PC. There is no software development on the PC.

Project Hardware

Figure 7.112 shows the circuit diagram of the project. The microcontroller is designed around a PIC18F45K22-type microcontroller chip, operating at 8 MHz. The Ethernet controller is based on the ENC28J60 chip, operating at 25 MHz. The interface between the microcontroller and the Ethernet chip is based on the SPI bus protocol, where SI, SO, and SCK pins of the Ethernet chip are connected to SPI pins (PORTC) of the microcontroller. The Ethernet controller chip operates at 3.3 V, and thus, its output pin SO cannot drive the microcontroller input pin without a voltage translator. In Figure 7.112, a 74HCT245-type buffer is used to boost the output level of pin SO. Other lower cost chips, such as 74HCT08 (AND gate), 74ACT125 (quad 3-state buffer) or other chips could also have been used.

The internal analog circuitry of the ENC28J60 chip requires that an external resistor be connected from RBIA to the ground. Some of the device's digital logic operates at 2.5 V,

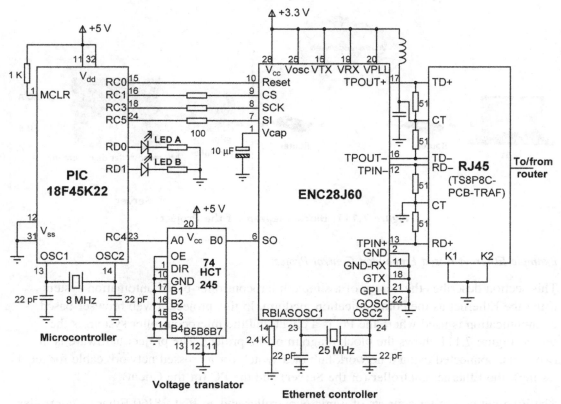

Figure 7.112: Circuit Diagram of the Project.

and an external filter capacitor should be connected from Vcap to ground. Transmit output pins of the Ethernet chip (TPOUT+ and TPOUT−) and the receive inputs (TPIN+ and TPIN−) are connected to an RJ45 network socket with an integrated Ethernet transformer (T58P8C-PCB-TRAF). Two LEDs on the Serial Ethernet board provide visual indication of the Link and Activity on the line (the RJ45 socket has a pair of built-in internal LEDs, but are not used in this project). A 5- to 3.3-V power supply regulator chip (e.g. MC33269DT-3.3) is used to provide power to the Ethernet chip. If the PC and the Ethernet controller are on the same network and close to each other, then the two can be connected together using a crossed network cable; otherwise, a hub or a switch may be required. If the PC and the Ethernet controller are located on different networks and are not close to each other, then routers may be required to establish connectivity between the two.

Two LEDs, LED A and LED B, are connected to pins RD0 and RD1 of the microcontroller, respectively. These LEDs are toggled under the control of a Web Browser command issued from the PC.

Figure 7.113: The Serial Ethernet Board.

The Construction

The project was constructed using the EasyPIC V7 development board and the mikroElektronika Serial Ethernet Board (Figure 7.113). This is a small board that plugs in directly to PORTC of the EasyPIC V7 development board via a 10-way IDC plug (Figure 7.114) simplifying the development of embedded Ethernet projects. The board is equipped with an EC28J60 Ethernet controller chip, a 74HCT245 voltage translation chip, three LEDs, a 5- to 3.3-V voltage regulator, and an RJ45 socket with an integrated transformer.

Figure 7.114: Connecting the Serial Ethernet Board to EasyPIC7 V7.

If you are using the EasyPIC V7 development board for this project, use a 10-way ribbon cable and make sure that you plug in one side of the cable to PORTC on the development board, and the other side to the bottom connector on the Serial Ethernet board (Figure 7.114).

Project PDL

Figure 7.115 shows the project PDL.

Project Software

mikroC Pro for PIC

The mikroC Pro for PIC program listing is given in Figure 7.116 (MIKROC-ETHER.C). At the beginning of the main program, PORTC and PORTD are configured as digital, and the SPI bus is initialized by calling built-in library function *SPI_Init*. Then, the Serial Ethernet module is initialized by calling built-in function *SPI_Ethernet_Init* and specifying the MAC address of the ethernet board, the IP address to be used, and the mode of operation as full duplex.

```
MAIN program

BEGIN
        Configure I/O ports
        Initialise SPI bus
        Initialise Serial Ethernet Library
        DO FOREVER
                Check for packets
        ENDDO
END


BEGIN/SPI_Ethernet_UserTCP
        IF a GET packet is received THEN
                IF two characters starting at index 6 are "TA"
                        Toggle LED A
                ELSE IF two characters starting at index 6 are "TB"
                        Toggle LED B
                ENDIF
        ENDIF
        Send HTML response page to the Client
END/SPI_Ethernet_UserTCP


BEGIN/SPI_Ethernet_UserUDP
END/SPI_Ethernet_UserUDP
```

Figure 7.115: Project PDL.

```
/********************************************************************************
                    WEB BROWSER BASED ETHERNET CONTROL PROJECT
                    =============================================
```

This project shows how the ETHERNET can be used in microcontroller based projects. In this project a Serial Ethernet Board (www.mikroe.com) is connected to the EasyPIC V7 development board.

The project uses the Web Browser method to establish Ethernet based communication between a PC and the microcontroller system.

The PC is the client and the microcontroller system is the server.

Two LEDs (LED A and LED B) are connected to the microcontroller system. These LEDs are toggled remotely by entering commands on the PC. The HTTP protocol is used in the project

```
Author:  Dogan Ibrahim
Date:    October, 2013
File:    MIKROC-ETHER.C
***** *************************************************************************/
const char HTTPheader[] = "HTTP/1.1 200 OK\nContent-type:";
const char HTTPMimeTypeHTML[] = "text/html\n\n";
const char HTTPMimeTypeScript[] = "text/plain\n\n";
//
// Define the HTML page to be sent to the PC
//
char StrtPage[] =
"<html><body>\
<form name=\"input\" method=\"get\"><table align=center width=500 \
bgcolor=Red border=4><tr><td align=center colspan=2><font size=7 \
color=white face=\"verdana\"><b>LED CONTROL</b></font></td></tr>\
<tr><td align=center bgcolor=Blue><input name=\"TA\" type=\"submit\" \
value=\"TOGGLE LED A\"></td><td align=center bgcolor=Green> \
<input name=\"TB\" type=\"submit\" value=\"TOGGLE LED B\"></td></tr>\
</table></form></body></html>";

//
// Ethernet NIC interface definitions
//
sfr sbit SPI_Ethernet_Rst at RC0_bit;
sfr sbit SPI_Ethernet_CS  at RC1_bit;
sfr sbit SPI_Ethernet_Rst_Direction at TRISC0_bit;
sfr sbit SPI_Ethernet_CS_Direction  at TRISC1_bit;
//
// Define Serial Ethernet Board MAC Address, and IP address to be used for the communication
//
unsigned char MACAddr[6] = {0x00, 0x14, 0xA5, 0x76, 0x19, 0x3f} ;
unsigned char IPAddr[4] = {192,168,1,15};
unsigned char getRequest[10];

typedef struct
```

Figure 7.116: Program Listing of the Project.

```
{
 unsigned canCloseTCP:1;
 unsigned isBroadcast:1;
}TethPktFlags;

//
// TCP routine. This is where the user request to toggle LED A or LED B are processed
//
//
unsigned int SPI_Ethernet_UserTCP(unsigned char *remoteHost,
                  unsigned int remotePort, unsigned int localPort,
                  unsigned int reqLength, TEthPktFlags *flags)
{
 unsigned int Len;
 for(len=0; len<10; len++)getRequest[len]=SPI_Ethernet_getByte();
 getRequest[len]=0;
 if(memcmp(getRequest,"GET /",5))return(0);

 if(!memcmp(getRequest+6,"TA",2))RD0_bit = ~ RD0_bit;
 else if(!memcmp(getRequest+6,"TB",2))RD1_bit = ~ RD1_bit;

 if(localPort != 80)return(0);
 Len = SPI_Ethernet_putConstString(HTTPheader);
 Len += SPI_Ethernet_putConstString(HTTPMimeTypeHTML);
 Len += SPI_Ethernet_putString(StrtPage);
 return Len;
}

//
// UDP routine. Must be declared even though it is not used
//
unsigned int SPI_Ethernet_UserUDP(unsigned char *remoteHost,
                  unsigned int remotePort, unsigned int destPort,
                  unsigned int reqLength, TEthPktFlags *flags)
{
 return(0);
}

//
// Start of MAIN program
//
void main()
{
  ANSELC = 0;                              // Configure PORTC as digital
  ANSELD = 0;                              // Configure PORTD as digital
  TRISD = 0;                               // Configure PORTD as output
  PORTD = 0;
  SPI1_Init();                             // Initialize SPI module
  SPI_Ethernet_Init(MACAddr, IPAddr, 0x01); // Initialize Ethernet module

  while(1)                                 // Do forever

  {
   SPI_Ethernet_doPacket();                // Process next received packet
  }
}
```

Figure 7.116
cont'd

```
<html>
<body>
<form name="input" method="get">
  <table align=center width=500 bgcolor=Red border=4>
   <tr>
     <td align=center colspan=2><font size=7 color=white face="verdana"><b>LED CONTROL</b></font></td>
   </tr>
   <tr>
     <td align=center bgcolor=Blue><input name="TA" type="submit" value="TOGGLE LED A"></td>
     <td align=center bgcolor=Green><input name="TB" type="submit" value="TOGGLE LED B"></td>
   </tr>
  </table>
 </form>
</body>
</html>
```

Figure 7.117: HTML Code Sent to the Web Browser.

The MAC address of the Serial Ethernet Board used in this project is set at factory to "0x00, 0x14, 0xA5, 0x76, 0x19, 0x3F". The program sets the IP address of the board to "192.168.1.15". The main program then enters an infinite loop where built-in library function *SPI_Ethernet_doPacket* is called to check for the arrival of packets and also to send any outstanding packets to their destinations. The Ethernet library requires both the UDP and TCP functions to be present in the program even though they may not be used. Only the TCP is used in this example as the Web Browser communication is based on TCP. Inside the TCP function, any received packets are checked, and the function continues if the packets are of type "GET/". Then, the command tail is checked and the LEDs are toggled as required. The transmit buffer is loaded with the HTML response and the length of the buffer is returned from the function which then sends the buffer to the client.

The array StrtPage at the beginning of the program defines the HTML page to be sent to the PC so that the PC can display it. This page is made up of the following commands. This HTML script basically displays a form (Figure 7.118) on the PC screen with two buttons. Clicking TOGGLE LED A toggles the state of LED A (if the LED is ON it turns OFF, and vice versa), and similarly, clicking TOGGLE LED B button toggles the state of LED B:

Figure 7.118: Form Displayed by the Web Browser on the PC.

```
char StrtPage[] =
"<html><body>\
<form name=\"input\" method=\"get\"><table align=center width=500 \
bgcolor=Red border=4><tr><td align=center colspan=2><font size=7 \
color=white face=\"verdana\">LED CONTROL</font></td></tr>\
<tr><td align=center bgcolor=Blue><input name=\"TA\" type=\"submit\" \
value=\"TOGGLE LED A\"></td><td align=center bgcolor=Green> \
<input name=\"TB\" type=\"submit\" value=\"TOGGLE LED B\"></td></tr>\
</table></form></body></html>";
```

The connectivity of the system can be checked by using the PING command to send packets from the PC to the Ethernet controller. If everything is working as expected, then PING replies should be displayed on the PC screen. To use the PING command, Click the START button (Windows 7) and type CMD, followed by the Enter key. Then enter the command

PING 192.168.1.15

You should get a response similar to the following lines:

```
Pinging 192.168.1.15 with 32 bytes of data:
Reply from 192.168.1.15: bytes = 32 time = 12 ms TTL = 128
Reply from 192.168.1.15: bytes = 32 time = 6 ms TTL = 128
Reply from 192.168.1.15: bytes = 32 time = 6 ms TTL = 128
Reply from 192.168.1.15: bytes = 32 time = 6 ms TTL = 128

Ping statistics for 192.168.1.15:
   Packets: Sent = 4, Received = 4, Lost = 0 (0% loss>
Approximate round trip time in milliseconds:
   Minimum = 6 ms, Maximum = 12 ms, Average = 7 ms
```

The operation of the system is described below in steps:

- Compile and load the program to the microcontroller. Connect the Serial Ethernet board to PORTC of the development board. Connect the PC and the Serial Ethernet board together using a hub, switch, or router, or using a crossed network cable for local testing.
- Open a web browser on the PC (e.g. Microsoft Internet Explorer or Firefox) and send an HTTP request by entering the following url: http://192.168.1.15
- Upon receipt of this request, the Ethernet controller sends the HTML code shown in Figure 7.117 to the Web Browser, together with the HTTP header. The *Content-Type* field is used by the browser to tell which format the document it receives is in. HTML is identified with "text/html", and ordinary text is identified with "text/plain".
- The Web Browser then displays the form shown in Figure 7.118.
- The user can toggle LED A or LED B by clicking on the appropriate buttons. Assuming that button LED A is clicked, the Web Browser sends the following command to the Ethernet controller:

GET /? TA = TOGGLE_LED + A

Similarly, if button B is pressed, the Web Browser sends the following command to the Ethernet controller:

GET /? TB = TOGGLE_LED + B

- The Ethernet controller checks the received command (inside function TCP) and toggles LED A or LED B as required.

Project 7.14—Using the Ethernet—UDP-Based Control

This is another project using the embedded Ethernet. In this project, communication is established between the PC and the microcontroller system using the UDP protocol. Eight LEDs are connected to PORTD of the microcontroller. A Graphical User Interface (GUI) program is developed on the PC using the Visual Studio, Visual Basic program (VB.NET). The user specifies which bits of PORTD should be turned ON by clicking the appropriate parts of a GUI form. The PC program establishes UDP communication with the microcontroller system and sends a packet about the PORTD bits that should be turned ON. The microcontroller system uses the UDP protocol to receive this packet and then turns ON the required bits of PORTD.

The Hardware

The circuit diagram of the project is similar to the one given in Figure 7.112, but here, eight LEDs are connected to PORTD instead of just 2.

The PC Program

The PC program is based on VB.NET. When the program is run, the form shown in Figure 7.119 is displayed. The form consists of 8 Textboxes (called TextBox0 to TextBox7) and 8 Buttons (called Button0 to Button7). The Textboxes correspond to the LEDs connected to the microcontroller, and they can take two values: ON or OFF. Clicking a button under a textbox toggles the contents of the Textbox from ON to OFF, or from OFF to ON. When in state ON, the corresponding LED is ON.

After selecting which bits of PORTD should be ON, the user should click the CONNECT button to establish UDP communication with the microcontroller system. After this, the SEND button should be clicked to send a packet to the microcontroller system so that the required LEDs can be turned ON. Byte array PortArray has elements as either 0 or 1, and it stores the required state of each individual bit. The UDP packet consists of 1 byte only which is the byte stored in variable PortValue, corresponding to

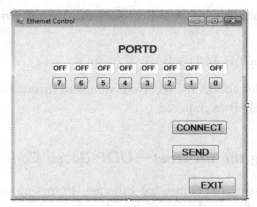

Figure 7.119: The PC Form.

the PORTD bits to be turned ON. Function ClientSocket.Send is used to send this byte to the microcontroller.

The PC program listing is shown in Figure 7.120.

The Microcontroller Program

mikroC Pro for PIC

The microcontroller program is shown in Figure 7.121 (MIKRO-UDP.C). At the beginning of the program, the connections between the Serial Ethernet board and the microcontroller are defined. Then the Ethernet board MAC address and the IP address to be used in the UDP communication are defined.

This program uses the UDP, but the function SPI_Ethernet_UserTCP must be declared with a return statement, even though the, TCP protocol is not used in the program.

The SPI_Ethernet_UserUDP is very simple. It checks the port number, and if the remote port number is invalid (i.e. is not 10001), then the function returns. Otherwise, the received packet is stored in array Txt using function SPI_Ethernet_getByte, and the contents of Txt[0], which is the byte corresponding to user's selection is sent to PORTD.

Project 7.15—Digital Signal Processing—Low Pass FIR Digital Filter Project

Digital filters are very important in many digital signal processing applications. The theory of digital filters is complex and is beyond the scope of this book. It is assumed that the

```
Imports System.Net.Sockets
Imports System.Text
Imports System.Threading

Public Class Form1
    Public Txt(1) As Byte
    Public PortArray(8) As Byte
    Public ClientSocket As New UdpClient

    Public ServerAddress As String = "192.168.1.15"     ' Set the IP address of the server
    Public PortNumber As Integer = 10001                ' Set port number

    Private Sub Button0_Click(sender As System.Object, e As System.EventArgs) Handles
Button0.Click
        If TextBox0.Text = "OFF" Then
            TextBox0.Text = "ON"
            PortArray(0) = 1
        Else
            TextBox0.Text = "OFF"
            PortArray(0) = 0
        End If
    End Sub

    Private Sub Button1_Click(sender As System.Object, e As System.EventArgs) Handles
Button1.Click
        If TextBox1.Text = "OFF" Then
            TextBox1.Text = "ON"
            PortArray(1) = 1
        Else
            TextBox1.Text = "OFF"
            PortArray(1) = 0
        End If
    End Sub

    Private Sub Button2_Click(sender As System.Object, e As System.EventArgs) Handles
Button2.Click
        If TextBox2.Text = "OFF" Then
            TextBox2.Text = "ON"
            PortArray(2) = 1
        Else
            TextBox2.Text = "OFF"
            PortArray(2) = 0
        End If
    End Sub

    Private Sub Button3_Click(sender As System.Object, e As System.EventArgs) Handles
Button3.Click
        If TextBox3.Text = "OFF" Then
            TextBox3.Text = "ON"
            PortArray(3) = 1
        Else
            TextBox3.Text = "OFF"
            PortArray(3) = 0
        End If
    End Sub

    Private Sub Button4_Click(sender As System.Object, e As System.EventArgs) Handles
Button4.Click
        If TextBox4.Text = "OFF" Then
```

Figure 7.120: The PC Program.

```
            TextBox4.Text = "ON"
            PortArray(4) = 1
        Else
            TextBox4.Text = "OFF"
            PortArray(4) = 0
        End If
    End Sub

    Private Sub Button5_Click(sender As System.Object, e As System.EventArgs) Handles
Button5.Click
        If TextBox5.Text = "OFF" Then
            TextBox5.Text = "ON"
            PortArray(5) = 1
        Else
            TextBox5.Text = "OFF"
            PortArray(5) = 0
        End If
    End Sub

    Private Sub Button6_Click(sender As System.Object, e As System.EventArgs) Handles
Button6.Click
        If TextBox6.Text = "OFF" Then
            TextBox6.Text = "ON"
            PortArray(6) = 1
        Else
            TextBox6.Text = "OFF"
            PortArray(6) = 0
        End If
    End Sub

    Private Sub Button7_Click(sender As System.Object, e As System.EventArgs) Handles
Button7.Click
        If TextBox7.Text = "OFF" Then
            TextBox7.Text = "ON"
            PortArray(7) = 1
        Else
            TextBox7.Text = "OFF"
            PortArray(7) = 0
        End If
    End Sub

    Private Sub ButtonConnect_Click(sender As System.Object, e As System.EventArgs)
Handles ButtonConnect.Click
        ClientSocket.Connect(ServerAddress, PortNumber)
    End Sub

    Private Sub ButtonSend_Click(sender As System.Object, e As System.EventArgs) Handles
ButtonSend.Click
        Dim PortValue As Byte
        PortValue = 0

        For i = 0 To 7
            PortValue = PortValue + PortArray(i) * 2 ^ i
        Next
        Txt(0) = PortValue
        ClientSocket.Send(Txt, 1)
    End Sub
```

Figure 7.120

cont'd

```
    Private Sub ButtonExit_Click(sender As System.Object, e As System.EventArgs) Handles
ButtonExit.Click
        ClientSocket.Close()
        End
    End Sub
End Class
```

Figure 7.120
cont'd

readers have a sufficient knowledge on finding the filter coefficients and the various digital filtering structures for a given design specifications. There are many books and references on the theory of digital filters that may be helpful.

In this project, we will be designing a low-pass FIR type digital filter with the following specifications:

Window Type	No Windowing
Cut-off frequency	50 Hz
Sampling frequency	1000 Hz
Filter order	10 (11 taps)

The block diagram of the project is shown in Figure 7.122. In this project, a Velleman PCSGU250 is used (Figure 7.123). This is a device that operates as a frequency generator, oscilloscope, transient recorder, spectrum analyzer, and frequency plotter (Bode plotter).

The analog signal generated by the PCSGU250 is applied to one of the analog channels of the microcontroller. The filtered signal is converted into the analog signal using a DAC chip and is then fed to the PCSGU250 for plotting the frequency response.

The Filter Structure

There are many software packages that could be used to find the filter coefficients. In this book, the highly popular ScopeFIR program is used.

The steps in finding the filter parameters using the ScopeFIR program are given below:

- Start the program.
- Create a new project by selecting the filter type as low-pass, Windowed Sinc, 11 taps, sampling frequency 100 Hz, and the cut-off frequency of 50 Hz (Figure 7.124).
- Click Design to design the filter. The frequency response, filter coefficients, and the impulse response of the filter to be designed will be displayed by the program. Figure 7.125 shows the frequency response and Figure 7.126 shows the required filter coefficients.

```
/*********************************************************************
                    UDP ETHERNET CONTROL PROJECT
                    ==============================
```

This project shows how the ETHERNET can be used in microcontroller based projects. In this project a Serial Ethernet Board (www.mikroe.com) is connected to the EasyPIC V7 development board.

The project uses the UDP method to establish Ethernet based communication between a PC and the microcontroller system.

The PC is the client and the microcontroller system is the server. The PC program is written using the Visual Studio, Visual Basic programming language (VB.NET). The PC sends a packet to the microcontroller system in the form of a UDP packet.

8 LEDs are connected to PORTD. The user interactively enters the LEDs to be turned ON using a GUI type window. This message is then passed to the microcontroller system via the UDP and then the required LEDs are turned ON by the microcontroller.

Port 10001 and IP address 192.168.1.15 are used in the project.

```
Author:  Dogan Ibrahim
Date:    October, 2013
File:    MIKROC-UDP.C
*********************************************************************/
//
// Ethernet NIC interface definitions
//
sfr sbit SPI_Ethernet_Rst at RC0_bit;
sfr sbit SPI_Ethernet_CS  at RC1_bit;
sfr sbit SPI_Ethernet_Rst_Direction at TRISC0_bit;
sfr sbit SPI_Ethernet_CS_Direction  at TRISC1_bit;
//
// Define Serial Ethernet Board MAC Address, and IP address to be used for the communication
//
unsigned char MACAddr[6] = {0x00, 0x14, 0xA5, 0x76, 0x19, 0x3f} ;
unsigned char IPAddr[4] = {192,168,1,15};
unsigned char getRequest[10];

typedef struct
{
  unsigned canCloseTCP:1;
  unsigned isBroadcast:1;
}TethPktFlags;

//
// TCP routine. This is where the user request to toggle LED A or LED B are processed
//
//
unsigned int SPI_Ethernet_UserTCP(unsigned char *remoteHost,
                unsigned int remotePort, unsigned int localPort,
                unsigned int reqLength, TEthPktFlags *flags)
```

Figure 7.121: mikroC Pro for PIC Program.

```
            {
              return (0);
            }

            //
            // UDP routine. Must be declared even though it is not used
            //
            unsigned int SPI_Ethernet_UserUDP(unsigned char *remoteHost,
                          unsigned int remotePort, unsigned int destPort,
                          unsigned int reqLength, TEthPktFlags *flags)
            {
              char Txt[10];
              char len=0;

              if(destport != 10001)return(0);                // Check that correct port is used
              while(reqLength--)
              {
                Txt[len++] = SPI_Ethernet_getByte();         // Extract the received bytes
              }

              PORTD = Txt[0];                                // Turn ON required LEDs
              return;
            }

            //
            // Start of MAIN program
            //
            void main()
            {
              ANSELC = 0;                                    // Configure PORTC as digital
              ANSELD = 0;                                    // Configure PORTD as digital
              TRISD = 0;                                     // Configure PORTD as output
              PORTD = 0;                                     // Clear PORTD to start with

              SPI1_Init();                                   // Initialize SPI module
              SPI_Ethernet_Init(MACAddr, IPAddr, 0x01);      // Initialize Ethernet module

              while(1)                                       // Do forever
              {
                SPI_Ethernet_doPacket();                     // Process next received packet
              }
            }
```

Figure 7.121

cont'd

The filter coefficients are

```
h[0] = h[10] = 0.0681
h[1] = h[9] = 0.0810
h[2] = h[8] = 0.0918
h[3] = h[7] = 0.1001
h[4] = h[6] = 0.1052
h[5] = 0.1070
```

The filter structure is shown in Figure 7.127.

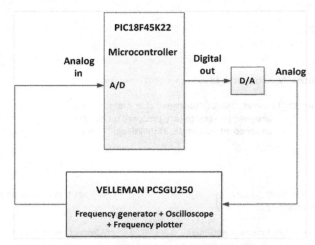

Figure 7.122: Block Diagram of the Project.

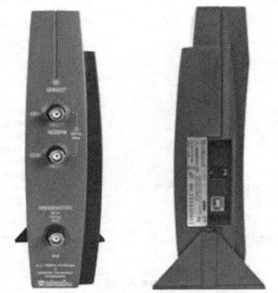

Figure 7.123: The Velleman PCSGU250 Device.

The Hardware

The circuit diagram of the project is shown in Figure 7.128. The MCP4921 12-bit serial DAC chip, controlled by the SPI bus, is used for the DAC.

Project PDL

The project PDL is shown in Figure 7.129.

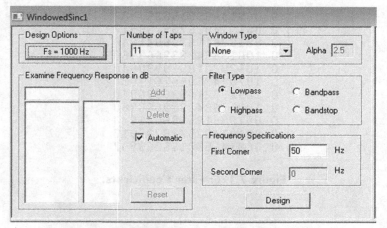

Figure 7.124: ScopeFIR Filter Design Program.

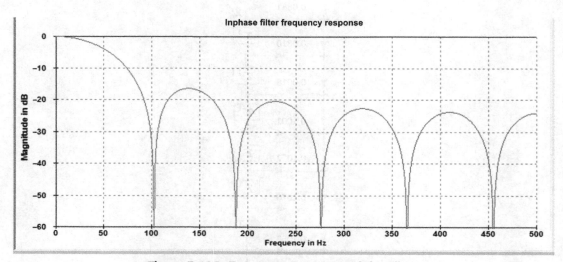

Figure 7.125: Frequency Response of the Filter.

Project Program

mikroC Pro for PIC

The mikroC Pro for the PIC program listing is given in Figure 7.130 (MIKROC-FIR.C). At the beginning of the program, the D/A converter chip select pin connection is defined. The 11 filter coefficients are then stored in a floating point array called *h*:

```
float h[N] = {0.0681, 0.0810, 0.0918, 0.1001, 0.1052, 0.1070, 0.1052, 0.1001, 0.0918, 0.0810, 0.0681};
```

```
0.068146550155253907
0.081014025735235917
0.091886195306041848
0.100138842939668960
0.105292210529990840
0.107044350667617030
0.105292210529990840
0.100138842939668960
0.091886195306041848
0.081014025735235917
0.068146550155253907
```

Figure 7.126: Filter Coefficients.

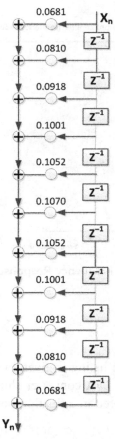

Figure 7.127: The 10th Order (11-tap) FIR Filter Structure.

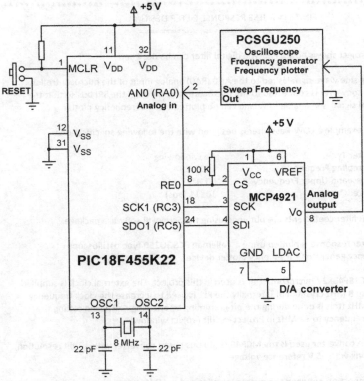

Figure 7.128: Circuit Diagram of the Project.

BEGIN/MAIN
      Configure the D/A chip select pin
      Store filter coefficients in an array
      Configure PORTC and PORTE as digital
      Configure AN0 (Channel 0) as analog input
      Initialize the SPI library
      Initialize A/D converter module
      Initialize Timer 1 for 1ms interrupts
      Wait for Timer 0 interrupts (TMR)
END/MAIN

BEGIN/TMR
      Get a new signal sample
      Calculate the output sample
      Send the output sample to the D/A converter
      Delay the input signals by one sample time
      Clear Timer 0 interrupt flag
END/TMR

Figure 7.129: Project PDL.

```
/*************************************************************************
                    FINITE IMPULSE RESPONSE FILTER DESIGN
                    ====================================
```

This project shows how a FIR type digital filter can be designed.

Analog sine wave signal is fed to the AN0 (RA0) analog input of the microcontroller.
A D/A converter is connected to the microcontroller through the SPI bus so that the
filtered signal is in analog form and can be plotted using a frequency plotter.

In this example a LOW-Pass filter is designed with the following specifications:

Filter Type:	No windowing
Sampling Frequency:	1000 Hz
Passband Upper Frequency:	50 Hz
Filter Order:	10 (11 taps)

The FIR filter coefficients are obtained using the ScopeFIR software package.

The filter response is plotted using a Velleman PCSGU250 type oscilloscope+
frequency generator + frequency plotter device.

The PIC18F45K22 microcontroller is used in this project. The external clock is supplied
using an 8 MHz crystal, but internally the PLL is used to increase the clock frequency
to 32 MHz (this is done during the programming by enabling 4xPLL and setting the
clock frequency to 32 MHz in Project -> Edit Project window).

The D/A converter used is the MCP4921 SPI bus based converter with 12-bit resolution,
operating with +5 V reference voltage.

The connection between the microcontroller and the D/A converter is as follows:

Author:	Dogan Ibrahim
Date:	October 2013
File:	MIKROC-FIR.C

```
*************************************************************************/
// DAC module connections
sbit Chip_Select at LATE0_bit;
sbit Chip_Select_Direction at TRISE0_bit;
// End DAC module connections

#define N 10                                // Filter order=10, having 11 taps

float Sample,xn, yn, x[N];
unsigned ADC;
unsigned char temp;
unsigned int DAC;
float h[N+1] = {0.0681, 0.0810, 0.0918, 0.1001, 0.1052, 0.1070, 0.1052, 0.1001, 0.0918, 0.0810,
0.0681};
//
// Timer 1 interrupt service routine. The program jumps here every 1000us (the sampling
```

Figure 7.130: mikroC Pro for PIC Program.

```
// frequency is 1 kHz, i.e. Period = 1ms = 1000us). Here, a new output is calculated and sent
// to the D/A converter
//
void interrupt()
{
    unsigned char i;

    TMR0L = 6;                                          // Re-load TMR0
    ADC = ADC_Get_Sample(0);                            // Get new input Sample from AN0
    x[0]=ADC;
    yn = 0.0;
//
// Calculate a new output yn
//
    for(i = 0; i <= N; i++)
    {
      yn = yn + h[i]*x[i];
    }
//
// Output the new Sample via the D/A converter
//
    DAC = yn;
    Chip_Select = 0;                                    // Select DAC chip

    // Send High Byte
    temp = (DAC >> 8) & 0x0F;                            // Store DAC[11..8] to temp[3..0]
    temp |= 0x30;                                        // Define D/A setting
    SPI1_Write(temp);                                   // Send high byte via SPI

    // Send Low Byte
    temp = DAC;                                          // Store DAC[7..0] to temp[7..0]
    SPI1_Write(temp);                                   // Send low byte via SPI

    Chip_Select = 1;                                    // Deselect D/A converter chip
//
// Shift the input samples for the delay action
//
    for(i = 0; i < N; i++)
    {
      x[N-i] = x[N-i-1];
    }
//
// Re-enable Timer 0 interrupts
//
    INTCON.TMR0IF = 0;                                  // Clear Timer 0 interrupt flag
}


//
//*********************** MAIN PROGRAM ***********************
// Start of MAIN program. In the main program the I/O ports are configured, SPI bus
```

Figure 7.130
cont'd

```
// library is initialized, and the A/D converter module library is initialized. In addition,
// Timer 0 is configured to interrupt at every one millisecond and interrupts are enabled.
//
void main()
{
    TRISA0_bit = 1;                        // AN0 (RA0) is input
    ANSELE = 0;                            // RE0 is digital I/O
    ANSELA = 1;                            // RA0 is analog I/O
    Chip_Select_Direction = 0;             // Configure CS pin as output
    Chip_Select = 1;                       // Disable D/A converter
    SPI1_Init();                           // Initialize SPI1
    ADC_Init();                            // Initialize A/D converter

//
// Configure Timer 0 for 1000us (1ms) interrupts
//
    T0CON = 0 x 44;                        // Disable TMR0, 8-bit, prescaler-32
    INTCON.TMR0IF = 0;                     // Clear Timer 0 interrupt flag
    INTCON.TMR0IE = 1;                     // Enable Timer 0 interrupts
    TMR0L = 6;                             // Load TMR0
    T0CON.TMR0ON = 1;                      // Enable Timer 0
    INTCON.GIE = 1;                        // Enable global interrupts

    for(;;)                                // Wait for Timer 0 interrupts
    {
    }
}
```

Figure 7.130
cont'd

Inside the main program, analog input AN0 (Channel 0, or pin RA0) is configured as an analog input, D/A converter is disabled, and the SPI library and A/D converter modules are initialized. Timer 0 is then configured to interrupt at every sampling time (1000 μs).

With a clock frequency of 32 MHz, the clock period is 0.03125 μs (note that the actual crystal frequency is 8 MHz, but the internal PLL module is used to multiply the external clock frequency by 4 to give an operating clock frequency of 32 MHz. This is done by enabling the 4xPLL option in the Project → Edit Project window before the microcontroller is programmed. In addition, the oscillator frequency should be selected as 32 MHz in this window). Since the instruction cycle time is four clock periods, the actual timer clock frequency is 8 MHz, or the actual timer clock period is 0.125 μs. The timer prescaler is set to 32, giving a value of 6 for the timer register TMR0L. Thus, when Timer 0 is loaded with 6 and timer and global interrupts are enabled, the timer will generate interrupts at every millisecond and the program will jump to the ISR declared by the programmer:

$$TMR0L = 256 - Delay/(Clock\ period \times prescaler\ value)$$

or,

$$TMR0L = 256 - 1000\ \mu s/(0.125\ \mu s \times 32) = 6$$

The main program then enters a loop and waits for timer interrupts to occur. The digital filtering operation is performed inside the timer ISR, which is entered every millisecond. Here, a new input sample is obtained from analog channel AN0 by calling function *ADC_Get_Sample* with channel number 0, and the output sample is calculated.

The program then sends the output sample to the D/A converter. The D/A converter is 12 bits wide and the high nibble (bits 8–11) is sent first, followed by the low byte (bits 0–7). Function *SPI1_Write* sends data to the D/A converter over the SPI bus. Note that the PIC18F45K22 microcontroller supports two SPI bus I/O pins, and in this project, only the first SPI bus is used. This is why the SPI statements are terminated with number 1. The following statements are used to send the processed data to the D/A converter:

```
Chip_Select = 0;              // Enable DAC chip
// Send High Byte
temp = (DAC >> 8) & 0x0F;     // Store DAC[11..8] to temp[3..0]
temp |= 0x30;                 // Define D/A setting
SPI1_Write(temp);             // Send high byte via SPI
// Send Low Byte
temp = DAC;                   // Store DAC[7..0] to temp[7..0]
SPI1_Write(temp);             // Send low byte via SPI
Chip_Select = 1;              // Disable D/A chip
```

The input samples are then shifted (delayed) by one sampling time using the following code:

```
for(i = 0; i < N; i++)
{
  x[N-i] = x[N - i - 1];
}
```

Just before exiting from the ISR, the timer interrupt flag TMR0IF is cleared so that further timer interrupts can be accepted by the processor.

Note that the A/D and D/A converters used in the design are unipolar (accept only positive voltages), and therefore, it is necessary to introduce the DC level to the input signal to shift it up so that it is always positive. This is done by clicking the *Offset* button at the bottom right-hand corner of the PCSGU250 screen.

In addition, the PCSGU250 device will not give an accurate graph in the stop band if the input signal is very small as may be the case in the stop band of low-pass filters.

Project 7.16—Automotive Project—Local Interconnect Network Bus Project

The Local Interconnect Network (LIN) is a serial network protocol developed for the automotive industry. The CAN bus was too expensive to implement for every electronic component in a car and the need for a cheap serial network arose as a result of this.

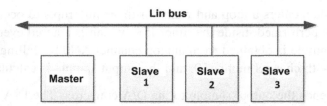

Figure 7.131: LIN Bus with One Master and Three Slaves.

The LIN bus is a low-cost serial protocol that can easily be implemented with microcontrollers having UART modules. The bus consists of a master and typically up to 16 slaves. All messages are initiated by the master with an identifier. The slave with the matching identifier replies to the message. As a result of this two-way communication, there is no collision on the bus. In typical applications, the master is a microcontroller requesting information or sending commands to a slave. The slaves are typically sensors or actuators that respond to the commands sent by the master.

Compared to the CAN bus, the LIB bus has the following advantages and disadvantages:

- The LIN bus has maximum data rate of 19,200 bps, while the CAN bus is much faster, up to 1 Mbps.
- The LIN bus is limited to 16 nodes (1 master and up to 15 slaves). The CAN bus can have up to 128 nodes.
- The LIN can is a single-wire bus (plus the chassis), while the CAN bus is a two-wire bus.
- The LIN bus is much cheaper to implement than the CAN bus.

The LIN bus can be up to 40 m long and is typically used in the following components of vehicles:

- Vehicle doors, central locking system, mirrors.
- Vehicle light sensor, sun roof, light control.
- Seat heater, occupancy sensor, seat motors.
- Small engine controls, such as cruise control, wiper, turn indicator, climate control, small motors, sensors, steering wheel.

Figure 7.131 shows a typical LIN bus implementation with one master and three slaves.

The LIN Protocol

The LIN protocol can be implemented with any microcontroller that supports a UART module. The protocol consists of frames, where each frame has two

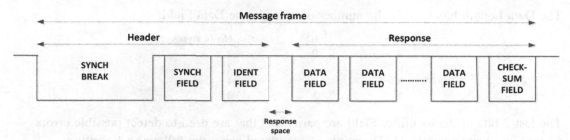

Figure 7.132: LIN Protocol.

parts: the Header, and the Response. The protocol consists of the following fields (Figure 7.132):

- Synch Break,
- Synch Field,
- Identifier Field,
- Data Field,
- Checksum Field.

Each byte, except the Synch Break, is transmitted (or received) in standard serial format with one start bit, eight data bits, and one stop bit.

Synch Break: The Synch Break is always initiated by the master, and it signifies the start of a frame. It is identified by a start bit, and at least 13 bits of 0s, followed by the stop bit.

Synch Field: This field is sent by the master as the Synch delimiter, and it allows the slaves to synchronize. The data sent are one start bit, eight data bits corresponding to hexadecimal number 0x55 (bit pattern 01010101), and one stop bit.

Identifier Field: This byte is sent by the master. and it consists of the start bit, eight identifier bits, and one stop bit in the following format:

- Bits 0 to 3—LIN ID,
- Bits 4 and 5—Data Length,
- Bits 6 and 7—Parity.

| ID0 | ID1 | ID2 | ID3 | ID4 | ID5 | P0 | P1 |

The LIN ID bits (ID0 to ID3) represent the identifier of the slave node who is to respond in the Response part of the frame.

The Data Length bits specify the number of bytes in the Data Field:

ID5	ID4	No of Bytes
0	0	2
0	1	2
1	0	4
1	1	8

The last 2 bits of the identifier Field are parity bits that are used to detect possible errors (there is no error correction). The parity is calculated using the following algorithm:

$$P0 = ID0 \oplus ID1 \oplus ID2 \oplus ID4$$
$$P1 = ID1 \oplus ID3 \oplus ID4 \oplus ID5$$

Identifiers in the range 0x00 to 0x3B are known as Unconditional frame identifiers, and there is only one sender of these frames. The identifiers, known as Command Frame Identifiers and having codes 0x60, 0x3C, and 0x3D are known as Diagnostic Frames and are reserved for diagnostic purposes. The Command Frame with the first byte set to 0x00 is used to put all slaves into the Sleep mode.

Data Field: the Data Field can contain 2, 4, or 8 bytes, each having one start bit, eight data bits, and one stop bit.

Checksum Field: This is the last byte in a frame. This byte contains the inverted modulo − 256 sum of all bytes within the Data Field. The sum is calculated by adding all data bytes with any carry bits and then inverting the answer (the property of inverted module-256 sum is that if the resultant number is added to the sum of all data bytes, the result will be 0xFF). For example, 0xFF + 0x01 = 0x01 and not 0x00.

Project Description

In this project, a master and a slave node are used. The slave node is connected to a temperature sensor. Temperature readings are sent to the master on request and are displayed on an LCD connected to the master.

Figure 7.133 shows the block diagram of the project.

Project Hardware

The circuit diagram of the project is shown in Figure 7.134. The MCP 201 LIN bus transceiver chip is used for the LIN bus to microcontroller interface. An LM35DZ-type temperature sensor is connected to the slave node.

The MCP 201 chip has the following features:

- Support up to 19,200 bps communication speed;
- Six- to 18-V supply voltage;

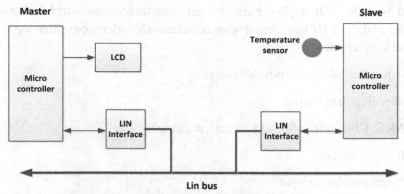

Figure 7.133: Block Diagram of the Project.

- Eight-pin DIL housing;
- Standard UART interface;
- Internal pull-up resistor and diode;
- A 40-mA current drive;
- Short-circuit current limit;
- Internal 5 V, 50-mA regulator.

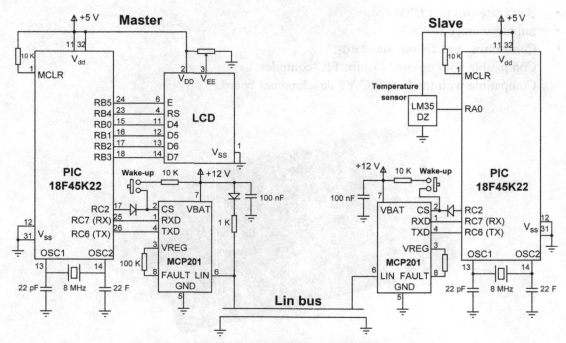

Figure 7.134: Circuit Diagram of the Project.

The MCP 201 provides half-duplex, bidirectional communications interface between a microcontroller and the LIN bus. The device translates the microcontroller logic levels to LIN logic, and vice versa.

The MCP 201 has the following pin definitions:

Pin 1, RXD: Receive data output.

Pin 2, CS/WAKE: Chip select (logic 1 to activate chip).

Pin 3, VREG: +5-V output.

Pin 4, TXD: Transmit data output.

Pin 5, VSS: Ground.

Pin 6, LIN: Lin bus connection.

Pin 7, VBAT: 6- to 18-V input.

Pin 8, FAULT: Fault detect output.

Two EasyPIC V7 development boards and two mikroElektronika LIN Bus Boards were used in the project development. These LIN Bus boards (Figure 7.135) have the following features:

- Baud rates up to 19,200 bps;
- Supply voltage 6−18 V;
- Conforming to LIN bus standards;
- Compatible with mikroC Pro for PIC compiler;
- Compatible with the EasyPIC V7 development board.

Figure 7.135: mikroElektronika LIN Bus Board.

If you are using the LIN Bus board together with the EasyPIC V7 development board, then plug in the board to PORTC connector at the edge of the development board and set the following jumpers on both the MASTER and the SLAVE boards (these jumper settings connect the UART pins RC6 and RC7 to the board. Also, the CS is connected to the RC2 pin of the microcontroller):

- Set DIL switch SW1 1, 4, and 7 ON.
- Leave J1 on the MASTER node.
- Remove J1 from the SLAVE node.
- Apply external a 6- to 18-V DC supply to the VBAT connectors.
- Establish the LIN bus between the MASTER and SLAVE by connecting LIN and GND connectors of both boards together.

Project PDL

The project PDL is shown in Figure 7.136.

Project Program

mikroC Pro for PIC
MASTER Node

The mikroC Pro for the PIC program listing of the MASTER node is shown in Figure 7.137 (MIKROC-LINMSTR.C). At the beginning of the program, the connections between the LCD and the microcontroller are defined. Also, the CS connection of the MCP 201 chip is defined as pin RC2 of the microcontroller. Symbols SYNCH_FIELD, SLAVE_NODE, and No_Of_Bytes are assigned values.

Inside the main program, PORTB and PORTC are configured as digital, MCP 201 chip is disabled, LCD is initialized, and message "LIN BUS PROJECT" is displayed for 2 s. The remainder of the program is executed in an endless loop, formed using a while statement. Inside this loop, the HEADER is sent to the LIN bus using function Send_Header, response is received from the slave using function Get_Response, and the received data (temperature in this project) are displayed on the LCD.

Send_Header Function

This function sends the BREAK sequence, SYNCH_FIELD, and the IDENT FIELD. The BREAK sequence is accomplished by forcing the UART to generate a frame error (i.e. the

MASTER:

<u>Main Program</u>

BEGIN

    Define connections between the LCD and microcontroller

    Configure PORTB and PORTC as digital

    Initialize LCD

    Send a message to the LCD

    Initialize UART

    DO FOREVER

        CALL Send_Header

        CALL Get_Response

        Clear LCD

        Display temperature

        Wait 1 second

    ENDDO

END

BEGIN/Send_Header

    Send BREAK sequence

    Send SYNCH FIELD

    Send IDENT FIELD

END/Send_Header

BEGIN/Get_Response

    Get response bytes from the slave

END/Get_Response

SLAVE:

<u>Main Program</u>

BEGIN

    Define connections between the LCD and microcontroller

    Configure RA0 as analog and PORTC as digital

    Initialize UART

    DO FOREVER

        Wait for BREAK sequence

        Get SYNC FIELD

        Get IDENT FIELD

        Extract identifier

        IF the request is for this node

            Read temperature from channel 0

            Send data to MASTER over the LIN bus

            CALL Calc_ Checksum

            Send Checksum to MASTER over the LIN bus

        ENDIF

    ENDDO

END

Begin/Calc_Checksum

    Add all data bytes including carry bit

    Invert the sum

    Return the sum to the caller

END/Calc_Checksum

Figure 7.136: The Project PDL

```
/*****************************************************************************
                        LIN BUS CONTROL PROJECT
                        =======================
```

This project shows how to use the Automotive LIN bus in microcontroller projects.

In this project 2 LIN bus nodes, named MASTER and SLAVE communicate with each other
at 9600 bps over the LIN BUS.

An LCD is connected to the MASTER node. An LM35DZ type temperature sensor is connected
to the SLAVE node. The MASTER node requests the temperature every second from the SLAVE
node. The SLAVE node reads the temperature and sends to the MASTER node which then displays
on the LCD.

This project is based on the EasyPIC V7 development board and the mikroElektronika
LIN bus boards. Two development boards are used, one as the MASTER, the other one as the
SLAVE. LIN bus boards are attached to PORTC of each development board and the two boards
are connected to each other with two cables (LIN bus and ground). The CS pins of the
LIN bus boards are connected to RC2 pin of the microcontroller.

There is one MASTER and one SLAVE in this project.

This is the MASTER node program. The program works with 8 MHz clock (clock PLL is disabled)

```
Author:  Dogan Ibrahim
Date:    October, 2013
File:    MIKROC-LINMSTR.C
*****************************************************************************/
// LCD module connections
sbit LCD_RS at RB4_bit;
sbit LCD_EN at RB5_bit;
sbit LCD_D4 at RB0_bit;
sbit LCD_D5 at RB1_bit;
sbit LCD_D6 at RB2_bit;
sbit LCD_D7 at RB3_bit;

sbit LCD_RS_Direction at TRISB4_bit;
sbit LCD_EN_Direction at TRISB5_bit;
sbit LCD_D4_Direction at TRISB0_bit;
sbit LCD_D5_Direction at TRISB1_bit;
sbit LCD_D6_Direction at TRISB2_bit;
sbit LCD_D7_Direction at TRISB3_bit;
// End LCD module connections

#define MCP201_CS PORTC.RC2              // MCP201 CS connection
#define SYNCH_FIELD 0x55                 // SYNCH FIELD
#define SLAVE_NODE_ID 0x01               // Slave node identifier
#define No_Of_Bytes 2                    // Number of bytes to receive
//
// This function sends HEADER to the SLAVE. The Header consists of:
// SYNCH BREAK, SYNCH FIELD, IDENT FIELD
```

Figure 7.137: mikroC Pro for PIC MASTER Program.

```
//
void Send_Header()
{
  bit P0, P1, ID0, ID1, ID2, ID3, ID4, ID5;
  unsigned char IDENT_FIELD, temp, Cnt, c;

//
// Enable BREAK sequence
//
  TXSTA1.TXEN = 1;                          // Set TXEN bit
  TXSTA1.SENDB = 1;                         // Set SENDB bit
  TXREG1 = 0x0;                             // send dummy data to start the sequence
  Uart1_Write(SYNCH_FIELD);                 // Send SYNCH FIELD character

//
// The Node Identifier is set to 1, and number of bytes is 2. Thus, the IDENT FIELD has the
// format P1 P0 00 0001. Find parity bits P1 and P0 and add to the IDENT field
  if(No_Of_Bytes == 2)                      // No of bytes
    Cnt = 0;                                // 2-bit field for the number of bytes
  else if(No_Of_Bytes == 4)                 // Cnt is 0,2, or 3 depending on byte count
    Cnt = 2;
  else if(No_Of_Bytes == 8)
    Cnt = 3;

  IDENT_FIELD = SLAVE_NODE_ID;
  Cnt = Cnt << 4;
  IDENT_FIELD = CNT | IDENT_FIELD;          // Add No of Bytes to IDENT field

  ID0 = IDENT_FIELD.F0;                     // Bit 0
  ID1 = IDENT_FIELD.F1;                     // Bit 1
  ID2 = IDENT_FIELD.F2;                     // Bit 2
  ID3 = IDENT_FIELD.F3;                     // Bit 3
  ID4 = IDENT_FIELD.F4;                     // Bit 4
  ID5 = IDENT_FIELD.F5;                     // Bit 5
  P0 = ID0 ^ ID1 ^ ID2 ^ ID4;               // Find P0
  P1 = ID1 ^ ID3 ^ ID4 ^ ID5;               // Find P1
  Temp = 0;
  Temp = P1 | P0;
  Temp = Temp << 6;
  IDENT_FIELD = IDENT_FIELD | Temp;         // Add the parity bits to IDENT field
  while(Uart1_Tx_Idle() == 0);              // Wait until ready
  Uart1_Write(IDENT_FIELD);                 // Send IDENT_FIELD
}

//
// This function reads data from the SLAVE node. The last byte received is the Checksum
//
void Get_Response(unsigned char c[])
{
  unsigned char i, Checksum;
```

Figure 7.137
cont'd

```
      for(i = 0; i < No_Of_Bytes; i++)            // Do to read all data
      {
        while(Uart1_Data_Ready() == 0);           // Wait until UART is ready
        c[i] = Uart1_Read();                       // get data from UART
      }
      c[i] = 0x0;                                  // Insert NULL terminator
      while(Uart1_Data_Ready() == 0);             // Wait until UART has data
      Checksum = Uart1_Read();                     // Read the Checksum
   }

//
// Start of MAIN program
//
void main()
{
   unsigned char Txt[3];

      ANSELB = 0;                                  // Configure PORTB as digital
      ANSELC = 0;                                  // Configure PORTC as digital
      MCP201_CS = 0;                               // Disable MCP201 to stat with
      TRISC.RC2 = 0;                               // Configure RC2 as output

      Lcd_Init();                                  // Initialize LCD
      Lcd_Cmd(_LCD_CURSOR_OFF);                    // LCD cursor OFF
      Lcd_Cmd(_LCD_CLEAR);                         // Clear LCD
      Lcd_Out(1,1,"LIN BUS PROJECT");              // Write message on row 1
      Delay_Ms(2000);                              // Wait to see the message

      Uart1_Init(9600);                            // Set UART baud rate to 9600bps
      MCP201_CS = 1;                               // Activate MCP201

      while(1)                                     // Do FOREVER
      {
        Send_Header();                             // Send LIN bus Header over the LIN bus
        Get_Response(Txt);                         // Get Response from the SLAVE
        Lcd_Cmd(_LCD_CLEAR);                       // Clear LCD
        Lcd_Out(2,1,Txt);                          // Display data (temperature)
        Delay_Ms(1000);                            // Wait 1 s
      }
}
```

Figure 7.137
cont'd

condition where the STOP bit is not received at the expected time). There are basically two ways that we can generate the BREAK sequence:

1. By sending a logic 0 (START bit) on pin RC6 and keeping the line low for at least 13 bit times, and then sending a STOP bit. For example, if the baud rate is 9600 bps, then

13 bits will take approximately 1.3 ms. The following statements will generate a BREAK sequence:

```
PORTC.RC6 = 0;    // Send START bit
Delay_Ms(2);      // Wait for 2 ms
PORTC.RC6 = 1;    // send STOP bit
```

2. Alternatively, we can force the UART module to send a BREAK sequence. The extended UART, found in most PIC18F series microcontrollers, supports sending the BREAK sequence. The following statements force the UART to send a BREAK sequence:

```
TXSTA1.TXEN = 1;    // Set TXEN bit
TXSTA1.SENDB = 1;   // Set send-break (SENDB) bit
TXREG1 = 0x0;       // Send dummy data to start the sequence
```

After sending the BREAK sequence, the SYNCH_FIELD character (0x55) is sent to the slave over the bus. Next, the IDENT field is formed by combining the slave node identifier, number of data bytes expected, and the parity bits P1 and P0, formed by Exclusive-OR'ing the appropriate bits.

Get_Response Function

This function reads the data bytes from the slave device, including the Checksum byte. Although the Checksum is received, it is not validated here for simplicity. The program finally displays the temperature on the LCD. The above process is repeated every second.

SLAVE Node

The mikroC Pro for the PIC program listing of the SLAVE node is shown in Figure 7.138 (MIKROC-LINSLAVE.C). At the beginning of the main program, PORTC is configured as a digital input, and RA0 is configured as an analog input. The UART is initialized to 9600 bps and the MCP 201 chip is activated. The remainder of the program is executed in an endless loop formed using a while statement. Inside this loop, the program waits until the BREAK sequence is received. There are several ways that the BREAK sequence can be detected:

1. By starting a timer when the START bit is detected and stopping the timer when the STOP bit is detected. If the timer value is equal or greater than 13 bit times (1.3 ms at 9600 bps), then it is assumed that the BREAK sequence is received and terminated.
2. By looking for the START bit and then a STOP bit, and assuming that the BREAK sequence is terminated when the STOP bit is received. Since when a slave device is waiting the only communication on the bus is the BREAK sequence, this is perhaps the simplest method of detecting the BREAK sequence.
3. By checking the framing error bit of the UART (RCSTA1.FERR).

```
/********************************************************************************
                        LIN BUS CONTROL PROJECT
                        =======================
```

This project shows how to use the Automotive LIN bus in microcontroller projects.

In this project 2 LIN bus nodes, named MASTER and SLAVE communicate with each other
at 9600 bps over the LIN BUS.

An LCD is connected to the MASTER node. An LM35DZ type temperature sensor is connected
to the SLAVE node. The MASTER node requests the temperature every second from the SLAVE
node. The SLAVE node reads the temperature and sends to the MASTER node which then displays
on the LCD.

This project is based on the easyPIC V7 development board and the mikroElektronika
LIN bus boards. Two development boards are used, one as the MASTER, the other one as
the SLAVE. LIN bus boards are attached to PORTC of each development board and the two boards
are connected to each other with two cables (LIN bus and ground). The CS pins of the
LIN bus boards are connected to RC2 pin of the microcontroller.

There is one MASTER and one SLAVE in this project.

This is the SLAVE node program. The program works with 8MHz clock (clock PLL is disabled)

```
Author:  Dogan Ibrahim
Date:    October, 2013
File:    MIKROC-LINSLAVE.C
********************************************************************************/
#define MCP201_CS PORTC.RC2                            // MCP201 CS bit
#define SLAVE_NODE_ID 0x01                             // Our Identifier
#define SYNCH_FIELD 0x55                               // SYNCH FIELD

//
// This function waits if UART is busy and then writes a character
//
void Write_Uart(unsigned char c)
{
    while(Uart1_Tx_Idle() == 0);                       // Wait if UART is busy
    Uart1_Write(c);                                    // Write the character
}


//
// This function checks if a character is ready in UART and then reads it
//
unsigned char Read_Uart()
{
    unsigned char c;

    while(Uart1_Data_Ready() == 0);                    // Wait to receive from the MASTER
    c = Uart1_Read();
    return c;
```

Figure 7.138: mikroC Pro for PIC SLAVE Program.

```
    }

//
// This function calculates the Checksum byte and returns to the calling program.
// The Checksum is calculated by adding all the data bytes (including carry bit)
// and then inverting the result
//
unsigned char Calc_Checksum(unsigned char N, unsigned char c[])
{
 unsigned char i, Checksum = 0;

 for(i = 0; i < N; i++)
 {
   Checksum = Checksum + c[i] - '0';              // Add data bytes (not ASCII)
   if(Checksum > 255)Checksum++;                  // Add carry (if any)
 }
 Checksum = ~Checksum;                            // Invert the data
 return Checksum;
}


// Start of MAIN program
//
void main()
{
    unsigned char c, i, Checksum, No_Of_Bytes, ID, MYID, Txt[7];
    unsigned int temp, mV;

    ANSELA = 1;                                   // RA0 is analog
    ANSELC = 0;                                   // Configure PORTC as digital
    TRISA.RA0 = 1;                                // RA0 is input
    MCP201_CS = 0;                                // Disable MCP201
    TRISC.RC2 = 0;                                // Configure RC2 as output


    Uart1_Init(9600);                             // Set UART baud rate to 9600 bps
    Delay_Ms(10);                                 // Wait until UART is settled
    MCP201_CS = 1;                                // Activate MCP201


//
// START OF LOOP
//
    while(1)                                      // Do FOREVER
    {
//
// Wait to receive the BREAK sequence. The BREAK sequence is identified when Framing
// error occurs
//
        TRISC.RC7=1;                              // Configure RC7 as input
        while(PORTC.RC7 == 1);                    // Wait for START bit
```

Figure 7.138
cont'd

```
        while(PORTC.RC7 == 0);                              // Wait for STOP bit

//
// Receive the SYNCH_FIELD byte
//
        c = Read_Uart();                                    // Read a character
        if(c == SYNCH_FIELD)                                // If SYNCH FIELD
        {
            ID = Read_Uart();                               // Read the IDENT FIELD
            MYID = ID & 0x0F;                               // Extract ID nibble

            if(MYID == SLAVE_NODE_ID)                       // If this node is requested
            {
                No_Of_Bytes = ID & 0x30;                    // Extract No Of Bytes to send
                No_Of_Bytes = No_Of_Bytes >> 4;

                if(No_Of_Bytes == 0 || No_Of_Bytes == 1)
                    No_Of_Bytes = 2;
                if(No_Of_Bytes == 2)
                    No_Of_Bytes = 4;
                if(No_Of_Bytes == 3)
                    No_Of_Bytes = 8;

                temp = ADC_READ(0);                         // Read the analog temperature
                mV = temp*5;                                // In millivolts (approximate)
                mV = mV / 10;                               // Temperature in Degrees C
                IntToStr(mV, Txt);                          // Convert to string
                Ltrim(Txt);                                 // Remove leading spaces

                for(i = 0; i < No_Of_Bytes; i++)            // Do for all requested bytes
                {
                    Write_Uart(Txt[i]);                     // Send temperature to MASTER
                }
                Checksum = Calc_Checksum(No_Of_Bytes, Txt); // Calculate the Checksum
                Write_Uart(Checksum);                       // Send the Checksum byte
            }
        }
    }
}
```

Figure 7.138
cont'd

In this program, option 2 is used to detect the end of the BREAK sequence:

```
TRISC.RC7 = 1;             // Configure RC7 as input
while(PORTC.RC7 == 1);     // Wait for START bit
while(PORTC.RC7 == 0);     // Wait for STOP bit
```

The program then reads a byte and checks to make sure that the received byte is actually a SYNCH byte (0x55). Then, the IDENT FIELD is read and the program extracts the ID to check if the request is for this node. If so, the temperature is read from analog channel 0 (RA0, AN0) of the microcontroller, converted into a string, and sent to the MASTER device over the LIN bus. Finally, the Checksum is calculated and sent over the bus. Although the Checksum is sent, it is not validated by the MASTER for simplicity. The program then waits for the next request.

Project 7.17—Automotive Project—Can Bus Project

CAN is a serial bus communications protocol developed by Bosch (an electrical equipment manufacturer in Germany) in the early 1980s. Thereafter CAN was standardized in ISO-11898 and ISO-11519, establishing itself as the standard protocol for in-vehicle networking in car industry. CAN defines an efficient communication protocol between sensors, actuators, controllers, and other nodes in real time applications. The early CAN development was mainly supported by the vehicle industry, and it was used in passenger cars, boats, trucks, and other types of vehicles. Today, the CAN protocol is also used in many other fields requiring networked embedded control, such as industrial automation, medical applications, building automation, weaving machines, and production machinery.

CAN is widely accepted for its simplicity, high performance, and reliability. In the early days of the automotive industry various actuators and electromechanical subsystems were controlled using standalone, localized controllers. By networking all the electronics in vehicles, it became possible to control them from a central point, the engine control unit, and this has increased the functionality, added modularity, and made it possible to carry out diagnostics more efficiently.

CAN is based on a bus topology, and only two wires are needed for communication over the bus. The bus has a multimaster structure where each device on the bus is capable of sending or receiving data. Only one device can send data at any time while all the other devices listen. If two or more devices attempt to send data at the same time, then the device with the higher priority is allowed to send its data while the others return into the receive mode.

The use of CAN in the automotive industry has caused the mass production of CAN controllers. Today, it is estimated that >400 million CAN modules are sold every year, and CAN controllers are integrated on many microcontrollers (e.g. PIC microcontrollers) and are available at a low cost.

Figure 7.139 shows a CAN bus with three nodes. The CAN protocol is based on CSMA/CD + AMP (Carrier Sense Multiple Access/Collision Detection with Arbitration on Message Priority) protocol, which is similar to the protocol used in an Ethernet LAN. When Ethernet detects a collision the sending nodes stop transmitting. They wait a random time before trying to send again. The CAN protocol solves the collision problem with the principle of arbitration where only the higher priority node is given the right to send its data.

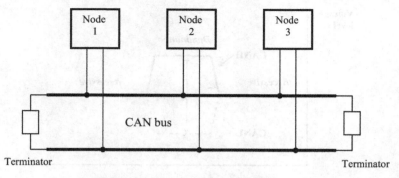

Figure 7.139: Example CAN Bus.

There are basically two types of CAN protocols: standard CAN 2.0A and CAN 2.0B. CAN 2.0A is the earlier standard with 11 bits of identifier, while Can 2.0B is the new extended standard with 29 bits of identifier. The 2.0B controllers are completely backward compatible with 2.0A controllers and can receive and transmit messages in either format. There are two types of 2.0A controllers: the first is capable of sending and receiving 2.0A messages only and reception of any 2.0B message will flag an error. The second type of 2.0A controller (known as 2.0B passive) can also send and receive 2.0A messages, but in addition, they will acknowledge receipt of 2.0B messages and then ignore them.

Some of the features of CAN protocol are as follows:

- CAN bus is a multimaster. When the bus is free, any device attached to the bus can start sending a message.
- CAN bus protocol is flexible. The devices connected to the bus have no addresses. This means that messages are not transmitted from one node to another node based on addresses. All the nodes in the system receive every message transmitted on the bus and it is up to each node in the system to decide whether the message received should be kept or discarded. A single message can be destined for one particular node to receive, or many nodes based on the way the system is designed. Another advantage of not having addresses is that, when a new device is added or an existing device is removed from the bus, there is no need to change any configuration data, that is, the bus is "hot pluggable".
- Another feature of the CAN protocol is the ability for a node to request information from other nodes on the bus. This is called **Remote Transmit Request** (RTR). Thus, instead of waiting for information to be sent continuously by a node, a request can be sent to the node to get its information. For example, in a vehicle, the engine temperature is an important parameter. The system can be designed such that the temperature can be sent periodically over the bus. A more elegant solution would be to request the temperature whenever required. This approach would minimize the bus traffic while still maintaining the integrity of the network.

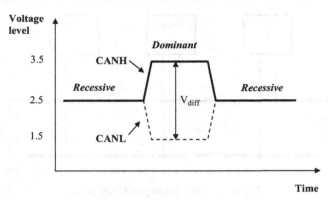

Figure 7.140: CAN Logic States.

- CAN bus communication speed is not fixed. Any communication speed can be set for the devices attached to a bus.
- All devices on the bus can detect an error. The device that has detected an error immediately notifies all other devices.
- Multiple devices can be connected to the bus at the same time. There are no logical limits to the number of devices that can be connected to the bus. In practice, the number of units that can be attached to the bus is limited by the delay time and electrical load of the bus.

The data on the CAN bus are differential, and there can be two states: **dominant** state and **recessive** state. Figure 7.140 shows the state of voltages on the bus. The bus defines a logic bit 0 as a dominant bit and a logic bit 1 as a recessive bit. When there is arbitration on the bus, a dominant bit state always wins arbitration over a recessive bit state. In the recessive state, the differential voltage CANH and CANL is less than the minimum threshold, that is, <0.5-V receiver input and <1.5-V transmitter output. In the dominant state, the differential voltage CANH and CANL is greater than the minimum threshold.

ISO-11898 specifies that a device on the CAN bus must be able to drive a 40-m cable at 1 Mb/s. A much longer bus length can usually be achieved by lowering the bus speed. A CAN bus is terminated to minimize signal reflections on the bus. ISO-11898 requires that the bus have a characteristic impedance of 120 Ω. The bus can be terminated in one of the following methods:

- Standard termination,
- Split termination,
- Biased split termination.

In standard termination, a 120-Ω resistor is used at each end of the bus as shown in Figure 7.141(a), and this is the most commonly used termination method. In split

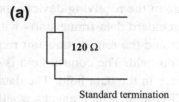

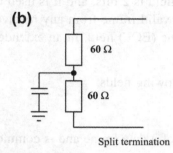

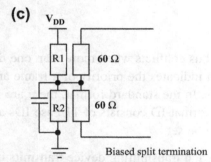

Figure 7.141: Bus Termination Methods.

termination, the ends of the bus is split and a single 60-Ω resistor is used as shown in Figure 7.141(b). Split termination provides reduced emission, and this method is gaining popularity. The biased split termination is similar to the split termination, but here, a voltage divider circuit and a capacitor are used at each end of the bus. This method increases the EMC performance of the bus (Figure 7.141(c)).

There are basically four message frames in CAN: **data**, **remote**, **error**, and **overload** frames. The data and remote frames need to be set by the user. Other frames are set by the CAN hardware.

Data Frame

The data frame is in two formats: standard and extended. The standard format has a 11-bit ID, and the extended format has a 29-bit ID. The data frame is used by the transmitting device to send data to the receiving device, and this is the most important frame handled by the user. A standard data frame starts with the SOF bit. It is then followed by an 11-bit identifier and the remote transmit request (RTR) bit. The identifier and the RTR form the arbitration field. The control field is 6 bits wide and it indicates how many bytes of data there are in the data field. The data field can be 0–8 bytes. The data field is followed by the CRC field, which checks whether or not the received bit sequence is corrupted. The ACK field is 2 bits, and it is used by the transmitter to receive an acknowledgement of a valid frame from any receiver. The end of the message is indicated by a 7-bit end-of-frame (EOF) field. In an extended data frame, the arbitration field is 29 bits wide.

The data frame consists of the following fields:

Start of Frame

This field indicates the beginning of a data frame and is common to both standard and extended formats.

Arbitration Field

Arbitration is used to resolve bus conflicts when more than one device starts sending a message on the bus. This field indicates the priority of a frame and differs between the standard and extended formats. In the standard format, there are 11 bits, and up to 2032 IDs can be set. The extended format ID consists of 11 base IDs and 18 extended IDs. Up to 2032×2^{18} discrete IDs can be set.

During the arbitration phase, each transmitting device transmits its identifier and compares it with the level on the bus. If the levels are equal, the device continues to transmit. If the device detects a dominant level on the bus while it was trying to transmit a recessive level, then it quits transmitting and becomes a receiving device. When the arbitration is over, there is only one transmitter left on the bus, and this transmitter continues to control field, data field etc.

Control Field

The control field is 6 bits wide, and it indicates the number of data bytes in a message to be transmitted. This field consists of two reserved bits and four data length code (DLC) bits. The control field is coded as shown in Table 7.19, where up to eight transmit bytes can be coded with 6 bits.

Data Field

This field indicates the actual content of data. The data size can vary from 0 to 8 bytes. The data are transmitted with the MSB first.

Table 7.19: Coding the Control Field.

No of Data Bytes	DLC3	DLC2	DLC1	DLC0
0	D	D	D	D
1	D	D	D	R
2	D	D	R	D
3	D	D	R	R
4	D	R	D	D
5	D	R	D	R
6	D	R	R	D
7	D	R	R	R
8	R	D or R	D or R	D or R

D: Dominant level, R: Recessive level.

CRC Field

The CRC field is used to check the frame for a transmission error. This field consists of a 15-bit CRC sequence and a 1-bit CRC delimiter. The CRC calculation includes the SOF, arbitration field, control field, and data field. The calculated CRC and the received CRC sequence are compared, and if they do not match, an error is assumed.

ACK Field

The ACK field indicates that the frame has been received normally. This field consists of 2 bits, one for ACK slot and one for ACK delimiter.

Remote Frame

This frame is used by the receive unit to request transmission of a message from the transmitting unit. The remote frame consists of six fields: **start of frame**, **arbitration field**, **control field**, **CRC field**, **ACK field**, and **end of frame field**. The remote field is the same as a data frame except that it does not have a data field.

Error Frame

The error frame is used to notify an error that has occurred during transmission. This field consists of an **error flag** and an **error delimiter**. Error frames are generated and transmitted by the CAN hardware. There are two types of error flags: active error flag, and passive error flag. Active error flag consists of six dominant bits. Passive error flag consists of six recessive bits. The error delimiter consists of eight recessive bits.

Overload Frame

The overload frame is used by the receive unit to notify that it is not ready to receive frames yet. This frame consists of an **overload flag** and an **overload delimiter**. The overload flag consists of six dominant bits, and it is structured the same way as the active error flag of the error frame. The overload delimiter consists of eight recessive bits and this field is structured the same way as the error delimiter of the error frame.

Bit Stuffing

CAN bus uses bit stuffing technique that is used to periodically synchronize transmit–receive operations to prevent timing errors between receive nodes. If five consecutive bits with the same level appear, 1 bit of inverted data is added to the sequence. During sending of a data or a remote frame, the same level occurs in five consecutive bits during the SOF to CRC sequence, an inverted level preceding 5 bits is inserted in the next (i.e. the sixth) bit. During receiving a data or a remote frame, the same level occurs in five consecutive bits during the SOF to CRC sequence, the next (sixth) bit is deleted from the received frame. If the deleted sixth bit is at the same level as the preceding fifth bit, an error (stuffing error) is detected.

Nominal Bit Timing

The CAN bus nominal bit rate (NMR) is defined as the number of bits transmitted every second without resynchronization. The inverse of the NMR is the nominal bit time. All devices on the CAN bus must use the same bit rate, even though each device can have its own different clock frequency. One message bit consists of four nonoverlapping time segments:

- Synchronization segment (Sync_Seg),
- Propagation time segment (Prop_Seg),
- Phase buffer segment 1 (Phase_Seg1),
- Phase buffer segment 2 (Phase_Seg2).

The **Sync_Seg** segment is used to synchronize various nodes on the bus, and an edge is expected to lie within this segment. The **Prop_Seg** segment compensates for the physical delay times within the network. The **Phase_Seg1** and **Phase_Seg2** segments are used to compensate for edge phase errors. These segments can be lengthened or shortened by synchronization. The sample point is the point in time where the actual bit value is located. The sample point is at the end of **Phase_Seg1**. A CAN controller can be configured to sample three times and use a majority function to determine the actual bit value.

Each segment is divided into units known as **Time Quantum**, or T_Q. A desired bit timing can be set by adjusting the number of T_Q's that comprise one message bit and the number of T_Q's that comprise each segment in it. The T_Q is a fixed unit derived from the oscillator period and the time quantum of each segment can vary from 1 to 8. The length of each time segment is

- **Sync_Seg** is one time quantum long.
- **Prop_Seg** is programmable to be one to eight time quanta long.
- **Phase_Seg1** is programmable to be one to eight time quanta long.
- **Phase_Seg2** is programmable to be two to eight time quanta long.

By setting the bit timing, it is possible to set a sampling point so that multiple units on the bus can sample messages with the same timing.

The nominal bit time is programmable from a minimum of eight time quanta to a maximum of 25 time quanta. By definition, the minimum nominal bit time is 1 μs corresponding to a maximum 1 Mb/s rate. The nominal bit time (T_{BIT}) is given by

$$T_{BIT} = T_Q{}^*(Sync\_Seg + Prop\_Seg + Phase\_Seg1 + Phase\_Seg2) \qquad (7.1)$$

and the NMR is

$$NBR = 1/T_{BIT}.$$

The time quantum is derived from the oscillator frequency and the programmable baud rate prescaler, with integer values from 1 to 64. The time quantum can be expressed as

$$T_Q = 2 * (BRP + 1)/F_{OSC},$$

Where

T_Q is in microseconds, and F_{OSC} is in megahertz, and BRP is the baud rate prescaler (0−63).

We can also write

$$T_Q = 2 * (BRP + 1) * T_{OSC},$$

Where T_{OSC} is in microseconds.

An example is given below for the calculation of the NMR.

Example 7.1

Assuming a clock frequency of 20 MHz, a baud rate prescaler value of 1, and a nominal bit time of $T_{BIT} = 8 * T_Q$, determine the NMR.

Solution 7.1
From the above equations,

$$T_Q = 2 * (1 + 1)/20 = 0.2 \text{ μs}.$$

Also,

$$T_{BIT} = 8 * T_Q = 8 * 0.2 = 1.6 \ \mu s$$

and

$$NBR = 1/T_{BIT} = 1/1.6 \ \mu s = 625,000 \ bits/s \ or \ 625 \ Kb/s.$$

To compensate for phase shifts between the oscillator frequencies of the nodes on the bus, each CAN controller must synchronize to the relevant signal edge of the received signal. Two types of synchronization are defined: *hard synchronization* and *resynchronization*. Hard synchronization is done only at the beginning of a message frame, when each CAN node aligns the **Sync_Seg** of its current bit time to the recessive or dominant edge of the transmitted SOF. According to the rules of synchronization, if a hard synchronization occurs, there will not be a resynchronization within that bit time.

With resynchronization, the **Phase_Seg1** may be lengthened or **Phase_Seg2** may be shortened. The amount of change of the phase buffer segments has an upper bound given by the **Synchronization Jump Width** (SJW). The SJW is programmable between 1 and 4, and its value is added to **Phase_Seg1**, or subtracted from **Phase_Seg2**.

PIC Microcontroller CAN Interface

PIC microcontrollers can be used in CAN bus-based projects. In general, any type of PIC microcontroller can be used, but some PIC microcontrollers (e.g. PIC18F258) have built-in CAN modules that simplify the design of CAN bus-based systems. Microcontrollers with no built-in CAN modules can also be used in CAN bus applications, but additional hardware and software are required, making the design costly and also more complex.

Figure 7.142 shows the block diagram of a PIC microcontroller-based CAN bus application. Here, a PIC microcontroller with no built-in CAN module is used.

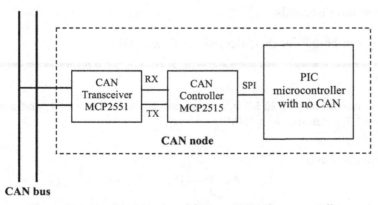

Figure 7.142: CAN Node with Any PIC Microcontroller.

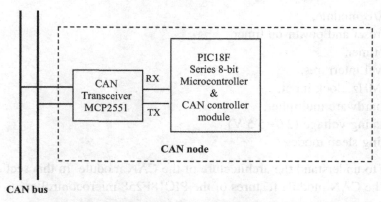

Figure 7.143: CAN Node with an Integrated CAN Module.

The microcontroller is connected to the CAN bus using an external MCP2515 CAN controller chip and an MCP2551 CAN bus transceiver chip. This configuration is suitable for a quick upgrade to an existing design using any PIC microcontroller.

For new CAN bus-based designs, it is easier to use a PIC microcontroller with a built-in CAN module. As shown in Figure 7.143, such devices include built-in CAN controller hardware on the chip. All that is required to make a CAN node is then to add a CAN transceiver chip.

In this project, the PIC18F258 microcontroller is used in a CAN bus-based project. The description and operating principles given in this section are in general applicable to other PIC microcontrollers with CAN modules.

PIC18F258 Microcontroller

PIC18F258 is a high performance 8-bit microcontroller with an integrated CAN module. The device has the following features:

- A 32 k flash program memory,
- A 1536-byte RAM data memory,
- A 256-byte EEPROM memory,
- Twenty two I/O ports,
- Five-channel 10-bit A/D converters,
- Three timers/counters,
- Three external interrupt pins,
- High current (25-mA) sink/source,
- Capture/compare/Pulse Width Modulation (PWM) module,
- SPI/I^{2}C module,

- CAN 2.0A/B module,
- Power-on reset and power-on timer,
- Watchdog timer,
- Priority level interrupts,
- DC to 40-MHz clock input,
- An 8×8 hardware multiplier,
- Wide operating voltage (2.0–5.5 V),
- Power saving sleep mode.

It is important to understand the architecture of the CAN module. In this section, we shall be looking at the CAN module features of the PIC18F258 microcontroller.

PIC18F258 microcontroller CAN module has the following features:

- Compatible with CAN 1.2, CAN 2.0A and CAN 2.0B,
- Supports standard and extended data frames,
- Programmable bit rate up to 1 Mbit/s,
- Double buffered receiver,
- Three transmit buffers,
- Two receive buffers,
- Programmable clock source,
- Six acceptance filters,
- Two acceptance filter masks,
- Loop-back mode for self-test,
- Low-power sleep mode,
- Interrupt capabilities.

The CAN module uses port pins RB3/CANRX and RB2/CANTX for CAN bus receive and transmit functions, respectively. These pins are connected to the CAN bus via an MCP2551 type CAN bus transceiver chip.

PIC18F258 microcontroller supports the following frame types:

- Standard data frame,
- Extended data frame,
- Remote frame,
- Error frame,
- Overload frame,
- Interframe space.

A node uses filters to decide whether or not to accept a received message. Message filtering is applied to the whole identifier field, and mask registers are used to specify which bits in the identifier are to be examined with the filters.

The CAN module in the PIC18F258 microcontroller has six modes of operation:

- Configuration mode,
- Disable mode,
- Normal operation mode,
- Listen-only mode,
- Loop-back mode,
- Error recognition mode.

Configuration Mode

The CAN module is initialized in the configuration mode. The module is not allowed to enter the configuration mode while a transmission is taking place. In the configuration mode, the module will not transmit or receive, the error counters are cleared, and the interrupt flags remain unchanged.

Disable Mode

In the disable mode, the module will not transmit or receive. In this mode, the internal clock is stopped unless the module is active. If the module is active, the module will wait for 11 recessive bits on the CAN bus, detect that condition as an IDLE bus, and then accept the module disable command. The WAKIF interrupt (wake-up interrupt) is the only CAN module interrupt that is active in the disable mode.

Normal Operation Mode

This is the standard operating mode of the CAN module. In this mode, the module monitors all bus messages and generates acknowledge bits, error frames, etc. This is the only mode that will transmit messages.

Listen-only Mode

This mode allows the CAN module to receive messages, including messages with errors. This mode can be used to monitor the bus activities or for detecting the baud rate on the bus. For autobaud detection, there must be at least two other nodes communicating with each other. The baud rate can be determined by testing different values until valid messages are received. No messages can be transmitted in the listen-only mode.

Loop-back Mode

In this mode, messages can be directed from internal transmit buffers to receive buffers, without actually transmitting messages on the CAN bus. This mode can be used during system developing and testing.

Error Recognition Mode

This mode is used to ignore all errors and receive all messages. In this mode, all messages, valid or invalid, are received and copied to the receive buffer.

CAN Message Transmission

PIC18F258 microcontroller implements three dedicated transmit buffers: TXB0, TXB1, and TXB2. Pending transmittable messages are in a priority queue. Prior to sending the SOF, the priority of all buffers that are queued for transmission is compared. The transmit buffer with the highest priority will be sent first. If two buffers have the same priorities, the buffer with the highest buffer number will be sent first. There are four levels of priority.

CAN Message Reception

Reception of a message is a more complex process. PIC18F258 microcontroller includes two receive buffers RXB0 and RXB1 with multiple acceptance filters for each (Figure 7.144). All received messages are assembled in Message Assembly Buffer (MAB). Once a message is received, regardless of the type of identifier and the number of data bytes received, the entire message is copied into MAB.

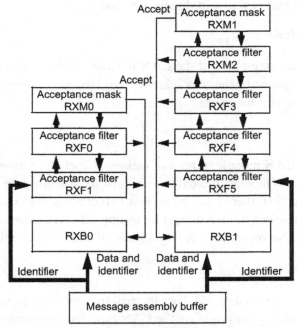

Figure 7.144: Receive Buffer Block Diagram.

Received messages have priorities. RXB0 is the higher priority buffer, and it has two message acceptance filters RXF0 and RXF1. RXB1 is the lower priority buffer, and it has four acceptance filters RXF2, RXF3, RXF4, and RXF5. There are also two programmable acceptance filter masks RXM0 and RXM1 available, one for each receive buffer.

The CAN module uses message acceptance filters and masks to determine if a message in the MAB should be loaded into a receive buffer. Once a valid message is received by the MAB, the identifier field of the message is compared to the filter values. If there is a match, that message will be loaded into the appropriate receive buffer. The filter masks are used to determine which bits in the identifier are examined with the filters.

Calculating the Timing Parameters

Setting the timing parameters of the nodes are very important for the reliable operation of the bus. Given the microcontroller clock frequency and the required CAN bus bit rate, we need to calculate the values of the following timing parameters:

- Baud rate prescaler value,
- Prop_Seg value,
- Phase_Seg1 value,
- Phase_Seg2 value,
- SJW value.

Correct timing requires that

- Prop_Seg + Phase_Seg1 ≥ Phase_Seg2,
- Phase_Seg2 ≥ SJW.

An example is given below to illustrate how the various timing parameters can be calculated.

Example 7.2

Assuming that the microcontroller oscillator clock rate is 20 MHz, and the required CAN bit rate is 125 kHz, calculate the timing parameters.

Solution 7.2

With a 20-MHz clock rate, the clock period is 50 ns. Choosing a baud rate prescaler value of 4, from Eqn (7.4),

$$T_Q = 2 * (BRP + 1) * T_{OSC}$$

gives a time quantum of $T_Q = 500$ ns. To obtain a NMR of 125 kHz, the nominal bit time must be

$$T_{BIT} = 1/0.125 \text{ MHz} = 8 \text{ μs, or } 16 T_Q.$$

The **Sync_Segment** is $1T_Q$. Choosing $2T_Q$ for the **Prop_Seg**, and $7T_Q$ for **Phase_Seg1** leaves $6T_Q$ for **Phase_Seg2** and places the sampling point at $10T_Q$ (at the end of **Phase_Seg1**).

By the rules given above, the SJW could be the maximum allowed (i.e. 4). However, a large SJW is only necessary when the clock generation of different nodes is not stable or accurate (e.g. if using ceramic resonators). Typically an SJW of 1 is enough. In summary, the required timing parameters are as follows:

```
Baud rate prescaler (BRP) = 4
Sync_Seg              = 1
Prop_Seg              = 2
Phase_Seg1            = 7
Phase_Seg2            = 6
SJW                   = 1
```

The sampling point is at $10T_Q$ that corresponds to 62.5% of the total bit time.

There are several tools available on the Internet for calculating the CAN bus timing parameters accurately. Interested readers should refer to the excellent book of the author on this topic, entitled "Controller Area Network Project, Elektor Int. Media, ISBN: 978-1-907920-04-2"

mikroC Pro for PIC CAN Functions

mikroC Pro for the PIC language provides two sets of libraries for CAN bus applications: the library for PIC microcontrollers with built-in CAN modules, and the library based on the use of SPI bus for PIC microcontrollers having no built-in CAN modules. In this project, we shall only be looking at the library functions available for PIC microcontrollers with built-in CAN modules. Similar functions are available for PIC microcontrollers with no built-in CAN modules.

mikroC CAN functions are supported only by PIC18XXX8 microcontrollers with MCP2551 or similar CAN transceivers. Both standard (11 identifier bits) and extended format (29 identifier bits) messages are supported.

The following mikroC functions are provided:

- CANSetOperationMode,
- CANGetOperationMode,
- CANInitialize,
- CANSetBaudRAte,
- CANSetMask,
- CANSetFilter,
- CANRead,
- CANWrite.

These functions are described below.

1. **CANSetOperationMode**

This function is used to set the CAN operation mode. The function prototype is

   void CANSetOperationMode(**char** mode, **char** wait_flag)

Parameter **wait_flag** is either 0 or 0xFF. If set to 0xFF, the function blocks and will not return until the requested mode is set. If 0, the function returns as a nonblocking call.

The mode can be one of the following:

*   _CAN_MODE_NORMAL   -   normal mode of operation,
*   _CAN_MODE_SLEEP   -   SLEEP mode of operation,
*   _CAN_MODE_LOOP   -   Loop-back mode of operation,
*   _CAN_MODE_LISTEN   -   Listen Only mode of operation,
*   _CAN_MODE_CONFIG   -   Configuration mode of operation,

2. **CANGetOperationMode**

This function returns the current CAN operation mode. The function prototype is

   char CANGetOperationMode(void)

3. **CAN_Initialize**

This function initializes the CAN module. All mask registers are cleared to 0 to allow all messages. Upon execution of this function, the Normal mode is set. The function prototype is

```
void CANInitialize(char SJW, char BRP, char PHSEG1,
             char PHSEG2,char PROPEG, char CAN_CONFIG_FLAGS)
```

where

SJW	is the Synchronization Jump Width,
BRP	is the Baud Rate prescaler,
PHSEG1	is the Phase_Seg1 timing parameter,
PHSEG2	is the Phase_Seg2 timing parameter,
PROPSEG	is the Prop_Seg.

CAN_CONFIG_FLAGS can be one of the following configuration flags:

Value	Meaning
_CAN_CONFIG_DEFAULT	Default flags
_CAN_CONFIG_PHSEG2_PRG_ON	Use supplied PHSEG2 value
_CAN_CONFIG_PHSEG2_PRG_OFF	Use a maximum of PHSEG1 or information processing Time whichever is greater

Continued

Value	Meaning
_CAN_CONFIG_LINE_FILTER_ON	Use the CAN bus line filter for wake-up
_CAN_CONFIG_FILTER_OFF	Do not use a CAN bus line filter
_CAN_CONFIG_SAMPLE_ONCE	Sample bus once at the sample point
_CAN_CONFIG_SAMPLE_THRICE	Sample bus three times prior to the sample point
_CAN_CONFIG_STD_MSG	Accept only standard identifier messages
_CAN_CONFIG_XTD_MSG	Accept only extended identifier messages
_CAN_CONFIG_DBL_BUFFER_ON	Use double buffering to receive data
_CAN_CONFIG_DBL_BUFFER_OFF	Do not use double buffering
_CAN_CONFIG_ALL_MSG	Accept all messages including invalid ones
_CAN_CONFIG_VALID_XTD_MSG	Accept only valid extended identifier messages
_CAN_CONFIG_VALID_STD_MSG	Accept only valid standard identifier messages
_CAN_CONFIG_ALL_VALID_MSG	Accept all valid messages

The above configuration values can be bitwise AND'ed to form complex configuration values.

4. CANSetBaudRate

This function is used to set the CAN bus baud rate. The function prototype is

```
void CANSetBaudRate(char SJW, char BRP, char PHSEG1, char PHSEG2,
                char PROPSEG, char CAN_CONFIG_FLAGS)
```

the arguments of the function are as in function *CANInitialize*.

5. CANSetMask

This function sets the mask for filtering of messages. The function prototype is

```
void CANSetMask(char CAN_MASK, long value, char
            CAN_CONFIGFLAGS)
```

CAN_MASK can be one of the following:

*   _CAN_MASK_B1—Receive Buffer 1 mask value,
*   _CAN_MASK_B2—Receive Buffer 2 mask value.

value is the mask register value. CAN_CONFIG_FLAGS can be either _CAN_CONFIG_XTD (extended message), or _CAN_CONFIG_STD (standard message).

6. CANSetFilter

This function sets filter values. The function prototype is

```
void CANSetFilter(char CAN_FILTER, long value, char
            CAN_CONFIG_FLAGS)
```

CAN_FILTER can be one of the following:

- _CAN_FILTER_B1_F1—Filter 1 for Buffer 1,
- _CAN_FILTER_B1_F2—Filter 2 for Buffer 1,
- _CAN_FILTER_B2_F1—Filter 1 for Buffer 2,
- _CAN_FILTER_B2_F2—Filter 2 for Buffer 2,
- _CAN_FILTER_B2_F3—Filter 3 for Buffer 2,
- _CAN_FILTER_B2_F4—Filter 4 for Buffer 2.

CAN_CONFIG_FLAGS can be either _CAN_CONFIG_XTD (extended message), or _CAN_CONFIG_STD (standard message).

7. **CANRead**

This function is used to read messages from the CAN bus. Zero is returned if no message is available. The function prototype is

```
char CANRead(long *id, char *data, char *datalen, char
            *CAN_RX_MSG_FLAGS)
```

id is the CAN message identifier. Only 11 or 29 bits may be used depending on message type (standard or extended). **data** is an array of bytes up to eight where the received data are stored. **datalen** is the length of the received data (1—8).

CAN_RX_MSG_FLAGS can be one of the following:

- _CAN_RX_FILTER_1—Receive Buffer Filter 1 accepted this message.
- _CAN_RX_FILTER_2—Receive Buffer Filter 2 accepted this message.
- _CAN_RX_FILTER_3—Receive Buffer Filter 3 accepted this message.
- _CAN_RX_FILTER_4—Receive Buffer Filter 4 accepted this message.
- _CAN_RX_FILTER_5—Receive Buffer Filter 5 accepted this message.
- _CAN_RX_FILTER_6—Receive Buffer Filter 6 accepted this message.

- _CAN_RX_OVERFLOW—Receive buffer overflow occurred.
- _CAN_RX_INVALID_MSG—Invalid message received.
- _CAN_RX_XTD_FRAME—Extended Identifier message received.
- _CAN_RX_RTR_FRAME—RTR frame message received.
- _CAN_RX_DBL_BUFFERED—This message was double-buffered.

The above flags can be bitwise AND'ed if desired.

8. **CANWrite**

This function is used to send a message to the CAN bus. A zero is returned if the message cannot be queued (buffer full). The function prototype is

```
char CANWrite(long id, char *data, char datalen, char
        CAN_TX_MSG_FLAGS)
```

id is the CAN message identifier. Only 11 or 29 bits may be used depending on the message type (standard or extended). **data** is an array of bytes up to eight where the data to be sent are stored. **datalen** is the length of the data (1–8).

CAN_TX_MSG_FLAGS can be one of the following:

- _CAN_TX_PRIORITY_0   - Transmit priority 0
- _CAN_TX_PRIORITY_1   - Transmit priority 1
- _CAN_TX_PRIORITY_2   - Transmit priority 2
- _CAN_TX_PRIORITY_3   - Transmit priority 3

- _CAN_TX_STD_FRAME   - Standard Identifier message
- _CAN_TX_XTD_FRAME   - Extended Identifier message

- _CAN_TX_NO_RTR_FRAME   - Non-RTR message,
- _CAN_TX_RTR_FRAME   - RTR message.

The above flags can be bitwise AND'ed if desired.

CAN Bus Programming

To operate the PIC18F258 microcontroller on the CAN bus, the following steps should be carried out:

- Configure CAN bus I/O port directions (RB2 and RB3).
- Initialize the CAN module (CANInitialize).
- Set CAN module to the CONFIG mode (CANSetOperationMode).
- Set Mask registers (CANSetMask).
- Set Filter registers (CANSetFilter).
- Set CAN module to the Normal mode (CANSetOperationMode).
- Write/Read data (CANWrite/CANRead).

CAN Bus Project Description—Temperature Sensor and Display

This is a simple two-node CAN bus based project. The block diagram of the project is shown in Figure 7.145. The system is made up of two CAN nodes. One node (called the **COLLECTOR** node) reads the temperature from an external semiconductor temperature sensor. The other node (called the **DISPLAY** node) requests the temperature every second and then displays it on an LCD. This process is repeated continuously.

The circuit diagram of the project is given in Figure 7.146. Two CAN nodes are connected together using a 2-m twisted pair cable, terminated with 120-Ω resistors at both ends.

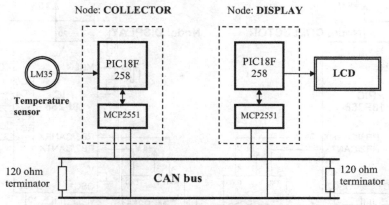

Node: **COLLECTOR**   Node: **DISPLAY**

Figure 7.145: Block Diagram of the Project.

The COLLECTOR Processor

The COLLECTOR processor consists of a PIC18F258 microcontroller with a built-in CAN module and an MCP2551 transceiver chip. The microcontroller is operated from an 8-MHz crystal. The MCLR input is connected to an external reset button. Analog input AN0 of the microcontroller is connected to a LM35DZ-type semiconductor temperature sensor. The sensor can measure temperature in the range 0–100 °C and generates an analog voltage directly proportional to the measured temperature, that is, the output is 10 mV/°C. For example, at 20 °C, the output voltage is 200 mV.

The CAN outputs (RB2/CANTX and RB3/CANRX) of the microcontroller are connected to the TXD and RXD inputs of an MCP2551-type CAN transceiver chip. The CANH and CANL outputs of this chip are connected directly to a twisted cable terminated CAN bus. MCP2551 is an eight-pin chip that supports data rates up to 1 Mb/s. The chip can drive up to 112 nodes. An external resistor connected to pin 8 of the chip controls the rise and fall times of CANH and CANL so that EMI can be reduced. For high-speed operation, this pin should be connected to the ground. A reference voltage equal to $V_{DD}/2$ is output from pin 5 of the chip.

The DISPLAY Processor

As in the COLLECTOR processor, the DISPLAY processor consists of a PIC18F258 microcontroller and an MCP2551 transceiver chip. The microcontroller is operated from an 8-MHz crystal. The MCLR input is connected to an external reset button. The CAN outputs (RB2/CANTX and RB3/CANRX) of the microcontroller are connected to the TXD and RXD inputs of the MCP2551. Pins CANH and CANL of the transceiver chip are

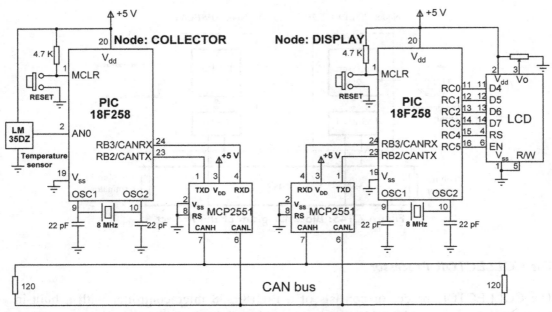

Figure 7.146: Circuit Diagram of the Project.

connected to the CAN bus. An LCD is connected to PORTC of the microcontroller to display the temperature values.

The program listing is in two parts: the COLLECTOR program and the DISPLAY program. The operation of the system is as follows:

- The DISPLAY processor requests the current temperature from the COLLECTOR processor over the CAN bus.
- The COLLECTOR processor reads the temperature, formats it, and sends to the DISPLAY processor over the CAN bus.
- The DISPLAY processor reads the temperature from the CAN bus and then displays it on the LCD.
- The above process is repeated every second.

DISPLAY Program

Figure 7.147 shows the program listing of the DISPLAY program, called CAN-DISPLAY.C. At the beginning of the program, PORTC pins are configured as outputs, RB3 is configured as input (CANRX), and RB2 is configured as output (CANTX). In this project, the CAN bus bit rate is selected as 100 kb/s. With a microcontroller clock frequency of 8 MHz, the timing parameters were calculated to be

```
/************************************************************************
                    CAN BUS EXAMPLE - NODE: DISPLAY
                    =================================
```

This is the DISPLAY node of the CAN bus example. In this project a PIC18F258 type microcontroller is used. An MCP2551 type CAN bus transceiver is used to connect the microcontroller to the CAN bus. The microcontroller is operated from an 8 MHz crystal with an external reset button.

Pin CANRX and CANTX of the microcontroller are connected to pins RXD and TXD of the transceiver chip respectively. Pins CANH and CANL of the transceiver chip are connected to the CAN bus.

An LCD is connected to PORT C of the microcontroller. The ambient temperature is read from another CAN node and is displayed on the LCD.

The LCD is connected to the microcontroller as follows:

Microcontroller	LCD
RC0	D4
RC1	D5
RC2	D6
RC3	D7
RC4	RS
RC5	EN

CAN speed parameters are:

Microcontroller clock:	8 MHz
CAN Bus bit rate:	100 Kb/s
Sync_Seg:	1
Prop_Seg:	6
Phase_Seg1:	6
Phase_Seg2:	7
SJW:	1
BRP:	1
Sample point:	65%

```
Author:   Dogan Ibrahim
Date:     October 2013
File:     CAN-DISPLAY.C
*************************************************************/
// LCD module connections
sbit LCD_RS at RC4_bit;
sbit LCD_EN at RC5_bit;
sbit LCD_D4 at RC0_bit;
sbit LCD_D5 at RC1_bit;
sbit LCD_D6 at RC2_bit;
sbit LCD_D7 at RC3_bit;
```

Figure 7.147: DISPLAY Program Listing.

```
        sbit LCD_RS_Direction at TRISC4_bit;
        sbit LCD_EN_Direction at TRISC5_bit;
        sbit LCD_D4_Direction at TRISC0_bit;
        sbit LCD_D5_Direction at TRISC1_bit;
        sbit LCD_D6_Direction at TRISC2_bit;
        sbit LCD_D7_Direction at TRISC3_bit;
        // End LCD module connections

        void main()
        {
            unsigned char temperature, dat[8];
            unsigned char init_flag, send_flag, dt, len, read_flag;
            char SJW, BRP, Phase_Seg1, Phase_Seg2, Prop_Seg, txt[4];
            long id, mask;

            TRISC = 0;                              // PORTC are outputs (LCD)
            TRISB = 0x08;                           // RB2 is output, RB3 is input
        //
        // CAN BUS Parameters
        //
            SJW = 1;
            BRP = 1;
            Phase_Seg1 = 6;
            Phase_Seg2 = 7;
            Prop_Seg = 6;

            init_flag = _CAN_CONFIG_SAMPLE_THRICE   &
                        _CAN_CONFIG_PHSEG2_PRG_ON   &
                        _CAN_CONFIG_STD_MSG         &
                        _CAN_CONFIG_DBL_BUFFER_ON   &
                        _CAN_CONFIG_VALID_XTD_MSG   &
                        _CAN_CONFIG_LINE_FILTER_OFF;

            send_flag = _CAN_TX_PRIORITY_0          &
                        _CAN_TX_XTD_FRAME           &
                        _CAN_TX_NO_RTR_FRAME;

            read_flag = 0;
        //
        // Initialise CAN module
        //
            CANInitialize(SJW, BRP, Phase_Seg1, Phase_Seg2, Prop_Seg, init_flag);
        //
        // Set CAN CONFIG mode
        //
            CANSetOperationMode(_CAN_MODE_CONFIG, 0xFF);

            mask = -1;
        //
        // Set all MASK1 bits to 1's
        //
```

Figure 7.147
cont'd

```
            CANSetMask(_CAN_MASK_B1, mask, _CAN_CONFIG_XTD_MSG);
//
// Set all MASK2 bits to 1's
//
            CANSetMask(_CAN_MASK_B2, mask, _CAN_CONFIG_XTD_MSG);
//
// Set id of filter B2_F3 to 3
//
            CANSetFilter(_CAN_FILTER_B2_F3,3,_CAN_CONFIG_XTD_MSG);
//
// Set CAN module to NORMAL mode
//
            CANSetOperationMode(_CAN_MODE_NORMAL, 0xFF);


//
// Configure LCD
//
            Lcd_Init();                                  // Initializd LCD
            Lcd_Cmd(_LCD_CLEAR);                          // Clear LCD
            Lcd_Out(1,1,"CAN BUS");                       // Display heading on LCD
            Delay_ms(1000);                              // Wait for 2 seconds

//
// Program loop. Read the temperature from Node:COLLECTOR and display
// on the LCD continuously
//
            for(;;)                                       // Endless loop
            {
                Lcd_Cmd(_LCD_CLEAR);                      // Clear LCD
                Lcd_Out(1,1,"Temp = ");                   // Display "Temp = "
                //
                // Send a message to Node:COLLECTOR and ask for data
                //
                dat[0] = 'T';                             // Data to be sent
                id = 500;                                 // Identifier
                CANWrite(id, dat, 1, send_flag);          // Send 'T'
                //
                // Get temperature from node:COLLECT
                //
                dt = 0;
                while(!dt)dt = CANRead(&id, dat, &len, &read_flag);
                if(id == 3)
                {
                    temperature = dat[0];
                    ByteToStr(temperature,txt);           // Convert to string
                    Lcd_Out(1,8,txt);                     // Output to LCD
                    Delay_ms(1000);                       // Wait 1 second
                }
            }

}
```

Figure 7.147
cont'd

```
SJW = 1
BRP = 1
Phase_Seg1 = 6
Phase_Seg2 = 7
Prop_Seg = 6
```

mikroC Pro for PIC CAN bus function CANInitialize is used to initialize the CAN module. The timing parameters and the initialization flag are specified as arguments in this function. The initialization flag was made up from the bitwise AND of

```
init_flag = _CAN_CONFIG_SAMPLE_THRICE &
            _CAN_CONFIG_PHSEG2_PRG_ON &
            _CAN_CONFIG_STD_MSG  &
            _CAN_CONFIG_DBL_BUFFER_ON &
            _CAN_CONFIG_VALID_XTD_MSG &
            _CAN_CONFIG_LINE_FILTER_OFF;
```

Where it was specified to sample the bus three times, standard identifier was specified, double buffering was turned on, and the line filter was turned off.

Then, the operation mode was set to CONFIGURATION, and the filter masks and filter values were specified. Both mask 1 and mask 2 were set to all 1's (-1 is a short hand of writing 0xFFFFFFFF, that is, setting all mask bits to 1's) so that matching of all filter bits was required.

Filter 3 for Buffer 2 was set to value 3 so that identifiers having values 3 will be accepted by the receive buffer.

The operation mode is then set to NORMAL. The program then configures the LCD and displays message "CAN BUS" for 1 s on the LCD.

The main program loop executes continuously and starts with a **for** statement. Inside this loop, the LCD is cleared and text "TEMP =" is displayed on the LCD. Then character "T" is sent over the bus with identifier equal to 500 (the COLLECTOR node filter is set accept identifier 500). This is a request to the COLLECTOR node to send the temperature reading. The program then reads the temperature from the CAN bus, converts it to a string in array **txt**, and then displays it on the LCD. The above process is repeated after a 1-s delay.

COLLECTOR Program

Figure 7.148 shows the program listing of the COLLECTOR program, called CAN-COLLECTOR.C. The initial part of this program is the same as the DISPLAY program. Here, the receive filter is set to 500 so that messages with identifier 500 can be accepted by the program.

```
/**************************************************************************
                    CAN BUS EXAMPLE - NODE: COLLECTOR
                    ===================================
```

This is the COLLECTOR node of the CAN bus example. In this project a PIC18F258 type
microcontroller is used. An MCP2551 type CAN bus transceiver is used to connect
the microcontroller to the CAN bus. The microcontroller is operated from an 8 MHz
crystal with an external reset button.

Pin CANRX and CANTX of the microcontroller are connected to pins RXD and TXD of the
transceiver chip respectively. Pins CANH and CANL of the transceiver chip are connected to
the CAN bus.

An LM35DZ type analog temperature sensor is connected to port AN0 of the microcontroller.
The microcontroller reads the temperature when a request is received and then sends the
temperature value as a byte to Node:DISPLAY on the CAN bus.

CAN speed parameters are:

Microcontroller clock:	8 MHz
CAN Bus bit rate:	100 Kb/s
Sync_Seg:	1
Prop_Seg:	6
Phase_Seg1:	6
Phase_Seg2:	7
SJW:	1
BRP:	1
Sample point:	65%

Author: Dogan Ibrahim
Date: October 2013
File: CAN-COLLECTOR.C
**/

void main()
{
    unsigned char temperature, dat[8];
    unsigned short init_flag, send_flag, dt, len, read_flag;
    char SJW, BRP, Phase_Seg1, Phase_Seg2, Prop_Seg, txt[4];
    unsigned int temp;
    unsigned long mV;
    long id, mask;

    TRISA = 0xFF;   // PORT A are inputs
    TRISB = 0x08;   // RB2 is output, RB3 is input
//
// Configure A/D converter
//
    ADCON1 = 0x80;
//
// CAN BUS Timing Parameters
```

**Figure 7.148: COLLECT Program Listing.**
```

```
//
    SJW = 1;
    BRP = 1;
    Phase_Seg1 = 6;
    Phase_Seg2 = 7;
    BRP = 1;
    Prop_Seg = 6;

    init_flag = _CAN_CONFIG_SAMPLE_THRICE    &
                _CAN_CONFIG_PHSEG2_PRG_ON &
                _CAN_CONFIG_STD_MSG          &
                _CAN_CONFIG_DBL_BUFFER_ON &
                _CAN_CONFIG_VALID_XTD_MSG &
                _CAN_CONFIG_LINE_FILTER_OFF;

    send_flag = _CAN_TX_PRIORITY_0        &
                _CAN_TX_XTD_FRAME          &
                _CAN_TX_NO_RTR_FRAME;

    read_flag = 0;
//
// Initialise CAN module
//
    CANInitialize(SJW, BRP, Phase_Seg1, Phase_Seg2, Prop_Seg, init_flag);
//
// Set CAN CONFIG mode
//
    CANSetOperationMode(_CAN_MODE_CONFIG, 0xFF);

    mask = -1;
//
// Set all MASK1 bits to 1's
//
    CANSetMask(_CAN_MASK_B1, mask, _CAN_CONFIG_XTD_MSG);
//
// Set all MASK2 bits to 1's
//
    CANSetMask(_CAN_MASK_B2, mask, _CAN_CONFIG_XTD_MSG);
//
// Set id of filter B2_F3 to 500
//
    CANSetFilter(_CAN_FILTER_B2_F3,500,_CAN_CONFIG_XTD_MSG);
//
// Set CAN module to NORMAL mode
//
    CANSetOperationMode(_CAN_MODE_NORMAL, 0xFF);

//
// Program loop. Read the temperature from analog temperature
// sensor
//
```

Figure 7.148
cont'd

```
for(;;)                                                    // Endless loop
{
    //
    // Wait until a request is received
    //
    dt = 0;
    while(!dt) dt = CANRead(&id, dat, &len, &read_flag);
    if(id == 500 && dat[0] == 'T')
    {
        //
        // Now read the temperature
        //
        temp = Adc_Read(0);                                // Read temp
        mV = (unsigned long)temp * 5000 / 1024;            // in mV
        temperature = mV/10;                               // in degrees C
        //
        // send the temperature to Node:Display
        //
        dat[0] = temperature;
        id = 3;                                            // Identifier
        CANWrite(id, dat, 1, send_flag);                   // Send temperature
    }
}
}
```

Figure 7.148
cont'd

Inside the program loop, the program waits until it receives a request to send the temperature. Here, the request is identified by the reception of character "T". Once a valid request is received, the temperature is read and converted into degrees centigrade (stored in variable **temperature**) and then sent to the CAN bus as a byte with identifier value equal to 3. The above process repeats forever.

Figure 7.149 summarizes the operation of both nodes.

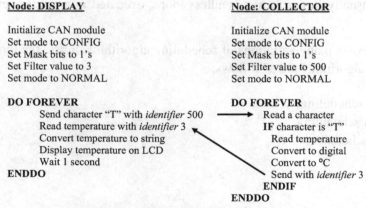

Node: DISPLAY	Node: COLLECTOR
Initialize CAN module	Initialize CAN module
Set mode to CONFIG	Set mode to CONFIG
Set Mask bits to 1's	Set Mask bits to 1's
Set Filter value to 3	Set Filter value to 500
Set mode to NORMAL	Set mode to NORMAL
DO FOREVER	**DO FOREVER**
Send character "T" with *identifier* 500 →	Read a character
Read temperature with *identifier* 3 ←	**IF** character is "T"
Convert temperature to string	Read temperature
Display temperature on LCD	Convert to digital
Wait 1 second	Convert to °C
ENDDO	Send with *identifier* 3
	ENDIF
	ENDDO

Figure 7.149: Operation of Both Nodes.

Project 7.18 Multitasking

Most complex real-time systems consist of a number of tasks running independently. This requires some form of scheduling and task control mechanisms. For example, consider an extremely simple real-time system that must flash an LED at required intervals and at the same time look for a key input from a keypad. One solution would be to scan the keypad in a loop at regular intervals while flashing the LED at the same time. Although this approach may work for simple systems, in most complex real-time systems, a real-time operating system (RTOS) or a multiprocessing approach are usually employed. Multiprocessing is beyond the scope of this project.

An RTOS is a program that manages system resources, scheduling the execution of various tasks in the system and provides services for intertask synchronization and messaging. There are many books and other sources of reference that describe the operation and principles of various RTOS systems.

Every RTOS consists of a kernel that provides the low-level functions, mainly the scheduling, creation of tasks and intertask resource management. Most complex RTOSs also provide file-handling services, disk input–output operations, interrupt servicing, network management, user management, and so on.

A task is an independent thread of execution in a multitasking system, usually with its own local set of data. A multitasking system consists of a number of independent tasks, each running its own code and communicating with each other to have orderly access to shared resources. The simplest RTOS consists of a scheduler that determines the order in which the tasks should run. This scheduler switches from one task to the next by performing a context switching where the context of the running task is stored and context of the next task is loaded so that execution can continue properly with the next task. Tasks are usually in the form of endless loops, executed in an order determined by the scheduler.

Although there exists many variations of scheduling algorithms in use, the three most commonly used algorithms are as follows:

- Cooperative scheduling,
- Round-robin scheduling,
- Preemptive scheduling.

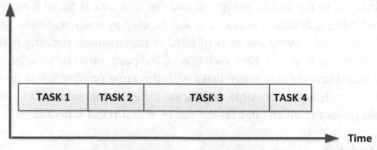

Figure 7.150: Cooperative Scheduling.

Cooperative Scheduling

This is perhaps the simplest algorithm (Figure 7.150) where tasks voluntarily give up the central processing unit (CPU) usage when they have nothing useful to do, or when they are waiting for some resources to become available (e.g. a key to be pressed and a timer to expire). This algorithm has the disadvantage that certain tasks can use excessive CPU times, and thus not allow some other important tasks to run when needed. Cooperative scheduling is used in simple multitasking systems with no time critical applications. A variation of the pure cooperative scheduling is to prioritize the tasks and run the highest priority computable task when the CPU becomes available. As we shall see in an example project later, cooperative scheduling can easily be implemented in microcontrollers using the *switch* statement.

Round-robin Scheduling

Round-robin scheduling (Figure 7.151) allocates each task an equal share of the CPU time. In its simplest form, tasks are in a circular queue and when a task's allocated CPU time

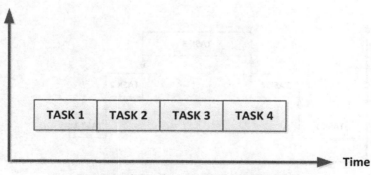

Figure 7.151: Round-robin Scheduling.

expires, the task is put to the end of the queue and the new task is taken from the front of the queue. Round-robin scheduling is not very satisfactory in many real-time applications where each task can have varying amounts of CPU requirements depending upon the complexity of processing required. One variation of the pure round-robin scheduling is to provide priority-based scheduling, where tasks with the same priority levels receive equal amounts of CPU time. It is also possible to allocate different maximum CPU times to each task. An example project is given later on the use of round-robin scheduling.

Preemptive Scheduling

This is the most commonly used and the most complex scheduling algorithm used in real-time systems. Here, the tasks are prioritized, and the task with the highest priority among all other tasks gets the CPU time (Figure 7.152). If a task with a priority higher than the currently executing task becomes ready to run, the scheduler saves the context of the current task and switches to the higher priority task by loading its context. Usually, the highest priority task runs to completion or until it becomes noncomputable (e.g. waiting for a resource to be available). Although the preemptive scheduling is very powerful, care is needed as an error in programming can place a high priority task in an endless loop and thus not release the CPU to other tasks. Some real-time systems employ a combination of round-robin and preemptive scheduling. In such systems, time critical tasks are usually prioritized and run under preemptive scheduling, whereas less time-critical tasks run under the round-robin scheduling, sharing the left CPU time among themselves.

There are many commercially available, shareware, and open-source RTOS software for the PIC microcontrollers. Brief details of some popular RTOS systems are given below.

Salvo (www.pumpkininc.com) is a low-cost, event driven, priority-based, multitasking RTOS designed for microcontrollers with limited data and program memories. Salvo

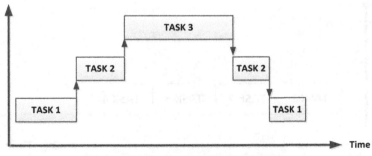

Figure 7.152: Preemptive Scheduling.

can be used for many microcontroller families and it supports large number of compilers, such as Keil C51, Hi-Tech 8051, Hi-Tech PICC-18, MPLAB C18, and many others. A demo version (Salvo Lite) is available for evaluation purposes. The Pro version is the top model aimed for professional applications, supporting unlimited number of tasks with priorities, event flags, semaphores, message queues, and many more features.

CCS compiler (www.ccsinfo.com) from Custom Computer services Inc supports a cooperative RTOS with a number of functions to start and terminate tasks, to send messages between tasks, to synchronize tasks using semaphores, and so on.

CMX-Tiny+ (www.cmx.com) supports large number of microcontrollers. This is a preemptive RTOS with a large number of features such as event-flags, cyclic timers, message queues, and semaphores. Although CMX-Tiny+ is a sophisticated RTOS, it has the disadvantage that the cost is relatively high.

PICos18 (www.picos18.com) is an open-source preemptive RTOS for the PIC18 microcontrollers. The full documentation and the source code are provided free of charge for people wishing to use the product.

MicroC/OS-II (http://micrium.com) is a preemptive RTOS, which has been ported to many microcontrollers, including the PIC family. This is a very sophisticated RTOS, providing semaphores, mailboxes, event-flags, timers, memory management, message queues, and many more.

FreeRTOS (www.freertos.org) is an open-source RTOS that can be used in microcontroller-based projects. This is a preemptive RTOS but can be configured for cooperative or hybrid operations.

Finally, OSA-RTOS (http://picosa.narod.ru) is freeware RTOS for PIC microcontrollers. The full source code and documentation are available and can be downloaded. OSA is a cooperative multitasking RTOS, offering many features such as semaphores, data queues, mutexes, memory pools, system services, and many more.

Project 1—Using Cooperative Multitasking

This is a simple project demonstrating how cooperative multitasking can be implemented easily using the C language. This is an example of an event counter with two-digit seven-segment LEDs. The project counts external events and displays the count on the LEDs. The following tasks are used in this project:

Task 1 (REFRESH_COUNT): This task refreshes the seven-segment LEDs every 3 ms and displays the current count.

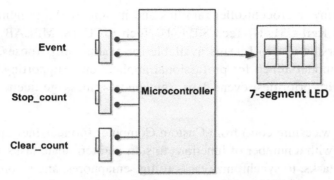

Figure 7.153: Block Diagram of the Project.

Task 2 (EVENT): This is the event counter task. An event is assumed to occur when a push-button switch goes from logic 1 to logic 0.

Task 3 (CLEAR_COUNT): This task clears the count. A 1 to 0 transition of a push-button switch clears the count.

Task 4 (STOP_COUNT): This task stops the count. A 1 to 0 transition of a push-button switch clears the count.

The block diagram of the project is shown in Figure 7.153.

Project Hardware

The circuit diagram of the project is shown in Figure 7.154. A two-digit seven-segment display is used to display the event count. Three push-button switches are used to initiate an event, clear the count, and to stop the count.

Project PDL

The project PDL is shown in Figure 7.155.

Project Program
mikroC Pro for PIC

Cooperative scheduling can easily be implemented using the *switch* statement. For example, a system with three tasks can be implemented as follows:

```
for(;;)
{
  state = 1;
  switch(state)
  {
```

```
case 1:
    Implement TASK 1
    state = 2;
    break;
case 2:
    Implement TASK 2
    state = 3;
    break;
case 3:
    Implement TASK 3
    state = 1;
    break;
  }
}
```

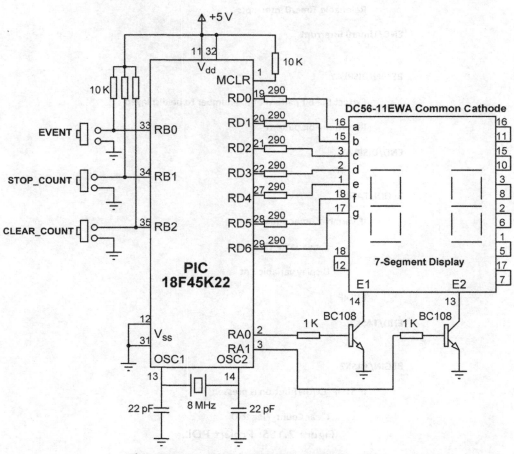

Figure 7.154: Circuit Diagram of the Project.

```
BEGIN
        Configure PORTA, PORTB, PORTD as digital
        Configure PORTA, PORTD as outputs, RB0:RB2 as inputs
        Disable the display
        Configure TIMER0 to interrupt at every millisecond
        DO FOREVER
                CALL TASK1
                CALL TASK2
                CALL TASK3
                CALL TASK4
        ENDDO
END

BEGIN/Timer0 Interrupt

        Re-load TMR0L

        Decrement variable Timer by 1

        Re-enable Timer0 interrupts

END/Timer0 Interrupt


BEGIN/DISPLAY

        Extract the bit pattern of the number to be displayed

        Return the bit pattern

END/DISPLAY


BEGIN/TASK1

        IF variable Timer equals 0

                Set variable Timer to 5

                Display variable Cnt

        ENDIF

END/TASK1


BEGIN/TASK2

        IF STOP_COUNT button is pressed

                Clear Count_Flag to 0
```

Figure 7.155: Project PDL.

```
                        ENDIF

                  END/TASK2


               BEGIN/TASK3

                     IF  Clear_Count button is pressed

                           Clear Cnt to 0

                     ENDIF

                  END/TASK3


               BEGIN/TASK4

                     IF  EVENT occurred

                           Wait until event is removed

                           Increment variable Cnt

                     ENDIF

                  END/TASK4
```

Figure 7.155
cont'd

The mikroC Pro for the PIC program listing is shown in Figure 7.156 (MIKROC-RTOS1.C). At the beginning of the program, PORTA, PORTB, and PORTD are configured as digital. PORTB and PORTD are configured as outputs and RB0, RB1, and RB2 are configured as inputs. Then, Timer 0 is configured to generate interrupts at every millisecond. The remaining part of the main program implements the multitasking loop where the tasks are executed in order one after the other.

TASK 1 refreshes the seven-segment display every 5 ms. Global variable Timer is loaded with five and is decremented by one every time a timer interrupt is generated. When the variable Timer is 0, the display is refreshed. First, the MSD digit is refreshed, followed by the LSD digit by enabling the corresponding digit enable bits.

```
/*******************************************************************************
                    Dual 7-SEGMENT DISPLAY EVENT COUNTER
                    =====================================
```

In this project two common cathode 7-segment LED displays are connected to PORTD of a
PIC18F45K22 microcontroller and the microcontroller is operated from an 8 MHz crystal.
Digit 1 (right digit) enable pin is connected to port pin RA0 and digit 2 (left digit) enable pin
is connected to port pin RA1 of the microcontroller.

The program is an event counter. 3 push-button switches are connected to PORTB as follows:

```
RB0   Event push button switch
RB1   Stop_Count push button switch
RB2   Clear_Count push button switch
```

Events are assumed to occur when RB0 goes from 1 to 0. When an event occurs, variable Cnt
is incremented by 1. The Display is refreshed every 5 ms and the count is displayed. Pressing
Stop Count button disables the counting. Pressing the Clear Count button clears the count to 0.

This program uses co-operative multitasking, implemented using a switch statement. There are
4 tasks in the program:

```
TASK1: Refreshes the Display every 5ms
TASK2: Disables counting
TASK3: Clears counting to zero
TASK4: Increments counting when an event is detected
```

```
Author:   Dogan Ibrahim
Date:     October 2013
File:     MIKROC-RTOS1.C
*******************************************************************************/
#define DIGIT1 PORTA.RA0                       // Display DIGIT 1 enable
#define DIGIT2 PORTA.RA1                       // Display DIGIT 2 enable
#define EVENT PORTB.RB0                        // Event input (push button)
#define STOP_COUNT PORTB.RB1                   // STOP_COUNT push button
#define CLEAR_COUNT PORTB.RB2                  // CLEAR_COUNT push button

unsigned char state = 1;                       // Initial value of state variable
unsigned char Count_Flag = 1;                  // Set when counting is enables
unsigned char refresh = 0;
unsigned char Cnt = 0;                         // Initial value of count
unsigned int Timer = 5;                        // Refreshing time (ms)
unsigned char Sbutton = 0;

//
// Generate Timer interrupts every milliseconds
//
void interrupt()
{
  TMR0L = 225;                                 // Re-load Timer0 for 1ms interrupts
  Timer--;                                     // Decrement variable Timer
```

Figure 7.156: mikroC Pro for the PIC Program.

```
      INTCON = 0x20;                                // Re-enable Timer0 interrupts
}

//
// This function finds the bit pattern to be sent to the port to display a number
// on the 7-segment LED. The number is passed in the argument list of the function.
//
unsigned char Display(unsigned char no)
{
    unsigned char Pattern;
    unsigned char SEGMENT[] = {0x3F,0x06,0x5B,0x4F,0x66,0x6D,
                    0x7D,0x07,0x7F,0x6F};

    Pattern = SEGMENT[no];                          // Pattern to return
    return (Pattern);
}

//
// This task refreshes the 7-segment LEDs every 5 milliseconds
//
void TASK1()
{
    unsigned char Msd, Lsd;

      if(Timer == 0)                                // If 5 ms has elapsed
      {
       Timer = 5;                                   // Re-load variable Timer
       if(refresh == 0)                             // Time to refresh MSD digit ?
       {
         refresh = 1;
         DIGIT1 = 0;
         Msd = Cnt / 10;                            // MSD digit
         PORTD = Display(Msd);                      // Send to PORTD
         DIGIT2 = 1;                                // Enable digit 2
       }
       else if(refresh == 1)                        // Time to refresh LSD digit ?
       {
         refresh = 0;
         DIGIT2 = 0;                                // Disable digit 2
         Lsd = Cnt % 10;                            // LSD digit
         PORTD = Display(Lsd);                      // Send to PORTD
         DIGIT1 = 1;                                // Enable digit 1
       }
      }
}

//
// This task stops the count
//
void TASK2()
```

Figure 7.156
cont'd

```
{
    if(STOP_COUNT == 0)Count_Flag = 0;              // Clear count flag to stop counting
}

//
// This task clears the count
//
void TASK3()
{
    if(CLEAR_COUNT == 0)Cnt = 0;                    // Clear the count to 0
}

//
// This task incrfements the event counter
//
void TASK4()
{
    if(EVENT == 0 && Count_Flag == 1)Sbutton = 1;   // If event and counting enabled
    if(EVENT == 1 && Sbutton == 1 && Count_Flag == 1)  // If event has been removed
    {
      Cnt++;                                        // Increment count
      Sbutton = 0;
    }

}


//
// Start of MAIN Program
//
void main()
{
    ANSELA = 0;                                     // Configure PORTA as digital
    ANSELB = 0;                                     // Configure PORTB as digital
    ANSELD = 0;                                     // Configure PORTD as digital
    TRISA = 0;                                      // Configure PORTA as outputs
    TRISB = 7;                                      // Configure RB0, RB1, RB2 as input
    TRISD = 0;                                      // Configure PORTD as outputs

    DIGIT1 = 0;                                     // Disable digit 1
    DIGIT2 = 0;                                     // Disable digit 2
//
// Set Timer0 to interrupt at every millisecond
//

    T0CON = 0xC5;                                   // Configure T0CON register
    TMR0L = 225;                                    // Load TMR0L register
    INTCON = 0xA0;                                  // Enable Timer0 interrupts
//
// Start of multitasking loop
//
```

Figure 7.156
cont'd

```
        while(1)
        {
          switch(state)
          {
            case 1:
              TASK1();                          // Do TASK1
              state = 2;                        // Next task is TASK2
              break;
            case 2:
              TASK2();                          // Do TASK2
              state = 3;                        // Next task is TASK3
              break;
            case 3:
              TASK3();                          // Do TASK3
              state = 4;                        // Next task is TASK4
              break;
            case 4:
              TASK4();                          // Do TASK4
              state = 1;                        // Next task is TASK1
              break;
          }
        }
      }
```

Figure 7.156
cont'd

TASK2 clears flag Count_Flag so that counting stops, that is, no count is generated when an event occurs.

TASK3 clears the count by clearing variable Cnt.

TASK4 checks the state of the EVENT button (RB0). This button is normally at logic 1. When the button is pressed, it goes to logic 0 and is back at logic 1 when the button is released. It is important that we generate only one count when the button is pressed and released. This task initially sets variable Sbutton to 1 when an event occurs (EVENT = 0) and also when the counting is enabled (Count_Flag = 1). At this point, the button is in the pressed state, and we have to wait until the button is released before incrementing the count; otherwise, the count increment while the button is kept pressed. The following program code detects when the button is pressed and then it increments the count only when the button is released:

```
if(EVENT == 0 && Count_Flag == 1)Sbutton = 1;
if(EVENT == 1 && Sbutton == 1 && Count_Flag == 1)
{
    Cnt++;
    Sbutton = 0;
}
```

Project 2—Using Round-Robin Multitasking With Variable CPU Time Allocation

In this project, we will be developing and using a round robin-type multitasking software with variable CPU time allocation. The software, called RTOS.C, can be included in multitasking programs. Each task is allocated maximum processing time selected by the user. The scheduler terminates a task when this time is reached, or when the task voluntarily gives-up CPU time by calling a function. The scheduler is interrupt driven and activates the tasks in order.

Each task in the user program is organized as a C function, running forever in a loop. The first thing a task does is to call scheduler function InitTask, which saves the task return address in an array called TStack. In addition, the maximum allocated duration of each task (in milliseconds) is also stored in array TTime. The program counter of a PIC18F microcontroller is 24 bits wide and is stored in three 8-bit stack registers TOSL, TOSH, and TOSU after a function call or an interrupt (Figure 7.157). These registers are accessed by the scheduler during the saving and restoring of task return addresses.

Figure 7.158 shows the program listing of the multitasking scheduler. In an application, the main program initially calls all the tasks in turn so that their return addresses can be saved. Then, function StartTasks is called. This function calls to SetUpTmrInt in order to configure timer TMR0 so that timer interrupts can be

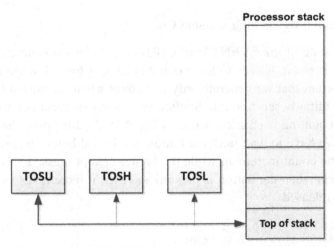

Figure 7.157: PIC18F Microcontroller Stack Structure.

generated every milliseconds for the scheduler. In addition, the return address of Task 0 is pushed onto the stack and a RETURN is executed so that task execution starts from Task 0. At the core of the scheduler, we have the timer ISR. The ISR determines the next task to run and performs the necessary context switching. The following operations are carried out within the ISR:

- Timer register TMR0 is reloaded for 1-ms interrupts.
- Current CPU registers W, STATUS, and BSR are saved.
- If the allocated duration of current task has not expired, then timer interrupts are reenabled and the ISR passes control back to the same task with no context changing.
- Otherwise, the return address of the current task is saved in array TStack.
- Task number of the next task is determined, and its return address is pushed onto processor stack.
- CPU registers W, STATUS, and BSR of the next task are restored.
- Timer interrupts are reenabled, and the ISR passes the CPU control to the next task.

Three timer registers called TimerA, TimerB, and TimerC are decremented every millisecond inside the ISR, and these registers can be used in task timing applications.

Project Description

This project is the same as the previous project where a two-digit seven-segment LED is used with four tasks:

Task 1 (REFRESH_COUNT): This task refreshes the seven-segment LEDs every 3 ms and displays the current count.

Task 2 (EVENT): This is the event counter task. An event is assumed to occur when a push-button switch goes from logic 1 to logic 0.

Task 3 (CLEAR_COUNT): This task clears the count. A 1—0 transition of a push-button switch clears the count.

Task 4 (STOP_COUNT): This task stops the count. A 1—0 transition of a push-button switch clears the count.

The block diagram of the project is as shown in Figure 7.153.

Project Hardware

The circuit diagram of the project is as given in Figure 7.154.

```
#pragma disablecontextsaving
#define freq 8                                              // Clock frequency
#define Prescale 64                                         // Timer prescaler
#define T 1000                                              // 1000 us (1 ms) timer
#define Timervalue 256-(T*freq/(4*Prescale))
#define StopTask while(1){Swapp = 1; INTCON.F2 = 1;}
#define SwapTask {Swapp = 1; INTCON.F2 = 1;}
unsigned char TMR0 = Timervalue;
unsigned char Temp, Twreg, Tstatus, Tbsr, Swapp = 0;
unsigned char Saved[MaxTsk][3];
unsigned char TaskNumber = 0;
unsigned char TStack[MaxTsk][4];
unsigned int TCount[MaxTsk];
unsigned int TTime[MaxTsk];
signed char TimerA, TimerB, TimerC;


//
// The program jumps here every 1 ms
//
void interrupt()
{
 TMR0L = TMR0;                                              // Reload timer register
 TimerA--;                                                  // Two general purpose timers
 TimerB--;
 TimerC--;

 Twreg = WREG; Tstatus = STATUS; Tbsr = BSR;                // Get current context

 TCount[TaskNumber]++;
 if((Swapp == 1) || (TCount[TaskNumber] >= TTime[TaskNumber]))
 {
   TCount[TaskNumber] = 0;
   if(Swapp == 1)Swapp = 0;

    Saved[TaskNumber][0] = Twreg;       Saved[TaskNumber][1] = Tstatus;
    Saved[TaskNumber][2] = Tbsr;
//
// Save return address of current task
//
    TStack[TaskNumber][0] = TOSL;       TStack[TaskNumber][1] = TOSH;
    TStack[TaskNumber][2] = TOSU;
    asm POP
//
// Get next task, and save its return address on TOS
//
    TaskNumber++;
    if(TaskNumber > MaxTsk - 1)TaskNumber = 0;
    asm PUSH
    Temp = TStack[TaskNumber][0];       TOSL = Temp;
    Temp = TStack[TaskNumber][1];       TOSH = Temp;
```

Figure 7.158: Scheduler (RTOS.C) Program Listing.

```
        Temp = TStack[TaskNumber][2];        TOSU = Temp;
//
// Restore task registers and return from interrupt
//
    INTCON = 0x20;
    Temp  = Saved[TaskNumber][1];        BSR = Saved[TaskNumber][2];
    WREG = Saved[TaskNumber][0];        STATUS = Temp;
    }
    else
    {
    INTCON = 0x20;                WREG = Twreg;
    BSR = Tbsr;                STATUS = Tstatus;
    }
    asm retfie 0
}


//
// Set up Timer0 interrupts every 100 microseconds
//
void SetUpTmrInt(void)
{
    T0CON = 0xC5;                                        // Prescaler = 64
    TMR0L = TMR0;
    INTCON = 0xA0;
}


//
// Store tasks return addresses on stack
//
void initTask(unsigned char TaskNo, unsigned int t)
{
    TStack[TaskNo][0] = TOSL;        TStack[TaskNo][1] = TOSH;
    TStack[TaskNo][2] = TOSU;        TTime[TaskNo] = t;
    asm POP
}


//
// Start tasks
//
void StartTasks(void)
{
    SetUpTmrInt();
    Temp = TStack[0][0];        TOSL = Temp;
    Temp = TStack[0][1];        TOSH = Temp;
    Temp = TStack[0][2];        TOSU = Temp;
    asm RETURN
}
```

Figure 7.158
cont'd

Project Program

mikroC Pro for PIC

The mikroC Pro for PIC program listing is shown in Figure 7.159 (MIKROC-RTOS2.C). At the beginning of the program, the number of tasks is specified using symbol MaxTsk, and multitasking scheduler file RTOS.C is included in the program.

The main program configures PORTA, PORTB, and PORTD as digital. Timer variable TimerA is set to 4 so that the LED refreshing time can be set to be every 4 ms in Task0. The tasks are then called one after the other and function StartTasks is called to start the tasks and pass control to the multitasking scheduler.

TASK0 refreshes the seven-segment display every 4 ms. Global variable TimerA is loaded with four and is decremented by one every time a timer interrupt is generated. When variable TimerA is ≤0, the display is refreshed. First, the MSD digit is refreshed, followed by the LSD digit by enabling the corresponding digit enable bits. Function call InitTask(0, 1) pushes the return address of Task0 on stack and allocates maximum processing time of 1 ms to TASK0.

TASK1 clears flag Count_Flag so that counting stops, that is, no count is generated when an event occurs. Function call InitTask(1, 1) pushes the return address of Task1 on stack and allocates maximum processing time of 1 ms to TASK1. Note that the following code is used for this task:

```
while(1)
{
    while(STOP_COUNT == 1);  // Wait until STOP_COUNT button is pressed
    Count_Flag = 0;          // Clear Count_Flag to stop counting
    SwapTask;                // Return to scheduler (Give up the CPU)
}
```

Note that since 1 ms is allocated to this task, the program will remain here for at least 1 ms if the button is not pressed. A quicker way of checking whether or not the button is pressed would be as in the following code. Here, the button is checked and if it is not pressed, the task releases the CPU by calling to function SwapTask:

```
while(1)
{
    if(STOP_COUNT == 0)Count_Flag = 0;
    SwapTask;                // Return to scheduler (Give up the CPU)
}
```

TASK2 clears the count by clearing variable Cnt.

TASK3 checks state of the EVENT button (RB0) and increments the event count Cnt when an event is detected and at the same time if the counting is enabled (Count_Flag = 1).

```
/*****************************************************************************
      Dual 7-SEGMENT DISPLAY EVENT COUNTER WITH ROUND-ROBIN MULTITASKING
      ====================================================================
```

In this project two common cathode 7-segment LED displays are connected to PORTD of a
PIC18F45K22 microcontroller and the microcontroller is operated from an 8 MHz crystal.
Digit 1 (right digit) enable pin is connected to port pin RA0 and digit 2 (left digit) enable pin is
connected to port pin RA1 of the microcontroller.

The program is an event counter. 3 push-button switches are connected to PORTB as follows:

```
   RB0  Event push button switch
   RB1  Stop Count push button switch
   RB2  Clear Count push button switch
```

Events are assumed to occur when RB0 goes from 1 to 0. When an event occurs the variable
Cnt is incremented by 1. The Display is refreshed every 5 ms and the count is displayed.
Pressing Stop Count button disables the counting. Pressing the Clear Count button clears
the count to 0.

This program uses round-robin multitasking algorithm with the modification that the allocated
CPU time to each task can be set by the user.

There are 4 tasks in the program:

```
TASK0: Refreshes the Display every 5ms
TASK1: Disables counting
TASK2: Clears counting to zero
TASK3: Increments counting when an event is detected
```

```
Author:   Dogan Ibrahim
Date:     October 2013
File:     MIKROC-RTOS2.C
*****************************************************************************/
#define MaxTsk 4                                    // Define number of tasks
#include "RTOS.C"                                    // Include RTOS.C file

#define DIGIT1 PORTA.RA0                             // Display DIGIT 1 enable
#define DIGIT2 PORTA.RA1                             // Display DIGIT 2 enable
#define EVENT PORTB.RB0                              // Event input (push button)
#define STOP_COUNT PORTB.RB1                         // STOP_COUNT push button
#define CLEAR_COUNT PORTB.RB2                        // CLEAR_COUNT push button

unsigned char refresh = 0;
unsigned char Count_Flag = 1;
unsigned char Cnt = 0;

//
// This function finds the bit pattern to be sent to the port to display a number
// on the 7-segment LED. The number is passed in the argument list of the function.
//
```

Figure 7.159: mikroC Pro for PIC Program.

```
unsigned char Display(unsigned char no)
{
    unsigned char Pattern;
    unsigned char SEGMENT[] = {0x3F,0x06,0x5B,0x4F,0x66,0x6D,
                0x7D,0x07,0x7F,0x6F};

    Pattern = SEGMENT[no];                          // Pattern to return
    return (Pattern);
}


//
// This task refreshes the 7-segment LEDs every 5 ms
//
void TASK0()
{
    unsigned char Msd, Lsd;

    InitTask(0,1);                                  // Allocate max 1 ms to TASK0
    while(1)
    {
     if(TimerA <= 0)                                // Time to refresh ?
     {
      TimerA = 4;                                   // Reload TimerA with 4 ms
      if(refresh == 0)                              // Time to refresh MSD digit ?
      {
       refresh = 1;
       DIGIT1 = 0;
       Msd = Cnt / 10;                              // MSD digit
       PORTD = Display(Msd);                        // Send to PORTD
       DIGIT2 = 1;                                  // Enable digit 2
      }
      else if(refresh == 1)                         // Time to refresh LSD digit ?
      {
       refresh = 0;
       DIGIT2 = 0;                                  // Disable digit 2
       Lsd = Cnt % 10;                              // LSD digit
       PORTD = Display(Lsd);                        // Send to PORTD
       DIGIT1 = 1;                                  // Enable digit 1
      }
     }
     SwapTask;                                      // Return to scheduler
    }
}



//
// This task stops the count
//
void TASK1()
{                                                   // Allocate maximun 1 ms to TASK1
    InitTask(1,1);
```

Figure 7.159
cont'd

```
            while(1)
            {
                while(STOP_COUNT == 1);                    // Wait until button is pressed
                Count_Flag = 0;                            // Clear count flag to stop counting
                SwapTask;                                  // Return to scheduler
            }
        }

        //
        // This task clears the count
        //
        void TASK2()
        {
            InitTask(2,1);                                 //Allocate maximum 1 ms to TASK2
            while(1)
            {
                while(CLEAR_COUNT == 1);                   // Wait until button is pressed
                Cnt = 0;                                   // Clear Cnt
                SwapTask;                                  // Return to scheduler
            }
        }

        //
        // This task increments the event counter
        //
        void TASK3()
        {
            InitTask(3,1);                                 // Allocate maximum 1 ms to TASK3
            while(1)
            {
                while(EVENT == 1);                         // Wait until button press -release
                while(EVENT == 0);
                if(Count_Flag == 1)Cnt++;                  // Increment Cnt
                SwapTask;                                  // Return to scheduler
            }

        }

        //
        // Start of MAIN Program
        //
        void main()
        {
            ANSELA = 0;                                    // Configure  PORTA as digital
            ANSELB = 0;                                    // Configure PORTB as digital
            ANSELD = 0;                                    // Configure PORTD as digital
            TRISA = 0;                                     // Configure PORTA as outputs
            TRISB = 7;                                     // Configure RB0, RB1, RB2 as input
            TRISD = 0;                                     // Configure PORTD  as outputs
```

Figure 7.159
cont'd

```
        DIGIT1 = 0;                        // Disable digit 1
        DIGIT2 = 0;                        // Disable digit 2
        TimerA = 4;                        // Set TimerA variable to   4ms

        TASK0();                           // Initialize TASK0
        TASK1();                           // Initialize TASK1
        TASK2();                           // Initialize TASK2
        TASK3();                           // Initialize TASK3
        StartTasks();                      // Start Tasks
    }
```

Figure 7.159
cont'd

Project 7.19—Stepper Motor Control Projects—Simple Unipolar Motor Drive

This project is about using stepper motors in microcontroller-based systems. This is an introductory project where a stepper motor is driven from a microcontroller.

Before going into the details of the project, it is worthwhile to look at the theory and operation of stepper motors briefly.

Stepper motors are commonly used in printers, disk drives, position control systems, and many more systems where precision position control is required. Stepper motors come in a variety of sizes, shapes, strengths, and precision. There are two basic types of stepper motors: unipolar and bipolar.

Unipolar Stepper Motors

Unipolar stepper motors have two identical and independent coils with center taps, and having five, six, or eight wires (Figure 7.160).

Unipolar stepper motors can be driven in three modes: One-phase full-step sequencing, two-phase full-step sequencing and two-phase half-step sequencing.

One-phase Full-step Sequencing

Table 7.20 shows the sequence of sending pulses to the motor. Each cycle consists of four pulses.

Two-phase Full-step Sequencing

Table 7.21 shows the sequence of sending pulses to the motor. The torque produced is higher in this mode of operation.

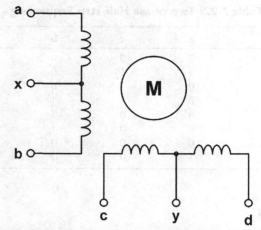

Figure 7.160: Unipolar Stepper Motor Windings.

Table 7.20: One-phase Full-step Sequencing.

Step	a	c	b	d
1	1	0	0	0
2	0	1	0	0
3	0	0	1	0
4	0	0	0	1

Two-phase Half-step Sequencing

Table 7.22 shows the sequence of sending pulses to the motor. This mode of operation gives more accurate control of the motor rotation, but it requires twice as many pulses for each cycle.

As we shall see later in the project, the motor can be connected to a microcontroller using power transistors or power MOSFET transistors.

Table 7.21: Two-phase Full-step Sequencing.

Step	a	c	b	d
1	1	0	0	1
2	1	1	0	0
3	0	1	1	0
4	0	0	1	1

Table 7.22: Two-phase Half-step Sequencing.

Step	a	c	b	d
1	1	0	0	0
2	1	1	0	0
3	0	1	0	0
4	0	1	1	0
5	0	0	1	0
6	0	0	1	1
7	0	0	0	1
8	1	0	0	1

Bipolar Stepper Motors

Bipolar stepper motors have two identical and independent coils and four wires, as shown in Figure 7.161.

The control of bipolar stepper motors is slightly more complex. Table 7.23 shows the driving sequence. The "+" and "−" signs denote the polarity of the voltage that should be given to the motor legs. Bipolar stepper motors are usually driven using H-bridge circuits.

Project Description

In this project, a unipolar stepper motor is used and the motor is rotated for 100 turns before it is then stopped.

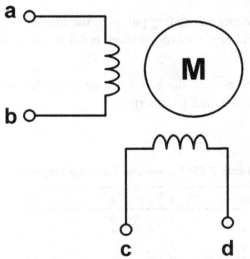

Figure 7.161: Bipolar Stepper Motor Windings.

Table 7.23: Bipolar Stepper Motor Driving Sequence.

Step	a	c	b	d
1	+	−	−	−
2	−	+	−	−
3	−	−	+	−
4	−	−	−	+

Project Hardware

The circuit diagram of the project is shown in Figure 7.162. In this project, an UAG2 type unipolar stepping motor is used. The motor is connected to RB0:RB3 pins of the microcontroller via IRLI520N-type power MOSFET transistors. UAG2 is a small stepper motor with an 18° stepping angle. Thus, a complete revolution requires 20 pulses. In this

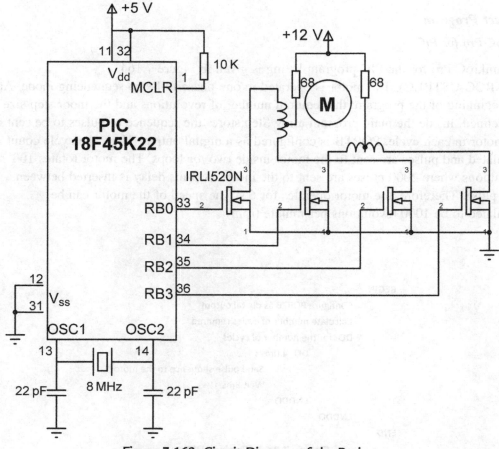

Figure 7.162: Circuit Diagram of the Project.

example, the motor rotates 100 turns (i.e. 2000 pulses are given), and then it stops. A 3-ms delay is inserted between each pulse to slow down the motor.

The pin connections of the UAG2 motor is as follows:

Pin	Description
1	Start of first coil
3	Middle connection of first coil
5	End of first coil
2	Start of second coil
4	Middle connection of second coil
6	End of second coil

Project PDL

The project PDL is shown in Figure 7.163.

Project Program

mikroC Pro for PIC

The mikroC Pro for the PIC program listing is given in Figure 7.164 (MIKROC-USTP1.C). The motor is operated in one-phase full-step sequencing mode. At the beginning of the program, the required number of revolutions and the moor step size are defined. Inside the main program array Step stores the sequence of pulses to be sent to the motor in each cycle. PORTB is configured as a digital output. Then, the cycle count is calculated and pulses are sent to the motor inside two *for* loops. The motor rotates 100 revolutions where 2000 pulses are sent to the motor. A 3-ms delay is inserted between each pulse. Therefore, the motor operates for 6 s. The speed of the motor can be calculated to be 1000 revolutions per minute (rpm).

```
BEGIN
        Configure PORTB as digital output
        Calculate number of cycles required
        DO  for the number of cycles
                DO  4 times
                        Send pulse sequence to the motor
                        Wait 3ms
                ENDDO
        ENDDO
END
```

Figure 7.163: Project PDL.

```
/************************************************************************
                    UNIPOLAR STEPPER MOTOR PROJECT
                    ================================
```

In this project an UAG2 model unipolar stepper motor is connected to PORTB of a PIC18F45K22 microcontroller through IRLI520N type MOSFET transistor switches.

The program rotates the motor 100 times and then stops. In total 2000 pulses are sent to the motor. 3ms delay is inserted between each pulse. Therefore, the motor rotates 100 revolutions in 6 s (2000 x 3 ms = 6 s) and then stops.

```
Author:  Dogan Ibrahim
Date:    October 2013
File:    MIKROC-USTP1.C
************************************************************************/
const unsigned int Req_Rev_Count = 100;                  // Required no of revs
const unsigned char Step_Size = 18;                      // Motor Step Size (degrees)

void main()
{
    unsigned char Step[4] = {1, 2, 4, 8};
    unsigned int One_Rev_Step, Step_Count, Cycle_Count, j;
    unsigned char i;

    ANSELB = 0;                                          // Configure PORTB as digital
    TRISB = 0;                                           // Configure PORTB as digital

    One_Rev_Step = 360/Step_Size;                        // No of steps for 1 revolution
    Step_Count = Req_Rev_Count*One_Rev_Step;             // Total no of steps required
    Cycle_Count = Step_Count/4;                          // No of cycles

    for(j = 0; j < Cycle_Count; j++)                     // Do for all cycles
    {
        for(i = 0; i < 4; i++)                           // Do for all steps
        {
            PORTB = Step[i];                             // Send pulses to the motor
            Delay_Ms(3);                                 // 3 ms delay between each pulse
        }
    }
    while(1);                                            // End. Wait here forever
}
```

Figure 7.164: mikroC Pro for PIC Program.

Project 7.20—Stepper Motor Control Projects—Complex Control Of A Unipolar Motor

In this project, a unipolar stepper motor is controlled in the following order:

- Turn 200 revolutions clockwise,
- Wait 5 s,
- Turn 50 revolutions anticlockwise,
- Wait 3 s,
- Turn 100 revolutions clockwise,
- Wait 1 s,
- Stop.

Project Hardware

The circuit diagram of the project is shown in Figure 7.165. In this project, a UCN5804B type stepper motor controller chip is used. This chip can drive small unipolar stepper motors up to +35 V and 1.25 A. The chip is connected to the microcontroller via its Step (pin 11) and Direction (pin 14) pins. The chip also has half-step (pin 10) and phase (pin 9) selection inputs. Depending upon the connection of pins 9 and 10, the chip can operate in one-phase full-step, two-phase full-step, or in two-phase half-step sequencing modes. In this project, both pins 9 and 10 are connected to ground to operate in two-phase full-step sequencing mode. The direction input controls direction of the motor. The motor rotates one step when a pulse is applied to the Step input. Notice that diodes are used at the output pins of the microcontroller to protect the pins from negative voltage.

Project Program

mikroC Pro for PIC

The mikroC Pro for PIC program listing is shown in Figure 7.166 (MIKROC-USTP2.C). The speed of a stepper motor depends upon the time delay between the step inputs. If the time delay between the steps is T, and the motor step constant is β degrees, then the motor rotates β/T steps in a second. Since a complete revolution is $360°$, the number of revolutions in a second is $\beta/360T$. The number of revolutions per minute, that is, the revolutions per minute of the motor is then given by

$$RPM = 60\beta/360T$$

or

$$RPM = \beta/6T.$$

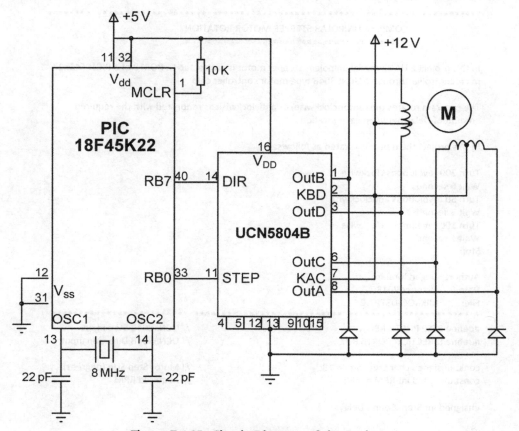

Figure 7.165: Circuit Diagram of the Project.

We can also write

$T = \beta/6 \times RPM$.

In this project, the motor has $\beta = 18°$ and therefore

$T = 3/RPM$ where T is in seconds

or

$T = 3000/RPM$ where T is in milliseconds.

If the required revolutions per minute is 500 rpm (assuming this is below the maximum revolutions per minute that can be provided by the motor) then the delay between each step is

$T = 3000/500 = 6$ ms

```
/********************************************************************
              COMPLEX UNIPOLAR STEPPER MOTOR ROTATION
              =========================================
```

In this project a UAG2 model unipolar   stepper motor is connected to PORTB of a PIC18F45K22
microcontroller through a UCN5804B type motor controller chip

The program rotates the motor clockwise, or anticlockwise as requested with the required
amount of Delay between each rotation.

In this project the motor is rotated as follows:

Turn 200 revolutions clockwise
Wait 5 seconds
Turn 50 revolutions anticlockwise
Wait 3 seconds
Turn 100 revolutions clockwise
Wait 1 second
Stop

```
Author:  Dogan Ibrahim
Date:    October 2013
File:    MIKROC-USTP2.C
********************************************************************/
#define STEP PORTB.RB0                              // UCN5804B Step input
#define DIRECTION PORTB.RB7                         // UCN5804B Direction input

const unsigned char Step_Size = 18;                 // Motor Step Size (degrees)
const unsigned int RPM = 500;                       // Required RPM

unsigned int Step_Count, Delay;

//
// This function operates the stepper motor as requested. "revcnt" is the rev count, "revdir"
// is the rev direction, and "gapdly" is the intergap Delay
// If the rev count is negative the motor rotates in the required direction continuously
// if rev is equal to 0 then the motor stops
//
void Stepper_Motor(int revcnt, unsigned char revdir, unsigned int gapdly)
{
  unsigned int k,p;

  if(revcnt < 0)
  {
    DIRECTION = revdir;
    for(;;)
    {
      STEP = 1;                                     // Send pulse to the motor
      STEP = 0;
      VDelay_Ms(Delay);
    }
```

Figure 7.166: mikroC Pro for PIC Program.

```
            }
         else
         {
            p = revcnt*Step_Count;
            DIRECTION = revdir;
            for(k = 0; k < p; k++)
            {
               STEP = 1;                                    // Send pulse to the motor
               STEP = 0;
               VDelay_Ms(Delay);
            }
            for(k = 0; k < gapdly; k++)Delay_Ms(1);          // Inter command delay
         }
      }


      //
      // This is the main program. The required Rev_Count, Rev_Direction, and Inter_Delay
      // must be specified here. In this example, the motor rotates 200 revs clockwise, then
      // waits 5s, rotates 50 revs anticlockwise, waits 3s, rotates 100revs clockwise, waits
      // 1s and then stops
      //
      void main()
      {
         int Rev_Count[] = {200, 50, 100, 0};               // 0 is the terminator
         unsigned char Rev_Direction[] = {0, 1, 0};         // 0=clockwise, 1=anticloc kwise
         unsigned char Inter_Delay[] = {5000, 3000, 1000};  // Delay in ms

         unsigned char j;

         ANSELB = 0;                                        // Configure PORTB as digital
         TRISB = 0;                                         // Configure PORTB as digital
         STEP = 0;                                          // Step = 0 to start with

         j = 0;
         Delay = 3000/RPM;                                  // Delay after each command
         Step_Count = 360/Step_Size;

         while(Rev_Count[j] != 0)
         {
            Stepper_Motor(Rev_Count[j], Rev_Direction[j], Inter_Delay[j]);
            j++;
         }

         while(1);                                          // End. Wait here forever
      }
```

Figure 7.166
cont'd

Three arrays are used in the program to specify the required number of revolutions, the direction of rotation, and the interstep gap:

Rev_Count: This array stores the required number of revolutions. A 0 entry specifies the end of the array. A negative value indicates that the motor is required to rotate continuously in the specified direction.

Rev_Direction: This array stores the corresponding motor rotation direction. 0 specifies clockwise and 1 specified anticlockwise rotation.

Inter_Delay: This array stores the required delay after each command.

For the example in this project, the arrays should have the following values:

```
Rev_Count[ ] = {200, 50, 100, 0};      // 0 is the terminator
Rev_Direction[ ] = {0, 1, 0};          // 0 = clockwise, 1 = anticlockwise
Inter_Delay[ ] = {5000, 3000, 1000};   // Delay is in ms
```

At the beginning of the program, symbols STEP and DIRECTION are assigned to port pins RB0 and RB7, respectively. Then, the above arrays are initialized, the step count is found, and a loop is formed. Inside this loop, the motor is activated by calling function Stepper_Motor. This array has three arguments: revolution count, revolution direction, and the delay after each command. If the revolution count is negative, then the motor rotates continuously in the specified direction. The rotation stops when the revolution is specified as 0. The motor is rotated by setting the required direction and then sending pulses to the STEP input of the UCN5804N controller chip.

Project 7.21—Stepper Motor Control Project—Simple Bipolar Motor Drive

The Bipolar Stepper motor is similar to the unipolar motor discussed in previous projects. The bipolar consists of two coils, but there is no center tap. As a result of this, the bipolar motor requires a controller where the current flow through the coils can be reversed. A bipolar motor is capable of a higher torque since entire coils may be energized, not just half of the coils. Bipolar stepper motors are usually controlled using H-bridge circuits where the current flow through the coils can easily be reversed.

Project Description

In this project, a bipolar stepper motor is used. The motor is rotated 10 revolutions in one direction, then stopped for 5 s, and then rotated 10 revolutions in the other direction, and is then stopped.

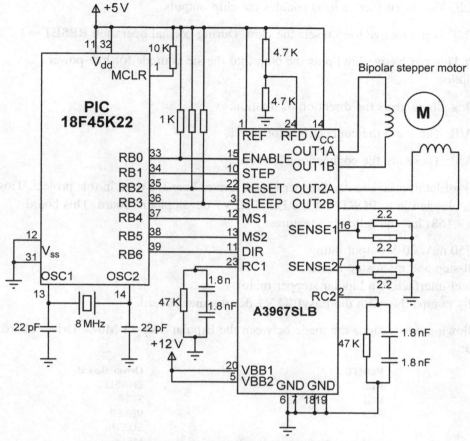

Figure 7.167: Circuit Diagram of the Project.

Project Hardware

The circuit diagram of the project is shown in Figure 7.167. In this project, A3967SLB type bipolar stepper motor controller chip is used. This chip can operate a bipolar stepper motor in the following modes, controlled by its MS1 and MS2 inputs:

- Full step,
- Half step,
- Quarter step,
- Eight microsteps.

In this project, the full step mode is used where MS1 = MS2 = 0. The other pins of interest are

STEP: A low-to-high transition on this pin rotates the motor by one step.

ENABLE: This input (active low) enables the chip outputs.

RESET: This pin (active low) resets the chip. During normal operation RESET = 1.

SLEEP: This pin (active low) puts the chip into the sleep mode for low-power consumption.

DIR: This pin controls the direction of rotation.

OUT1A/B: These are the connections for coil 1.

OUT2A/B: These are the connections for coil 2.

The mikroElektronika Bipolar Stepper Motor Driver board is used in this project. This board is plugged in to PORTB of the EasyPIC V7 development board. This board (Figure 7.168) has the following features:

- A 750 mA, 30-V output rating,
- Full-step and microstep modes,
- Direct interface to a bipolar stepper motor,
- Fully compatible with the EasyPIC V7 development board.

The following connections are made between the Bipolar Stepper Motor Driver board and PORTB:

PORTB	Driver Board
RC0	ENABLE
RC1	STEP
RC2	RESET
RC3	SLEEP
RC4	MS1
RC5	MS2
RC6	DIR

Figure 7.168: mikroElektronika Bipolar Stepper Motor Driver Board.

Figure 7.169: The 39HS02 Bipolar Stepper Motor.

The type of motor used in this project is the 39HS02 (Figure 7.169). This motor has the following features:

- A 1.8° step angle (200 steps for a complete revolution),
- A ±5% step angle accuracy,
- A 0.6-A phase current,
- Four leads (Coil 1: brown + gray, Coil 2: orange + green).

Project Program

mikroC Pro for PIC

The mikroC Pro for PIC program listing is shown in Figure 7.170 (MIKROC-BSTP.C). At the beginning of the program, the controller pins are defined by symbols. In the main program, the controller chip is enabled. Then, a loop is formed to send 2000 steps (10 revolutions) to the motor in one direction (DIR = 0). After a 5-s delay, the direction is changed (DIR = 1) and another 2000 steps are sent to the motor to rotate in the opposite direction. Note that a 1-ms delay is inserted between each step so that the speed of rotation is RPM = $1.8°/6 \times 1 \times 10^{-3}$ s = 300.

MPLAB XC8

The MPLAB XC8 program is shown in Figure 7.171 (XC8-BSTP.C). The program is similar to the mikroC version.

```
/*****************************************************************************
                      BIPOLAR STEPPER MOTOR ROTATION
                      ===============================
```

In this project a 39HS model bipolar stepper motor is connected to PORTB of a PIC18F45K22 microcontroller through an A3967SLB type bipolar motor controller chip

The program rotates the motor clockwise 10 revolutions, stops for 5 s, and then rotates anticlockwise for 10 turns and stops.

39HS motor has step angle of 1.8 degrees. thus, a complete revolution requires 200 pulses (steps) to be sent to the motor. For 10 revolutions, we have to send 2000 pulses. The DIR input controls the direction of rotation.

```
Author:   Dogan Ibrahim
Date:     October 2013
File:     MIKROC-BSTP.C
*****************************************************************************/
#define ENABLE PORTB.RB0                   // A3967SLB ENABLE input
#define STEP PORTB.RB1                      // A3967SLB STEP input
#define RST PORTB.RB2                       // A3967SLB RESET input
#define SLP PORTB.RB3                       // A3967SLB SLEEP input
#define MS1 PORTB.RB4                       // A3967SLB MS1 input
#define MS2 PORTB.RB5                       // A3967SLB MS2 input
#define DIR PORTB.RB6                       // A3967SLB DIR input
//
// This is the main program. The motor turns 10 revolutions clockwise with 1ms delay
// between each step. Then the motor stops for 5 seconds. The motor then rotates
// anticlockwise for 10 revolutions and then stops.
//
void main()
{
    unsigned int i;

    ANSELB = 0;                             // Configure PORTB as digital
    TRISB = 0;                              // Configure PORTB as digital
//
// A3967SLB chip configuration
//
    ENABLE = 0;                             // Enable the controller chip
    RST = 1;                                // Disable RESET
    SLP = 1;                                // Disable SLEEP
    MS1 = 0;                                // Full-step sequence
    MS2 = 0;
    STEP = 0;                               // Step = 0 to start with
//
// First rotate clockwise 10 revolutions
//
    DIR = 0;
    for(i = 0; i < 2000; i++)
    {
```

Figure 7.170: mikroC for the PIC Program.

```
            STEP = 1;                                // Send STEP pulses
            STEP = 0;
            Delay_Ms(1);
        }

        Delay_Ms(5000);                              // Wait 5 s
    //
    // Now rotate anticlockwise 10 revolutions
    //
        DIR = 1;
        for(i = 0; i < 2000; i++)
        {
            STEP = 1;                                // Send STEP pulses
            STEP = 0;
            Delay_Ms(1);
        }
        while(1);                                    // End. Wait here forever
    }
```

Figure 7.170
cont'd

Project 7.22—DC Motor Control Projects—Simple Motor Drive

DC motors are used in many industrial, commercial, and domestic applications. In this project, a DC motor is controlled by rotating it in either the clockwise or anticlockwise direction.

A simplified electrical circuit diagram of a DC motor is shown in Figure 7.172.

The motor torque is proportional to the current through the motor:

$$T = K_T i. \tag{7.2}$$

The back emf is given by

$$V_e = K_e w, \tag{7.3}$$

Where w is the motor angular speed (radians per second). We can write the following formula for the motor circuit:

$$V - V_e = Ri + L\, di/dt \tag{7.4}$$

or

$$V = Ri + L\, di/dt + K_e w. \tag{7.5}$$

At the same time,

$$T = J\, dw/dt. \tag{7.6}$$

```
/*******************************************************************************
                        BIPOLAR STEPPER MOTOR ROTATION
                        ===============================

In this project a 39HS model bipolar stepper motor is connected to PORTB of a PIC18F45K22
microcontroller through an A3967SLB type bipolar motor controller chip

The program rotates the motor clockwise 10 revolutions, stops for 5 s, and then
rotates anticlockwise for 10 turns and stops.

39HS motor has step angle of 1.8 degrees. thus, a complete revolution requires 200 pulses
(steps) to be sent to the motor. For 10 revolutions, we have to send 2000 pulses. The DIR
input controls the direction of rotation.

Author:  Dogan Ibrahim
Date:     October 2013
File:     XC8-BSTP.C
*******************************************************************************/
#include <xc.h>
#pragma config MCLRE = EXTMCLR, WDTEN = OFF, FOSC = HSHP
#define _XTAL_FREQ 8000000

#define ENABLE PORTBbits.RB0                    // A3967SLB ENABLE input
#define STEP PORTBbits.RB1                      // A3967SLB STEP input
#define RST PORTBbits.RB2                       // A3967SLB RESET input
#define SLP PORTBbits.RB3                       // A3967SLB SLEEP input
#define MS1 PORTBbits.RB4                       // A3967SLB MS1 input
#define MS2 PORTBbits.RB5                       // A3967SLB MS2 input
#define DIR PORTBbits.RB6                       // A3967SLB DIR input
//
// This function creates seconds delay. The argument specifies the delay time in seconds
//
void Delay_Seconds(unsigned char s)
{
    unsigned char i,j;

    for(j = 0; j < s; j++)
    {
        for(i = 0; i <  100; i++)__delay_ms(10);
    }
}


//
// This is the main program. The motor turns 10 revolutions clockwise with 1ms delay
// between each step. Then the motor stops for 5 seconds. The motor then rotates
// anticlockwise for 10 revolutions and then stops.
//
void main()
{
    unsigned int i;
```

Figure 7.171: MPLAB XC8 Program.

```
                ANSELB = 0;                           // Configure PORTB as digital
                TRISB = 0;                            // Configure PORTB as digital
            //
            // A3967SLB chip configuration
            //
                ENABLE = 0;                           // Enable the controller chip
                RST = 1;                              // Disable RESET
                SLP = 1;                              // Disable SLEEP
                MS1 = 0;                              // Full-step sequence
                MS2 = 0;
                STEP = 0;                             // Step = 0 to start with
            //
            // First rotate clockwise 10 revolutions
            //
                DIR = 0;
                for(i = 0; i < 2000; i++)
                {
                    STEP = 1;                         // Send STEP pulses
                    STEP = 0;
                    __delay_ms(1);
                }

                Delay_Seconds(5);                     // Wait 5 s
            //
            // Now rotate anticlockwise 10 revolutions
            //
                DIR = 1;
                for(i = 0; i < 2000; i++)
                {
                    STEP = 1;                         // Send STEP pulses
                    STEP = 0;
                    __delay_ms(1);
                }
                while(1);                             // End. Wait here forever
            }
```

Figure 7.171
cont'd

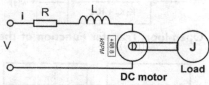

DC motor

Figure 7.172: Simplified Electrical Circuit of a DC Motor, Where, R is the Motor Resistance; L is the Motor Inductance; J is the Motor Inertia; V_e is the Back Electromotive Force (emf).

Thus, from Eqns (7.2) and (7.6),

$$I = J/K_T dw/dt. \qquad (7.7)$$

Combining Eqns (7.5) and (7.7),

$$V = \frac{RJ}{K_T}\frac{dw}{dt} + \frac{LJ}{K_T}\frac{d^2w}{dt^2} + K_e w. \qquad (7.8)$$

The inductance is negligible in small motors. If we remove the second-order inductance term from Eqn (7.8) we get

$$V = \frac{RJ}{K_T}\frac{dw}{dt} + K_e w. \qquad (7.9)$$

Taking the Laplace transform of both sides, the relationship between the motor speed and applied voltage is simply given by

$$\frac{w(s)}{V(s)} = \frac{K_T}{K_T K_e + RJs}. \qquad (7.10)$$

We can show the open-loop transfer function of the DC motor as in Figure 7.173.

The closed-loop motor transfer function is used in speed control applications. Figure 7.174 shows the closed-loop transfer function where the speed of the motor is sensed using either an optical encoder or a tachogenerator and is compared with the desired speed. The digital controller is usually a microcontroller or a PC that generates the control signals to drive the motor to obtain the required speed.

Project Description

This project is about controlling the direction of rotation of a DC motor. Two push-button switches are used: one to control the direction and another one to start/stop the motor.

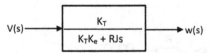

Figure 7.173: Open-loop Transfer Function of the DC Motor.

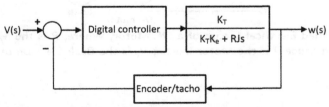

Figure 7.174: Closed-loop DC Motor Speed Control.

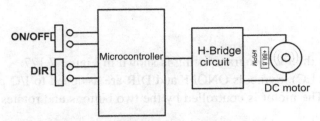

Figure 7.175: Block Diagram of the Project.

Figure 7.175 shows the block diagram of the project.

Project Hardware

The circuit diagram of the project is shown in Figure 7.176. An H-bridge circuit is
constructed from four MOSFET transistors, connected to PORTB of the microcontroller.
The motor is controlled as shown in the following table:

ON/OFF Button	Direction Button	RB3	RB2	RB1	RB0	Motor State
0	X	0	0	0	0 (0x00)	OFF
1	0	0	1	0	1 (0x05)	Clockwise
1	1	1	0	1	0 (0x0A)	Anticlockwise

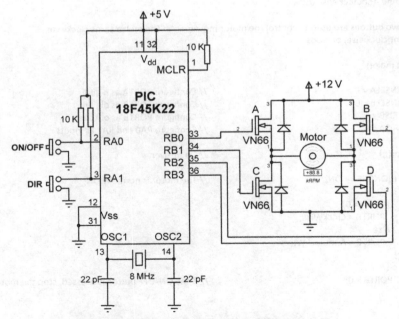

Figure 7.176: Circuit Diagram of the Project.

Project Program

mikro Pro for PIC

The mikroC Pro for the PIC program listing is shown in Figure 7.177
(MIKROC-DCMTR1.C). Symbols ONOFF and DIR are assigned to I/O ports RA0 and
RA1, respectively. The motor is controlled by the two buttons and rotates clockwise,
anticlockwise, or stops.

```
/********************************************************************************
                              DC MOTOR CONTROL
                              =================

In this project a DC motor is connected to PORTB of a microcontroller via an H-bridge circuit,
constructed from 4 MOSFET transistors.

Two push-button switches are used to control the motor. Normally the motor rotates
anticlockwise. Pressing the DIR button changes the direction of rotation. Pressing the ON/OFF
button stops the motor.

Author:  Dogan Ibrahim
Date:    October 2013
File:    MIKROC-DCMTR1.C
********************************************************************************/
#define ONOFF PORTA.RA0
#define DIR PORTA.RA1
#define ON 1
#define ClockWise 0x05
#define AntiClockWise 0x0A
//
// Two buttons are used to control the motor movements: The motor turns clockwise,
// anticlockwise, or stops
//
void main()
{
   ANSELA = 0;                        // Configure PORTA as digital
   ANSELB = 0;                        // Configure PORTB as digital
   TRISB = 0;                         // Configure PORTB as output
   TRISA = 3;                         // Configure RA0 and RA1 as inputs

   for(;;)                            // Do forever
   {
     if(ONOFF == ON)                  // The motor is normally ON
     {
       if(DIR == 0)
         PORTB = ClockWise;
       else
         PORTB = AntiClockWise;
     }
     else
       PORTB = 0;                     // If the ON/OFF button is pressed, stop the motor
   }
}
```

Figure 7.177: mikroC Pro for PIC Program.

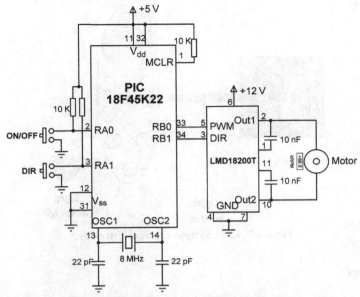

Figure 7.178: Using the LMD18200T for DC Motor Control.

Instead of using four transistors, we could have used a motor controller chip, for example, LMD18200T (Figure 7.178). This chip is connected directly to the motor. For fixed speed applications, we can apply logic 1 to the PWM input. The DIR input controls the motor direction.

Project 7.23—A Homemade Optical Encoder For Motor Speed Measurement

Optical encoders are used as sensors in motor speed and position control applications. There are several types of optical sensors available, such as incremental sensors, absolute sensors, and linear sensors. The encoder technology is not covered in this book and interested readers can search the Internet for further information and manufacturers' datasheets.

In this project, we will look at the design of an optical sensor to measure and display the speed of a motor. In the design, a round-shaped plate is attached to the motor shaft whose speed is to be measured. Holes are made on this plate at equal distances around the corner of the plate (Figure 7.179). A light source (e.g. infrared) emits light that passes through the holes as the plate rotates. At the other side of the plate, a light detector senses when a hole passes in front of it. This information is then passed to a microcontroller that counts the number of holes passing in front of the detector at a given time interval. This reading is then converted into motor speed.

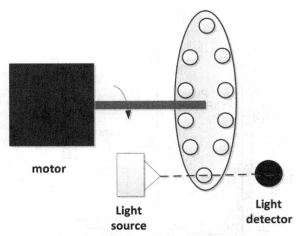

Figure 7.179: Simple Optical Encoder.

Project Hardware

In this project, a small plate with eight holes is attached to the motor shaft (Figure 7.180). The accuracy is increased when the number of holes is increased. Most commercially available encoders provide ≥200 holes. An infrared light source and light detector pair are

Figure 7.180: Homemade Optical Encoder Attached to a DC Motor.

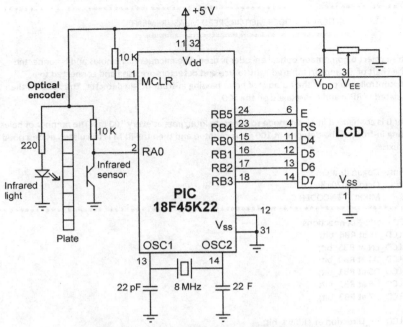

Figure 7.181: Circuit Diagram of the Project.

used in this project as shown in Figure 7.181. The output of the light detector is connected to pin RA0 of the microcontroller. The motor speed is displayed on the LCD.

Project Program

mikroC Pro for PIC

In this project, Timer0 of the microcontroller is configured to generate interrupts at every 100 ms and the number of holes passing in-front of the detector are counted within this time interval. If n is the number of holes passing in 100 ms, then we have 10n holes passing every second. If N is the number of holes on the encoder plate, then the plate makes 10n/N revolutions every second, that is,

Motor speed = 10n/N revolutions per second

In this project, N = 8 and therefore

Motor speed = 10n/8 = 1.25n revolutions per second

The speed is usually measured in revolutions per minute. Thus, multiplying the right-hand side by 60 we get

Motor Speed (revolutions per minute) = $60 \times 1.25n = 75n$.

```
/*******************************************************************************
                    OPTICAL ENCODER MOTOR SPEED MEASUREMENT
                    ===========================================
```

In this project a homemade optical encoder is used. The encoder has 8 holes and is connected
to the shaft of DC motor. Infrared light source and detectors are used and connected to a
microcontroller to count the number of holes passing in-front of the detector. The speed is then
calculated in RPM and is displayed on the LCD.

Timer 0 is configured in 16-bit mode to generate interrupts at every 100 ms. The number of holes
passing in-front of the detector in 100 ms is counted and then the RPM is calculated as described in
the text.

```
Author:  Dogan Ibrahim
Date:    October 2013
File:    MIKROC-ENCODER.C
*******************************************************************************/
// LCD module connections
sbit LCD_RS at RB4_bit;
sbit LCD_EN at RB5_bit;
sbit LCD_D4 at RB0_bit;
sbit LCD_D5 at RB1_bit;
sbit LCD_D6 at RB2_bit;
sbit LCD_D7 at RB3_bit;

sbit LCD_RS_Direction at TRISB4_bit;
sbit LCD_EN_Direction at TRISB5_bit;
sbit LCD_D4_Direction at TRISB0_bit;
sbit LCD_D5_Direction at TRISB1_bit;
sbit LCD_D6_Direction at TRISB2_bit;
sbit LCD_D7_Direction at TRISB3_bit;
// End LCD module connections

#define ENCODER PORTA.RA0
unsigned int RPM, Cnt = 0;
unsigned char Txt[] = "RPM=     ";
unsigned char DsplyCnt = 0;

void interrupt(void)
{
    TMR0H = 0x3C;                              // Reload TMR0 for 100ms interrupts
    TMR0L = 0xB0;
    RPM = 75*Cnt;                             // Speed in RPM
    IntToStr(RPM, Txt+4);                     // Convert to string
    Cnt = 0;
    DsplyCnt++;
    if(DsplyCnt == 10)                        // 1 s ?
    {
      DsplyCnt = 0;
      Lcd_Cmd(_LCD_CLEAR);                    // Clear LCD
      Ltrim(Txt+4);                           // Remove leading spaces
```

Figure 7.182: mikroC Pro for PIC Program.

```
        Lcd_Out(1,1,Txt);                          // Display speed every second
      }
      INTCON.T0IF = 0;                             // Re-enable TMR0 interrupts
}

//
// Two buttons are used to control the motor movements: The motor turns clockwise,
// anticlockwise, or stops
//
void main()
{
    ANSELA = 0;                                   // Configure PORTA as digital
    ANSELB = 0;                                   // Configure PORTB as digital
    TRISB = 0;                                    // Configure PORTB as output
    TRISA = 1;                                    // Configure RA0 as input

    Lcd_Init();                                   // Initialize LCD
//
// Configure TMR0 is 16 bit mode to generate interrupts every 100 ms
//
    T0CON = 0x81;                                 // 16-bit mode, prescaler = 4
    TMR0H = 0x3C;                                 // Load 0x3CB0 for 100 ms interrupts
    TMR0L = 0xB0;
    INTCON = 0xA0;                                // Enable TMR0 and global interrupts

    while(1)
    {
      while(ENCODER == 0);                        // Wait if 0
      Cnt++;                                      // Increment encoder count
      while(ENCODER == 1);                        // Wait if 1
    }
}
```

Figure 7.182
cont'd

Therefore, by counting the number of holes passing in-front of the detector in 100 ms we can easily find the motor speed in revolutions per minute.

In some applications, the motor speed is measured in radians per second. Since each rotation is 2π radians, the motor rotation in 1 s will be $10n \times 2\pi/N$. Thus, the motor speed in radians per second is given by

$$\text{Motor speed (radians/second)} = 10n \times 2\pi/N.$$

In this project, $N = 8$ and therefore,

$$\text{Motor speed (radians per second)} = 7.85n.$$

The mikroC Pro for PIC program listing is shown in Figure 7.182. The main program initializes the LCD, configures the timer, and then enters an endless loop. Inside this loop, the number of holes passing in front of the detector are counted and stored in variable Cnt.

Timer 0 is configured in the 16-bit mode to generate interrupts at every 100 ms. The value to be loaded into the timer registers for 100 ms (100,000 µs) interrupt is found as follows:

TMR0H:TMR0L = 65536 − 100,000/(4xTxPre-scaler).

With a clock frequency of 8 MHz (T = 0.125 µs) and the Prescaler value of 4,

TMR0H:TMR0L =− 15,536 which is 0x3CB0 in hexadecimal

that is, TMR0H = 0x3C and TMR0L = 0xB0.

Inside the ISR, the motor speed is calculated in revolutions per minute. The speed is displayed every second (every 10 times we enter the ISR) on the LCD. The motor used in this project had a speed of 5800 RPM, which is displayed as follows:

RPM = 5800

Project 7.24—Closed-Loop DC Motor Speed Control—On/Off Control

This project is about closed-loop speed control of a DC motor using a microcontroller. In the project, an optical encoder is used to detect the motor speed and feedback is applied to achieve the desired speed. The motor control signal is in the form of a PWM waveform.

PWM waveform is frequently used in power control applications, such as motor control, pump control, and heating control. By changing the duty cycle (ON time), we can control the average voltage (or power) applied to the load.

If V_i is the amplitude of the PWM signal, the average voltage supplied to the load is given by

$$\text{Average voltage} = \frac{t_{ON}}{t_{ON} + t_{OFF}} V_i = \frac{t_{ON}}{T} V_i.$$

The relationship between t_{ON} and the average load voltage is linear, and as t_{ON} changes from 0 to T (0−100% duty cycle), the average voltage delivered to the load changes from 0 to $+V_i$. Thus, we can control t_{ON} time to control the motor voltage and hence the motor speed.

In this project, the built-in PWM module of the microcontroller is used to generate the PWM signal. We can choose the PWM frequency from a few kilohertz to $\geq$10 kHz. Here, we chose 10 kHz (T = 0.1 ms). We can find the register values to generate the required PWM waveform. With an 8-MHz clock:

$$PR2 = \frac{\text{PWM period}}{\text{TMR2 prescale value} * 4 * T_{OSC}} - 1$$

or

$$PR2 = \frac{0.1 \times 10^{-3}}{4 * 4 * 0.125 \times 10^{-6}} - 1 = 49 \text{ or } 0x31 \text{ in hexadecimal.}$$

Also,

$$CCPR1L : CCP1CON\langle 5:4 \rangle = \frac{PWM \text{ pulse width}}{TMR2 \text{ prescale value} * T_{OSC}}$$

or

$$CCPR1L : CCP1CON\langle 5:4 \rangle = \frac{Duty \text{ cycle} \times 10^{-3}}{4 * 0.125 \times 10^{-6}} = 2000 \times Duty \text{ cycle},$$

where the Duty cycle is in milliseconds.

Thus, for a Duty cycle of 0.001 ms, CCPR1L:CCP1CON<5:4> = 2, which gives the average voltage as

$$Average \text{ voltage} = \frac{0.001}{0.1} V_i = 0.01 \, V_i$$

For 100% Duty cycle, Duty cycle = 0.1 ms and

CCPR1L:CCP1CON<5:4> = 200 which gives the average voltage equal to V_i.

As a summary, as we change the register value from 2 to 200, the average voltage applied to the motor will change linearly from $0.01V_i$ to V_i. Combining the above equations, we can write an expression for the average load voltage as follows:

$$Average \text{ voltage} = \frac{CCPR1L : CCP1CON < 5:4 >}{2000 \times 0.1} V_i$$

or

$$Average \text{ voltage} = \frac{CCPR1L : CCP1CON\langle 5:4 \rangle}{200} V_i.$$

The value loaded into CCPR1L:CCP1CON<5:4> changes from 2 to 200.

The CCPR1L:CCP1CON<5:4> register combination is 10 bits wide. Assuming that the number to be loaded into the register pair is integer N, the value to be loaded into CCPR1L can be found by right shifting N by 2 bits. The number to be loaded into bits 5 and 4 of CCP1CON can be found by taking the two least significant bits of N and left shifting by 4 bits. Bits 2 and 3 of CCP1CON are then set to 1 for

PWM operation. The following program code shows how to load the PWM register pair:

```
CCPR1L = N >> 2;
Temp = N & 0x03;
Temp = Temp << 4;
CCP1CON = Temp | 0x0C;
```

In this project, the homemade optical encoder (with eight holes) described in the previous project is used. The encoder data are read every 10 ms. If n is the number of holes passing in front of the detector in 0.01 s, then 100n holes pass in 1 s. If N is the number of holes on the encoder, the motor speed is given by

Motor speed = 100n/N revolutions per second

With $N = 8$,

Motor speed = 100n/8 = 12.5 revolutions per second

or

Motor speed (revolutions per minute) = 750n

The timer interrupt TMR0 is set to generate interrupts every 10 ms, and the motor speed is calculated inside the ISR.

Assuming a prescaler value of 128- and 8-MHz clock ($T = 0.125$ μs), the TMR0 register value for 10-ms interrupts can be found from

TMR0L = 256−10000/(4 × 0.125 × 128) = 100.

Project Hardware

The circuit diagram of the project is shown in Figure 7.183. The speed of the motor is fed back via the optical encoder to RA0 input of the microcontroller. The PWM output of the microcontroller drives the motor through a MOSFET transistor. An LCD is connected to PORTB to display the actual motor speed in revolutions per minute.

Project Program
mikroC Pro for PIC

The mikroC Pro for the PIC program listing is shown in Figure 7.184 (MIKROC-MTRONOFF.C). At the beginning of the main program, I/O ports are configured, and the LCD is initialized. Then, the PWM module is configured to generate pulses with a period of 9.1 ms. Timer 2 is used to provide clock to the PWM module, and Timer 0 is configured to generate interrupts at every 10 ms. The desired speed is set to

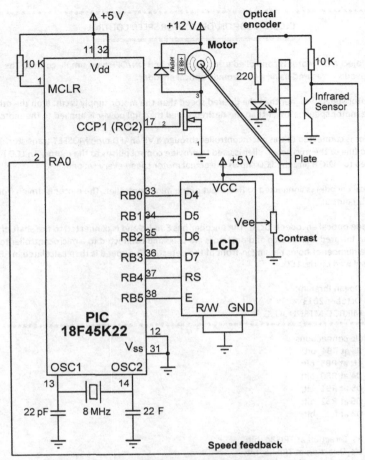

Figure 7.183: Circuit Diagram of the Project.

2000 RPM with a dead band of 100, that is, it is required to keep the speed between 2000 ± 100 RPM. The remainder of the main program counts the encoder pulses.

Inside the ISR the timer register TMR0L is reloaded, the motor speed is calculated and the ON/OFF control action is applied using the following program code:

```
/* IF SLOW */
if(RPM < (DesiredRPM — Deadband))
{
SetSpeed(200);      // Rotate motor FAST
}
/* IF FAST */
if(RPM > (DesiredRPM + Deadband))
{
SetSpeed(2);        // Rotate motor SLOW
}
```

```
/*************************************************************************
                    CLOSED-LOOP ON/OFF MOTOR SPEED CONTROL
                    =========================================
```

In this project a DC motor is controlled using feedback and ON/OFF type simple control. The motor speedis measured using a homemade optical encoder.

If the motor speed is higher than the desired speed than the motor supply is cut. If on the other hand the motor speed is lower than the desired speed than full power is applied to the motor.

The motor is connected to the microcontroller through a VN66AFD type MOSFET transistor. The PWM module of the microcontroller is used to provide control pulses to the motor. An LCD is connected to PORTB and the LCD shows the actual motor speed every second.

The optical encoder is connected to RA0 input of the microcontroller. The motor is driven from the CCP1 (RC2) output.

Homemade optical encoder is used. The encoder has 8 holes and is connected to the shaft of DC motor. Infrared light source and detectors are used and connected to a microcontroller to count the number of holes passing in-front of the detector. The speed is then calculated in RPM and is displayed on the LCD.

```
Author:   Dogan Ibrahim
Date:     October 2013
File:     MIKROC-MTRONOFF.C
*************************************************************************///
LCD module connections
sbit LCD_RS at RB4_bit;
sbit LCD_EN at RB5_bit;
sbit LCD_D4 at RB0_bit;
sbit LCD_D5 at RB1_bit;
sbit LCD_D6 at RB2_bit;
sbit LCD_D7 at RB3_bit;

sbit LCD_RS_Direction at TRISB4_bit;
sbit LCD_EN_Direction at TRISB5_bit;
sbit LCD_D4_Direction at TRISB0_bit;
sbit LCD_D5_Direction at TRISB1_bit;
sbit LCD_D6_Direction at TRISB2_bit;
sbit LCD_D7_Direction at TRISB3_bit;
// End LCD module connections

#define ENCODER PORTA.RA0
unsigned int Cnt = 0;
unsigned int RPM, DesiredRPM, DeadBand;
unsigned char Txt[] = "RPM=     ";
unsigned char DsplyCnt = 0;


//
// This function sets the motor speed
//
```

Figure 7.184: mikroC Pro for the PIC Program.

```c
void SetSpeed (unsigned char N)
{
  unsigned char Temp;

  CCPR1L = N >> 2;
  Temp = N & 0x03;
  Temp = Temp << 4;
  CCP1CON = Temp | 0x0C;
}

//
// This is the Timer0 interrupt service routine. The program jumps here every 10ms.
// The actual motor speed is compared with the desired speed and either the motor is
// operated with full voltage (maximum speed), or with low voltage (minimum speed). The
// motor speed is displayed on the LCD.
//
void interrupt(void)
{
  TMR0L = 100;                            // Reload Timer register
  RPM = 750*Cnt;                          // Actual motor speed (in RPM)
  Cnt = 0;
  /* IF SLOW */
  if(RPM < (DesiredRPM - Deadband))
  {
    SetSpeed(200);                        // Rotate motor FAST
  }

  /* IF FAST */
  if(RPM > (DesiredRPM + Deadband))
  {
    SetSpeed(2);                          // Rotate motor SLOW
  }
  DsplyCnt++;
  if(DsplyCnt == 100)                     // 1 second (10msx100=1s) ?
  {
    DsplyCnt = 0;
    Lcd_Cmd(_LCD_CLEAR);                  // Clear LCD
    IntToStr(RPM,Txt+4);                  // Convert into string
    Ltrim(Txt+4);                         // Remove leading spaces
    Lcd_Out(1,1,Txt);                     // Display speed every second
  }

  INTCON.T0IF = 0;                        // Re-enable TMR0 interrupts
}

//
// The desired RPM is set to 2000 and the DeadBand is set to 100 RPM
//
void main()
{
  ANSELA = 0;                             // Configure PORTA as digital
```

Figure 7.184
cont'd

```
            ANSELB = 0;                              // Configure PORTB as digital
            ANSELC = 0;                              // Configure PORTC as digital
            TRISA = 1;                               // Configure RA0 as input
            TRISB = 0;                               // Configure RB0 as input

            LCD_Init();                              // Initialize LCD
//
// Configure the PWM module
//
            T2CON = 0x05;                            // Set Timer 2 ON and prescaler to 4
            PR2 = 0x31;                              // Load Timer 2 PR2 register with 49
            CCPR1L = 0;                              // Motor OFF to start with
            TRISC = 0;                               // Configure CCP1 (RC2) as output
            CCP1CON = 0x0C;                          // Enable PWM module
            CCPTMRS0 = 0;                            // Use Timer 2 for the PWM module (CCP1)
//
// Configure TMR0 in 8 bit mode to generate interrupts every 10ms
//
            T0CON = 0xC6;                            // 8-bit mode, prescaler = 128
            TMR0L = 100;                             // Timer value for 10ms timer overflow
            INTCON = 0xA0;                           // Enable TMR0 interrupts

            DesiredRPM = 2000;                       // Desired RPM
            DeadBand = 100;                          // Deadband range

            while(1)
            {
              while(ENCODER == 0);                   // Wait if 0
              Cnt++;                                 // Increment encoder count
              while(ENCODER == 1);                   // Wait if 1
            }
}
```

Figure 7.184
cont'd

Function SetSpeed loads the CCPR1L and CCp1CON register of the PWM module accordingly. The LCD displays the actual motor speed every second.

Index